How Computers Do

UNIVERSITY *of* LIMERICK

D1513834

THE DEFINITIVE GUIDE TO

HOW COMPUTERS DO MATH

Featuring
The Virtual DIY Calculator

Clive "MAX" Maxfield
Alvin Brown

WILEY-INTERSCIENCE

A JOHN WILEY & SONS, INC., PUBLICATION

Copyright © 2005 by John Wiley & Sons, Inc. All rights reserved.

Published by John Wiley & Sons, Inc., Hoboken, New Jersey.
Published simultaneously in Canada.

For general information on our other products and services please contact our Customer Care
Department within the U.S. at 877-762-2974, outside the U.S. at 317-572-3993 or fax 317-572-4002.

Wiley also publishes its books in a variety of electronic formats. Some content that appears in print,
however, may not be available in electronic format.

Library of Congress Cataloging-in-Publication Data is available.

ISBN-13 978-0471-73278-5
ISBN-10 0-471-73278-8

Printed in the United States of America.

10 9 8 7 6 5 4 3 2 1

To all our friends
who make the world
such a wonderful place!

0`10101000`1011101010000101010110101001001001000110110000010101101000110101

Contents

Laboratories

Do You Speak Martian?

There are an abundance of books on computer architectures, computer logic, and computer mathematics, and most of these works discuss various techniques for representing and manipulating numbers inside computers. Sadly, however, it appears that the majority of these tomes are written by visitors from the planet Mars, whose keen understanding of higher mathematics is somewhat offset by their limited grasp of the English language.

"*Say it's not so!*" you cry, but the proof is irrefutable. When was the last time you waded through a book on computer mathematics without your brain overheating? Much like reading *Being and Nothingness* by the famous French philosopher Jean Paul Sartre, one could mull over many of these cryptic masterpieces until the end of time without gaining so much as the faintest clue as to what was in their authors' minds.

This is why we have been moved to write this modest attempt at introducing the basics of computer arithmetic in words we can all understand. An impossible task, some may say, but we dare to fly in the face of conventional wisdom. Now, read on. . . .

CHAPTER 0

WHY THIS BOOK IS SO COOL

The simplest schoolboy is now familiar with facts for which Archimedes would have sacrificed his life.

ERNEST RENAN (1823–1892) in *Souvenirs D'enfance et de Jeunesse* (1887)

In this chapter we will learn

- The sock color of choice for the discerning Viking warrior
- Why this book is so cool
- That there are jobs awaiting time-travelers
- Some "stuff" about calculators
- Why this chapter is numbered "0"

Fearsome Warriors or Slaves to Fashion?

Many of us are used to thinking of the Vikings as fearsome warriors (this isn't your Mother's math book!) who descended from the northlands and rampaged and pillaged across Europe. These formidable fighters are popularly believed to have laughed at danger and scoffed at the elements, so the recent archeological discovery that many Vikings wore red woolly socks is, to many of us, somewhat disconcerting. However, we digress. . .

This Book Is Cool Because . . .

This book really is cool because, together, we are going to discover all sorts of interesting snippets of knowledge, tidbits of data, and nuggets of trivia, all bundled together with a tremendous amount useful information as to how computers and calculators perform their magic; and it isn't going to make our heads hurt at all!

Experts (the authors' mothers in this case) agree that one of the best ways to learn something and remember it afterward is by means of hands-on experience, which you are poised to gain in huge dollops by using the virtual microcomputer/calculator that you'll find on the CD-ROM accompanying this book.

As one rocket scientist[1] who reviewed this manuscript told the authors: "The combination of this book and its associated virtual computer is fantastic! Experience over the last 50 years has shown me that there's only one way to truly understand how computers work, and that is to learn one computer and its instruction set, no matter how simple or primitive, from the ground up. Once you fully comprehend how that simple computer functions, you can easily extrapolate to more complex machines." However, once again, we digress. . .

Jobs Abound for Time-Travelers

Today, most of us are extremely familiar with using numbers to perform simple tasks such as addition, subtraction, multiplication, and division.

---

[1]Honestly. Huntsville, Alabama, U.S.A. (where the authors live) is known as the Space Capital of America, and it's difficult to take even a short strolling without bumping into at least one rocket scientist.

Due to the fact that we are so intimate with these concepts, we tend to forget the tremendous amounts of mental effort that have been expended by so many folks over the millennia to raise us to our present level of understanding.

In the days of yore, when few people even knew how to count beyond the number of fingers on their hands,[2] anyone who was capable of performing relatively rudimentary mathematical operations could easily achieve a position of power and standing in the community.

If you could predict an eclipse, for example, you were obviously someone to be reckoned with (especially if it actually came to pass). Similarly, if you were a warrior chieftain, it would be advantageous to know how many fighting men and women you had at your command, and the person who could provide you with this vital information would obviously rank high on your summer-solstice card list.[3]

So, we can all rest easy in our beds at night, secure in the knowledge that, should we ever be presented with the occasion to travel back through time, there would be numerous job opportunities awaiting our arrival. However (you guessed it), we digress. . .

Calculators Then and Now

Every now and then, strange and wonderful mechanisms from antiquity are discovered. In 1900, for example, a device of unknown purpose containing numerous gear wheels forming a sophisticated mechanism dating from 2200 BC was discovered in a shipwreck close to the tiny Greek island of Antikythera. This contraption, which is now known as the *Antikythera Mechanism* or the *Antikythera Calculator,* was created during the early years of the Hellenistic Period, a golden age during which science and art flourished in ancient Greece.

In many cases, objects like this prompt speculation that our antediluvian ancestors were the creators of complex mechanical calculators with which they could perform mathematical operations such as addition, subtraction, multiplication, and division. In reality, however, the concept of zero (0) as representing a true quantity in a place-value num-

[2]Sometimes, fingers and toes in the case of societies that employed vigesimal (base-20) numbering systems.
[3]You wouldn't have a Christmas card list, because Christmas cards weren't invented until 1843. (Try finding this tasty morsel of trivia in another computer book!)

> **Note**   The invention of zero and negative numbers, the use of tally sticks and the abacus, the origin of logic machines, calculators, and computers; and a wealth of other topics are discussed in more detail in "The History of Calculators, Computers, and Other Stuff" provided on the CD-ROM accompanying this book. See Appendix D for more details on additional resources, including this history.

ber system didn't appear until around 600 AD in India. Without the notion of zero in this context, it is really not possible to create a mechanical calculator in any form we would recognize.

This is not to say that these ancient mechanisms were not incredibly cunning and refined. However, such instruments were probably designed to measure things or to track time in one way or another; for example, to help in predicting the seasons and the activities of celestial objects like the sun, moon, planets, and constellations.

On the other hand, we might well wonder why, after the idea of zero as an actual number entered the scene, it took so long to actually invent a true mechanical calculator. Of course it's fair to say that Europe was undergoing a period of stagnation in art, literature, and science called the Dark Ages.[4] However, there were many brilliant minds in the Byzantine Empire and the Arabic, Chinese, Indian, and Persian cultures (to name but a few) that would almost certainly have been up to the task.

Be this as it may, the first true mechanical calculators of which we are aware were the Calculating Clock (1623), which was created by the German astronomer and mathematician Wilhelm Schickard (1592–1635); the Pascaline or Arithmetic Machine (1642), which was the inspiration of the French mathematician, physicist, and theologian Blaise Pascal (1623–1662); and the Step Reckoner (1694), which was the brainchild of a German Baron called Gottfried von Leibniz (1646–1716).

Countless mechanical calculating machines emerged to see the light of day over the next few hundred years, but these were all largely based on the underlying principles established by Schickard, Pascal, and Leibniz. However, "the times they were a'changin," as they say. The invention of the transistor in 1947 and the integrated circuit (silicon chip) in 1958 paved the way for an entirely new class of calculating devices.

---

[4]Some pundits equate the Dark Ages with the Middle Ages (the period in European history between Antiquity and the Renaissance, often dated from 476 AD to 1453 AD), whereas others regard the Dark Ages as encompassing only the early portion of the Middle Ages.

The first experimental model of an electronic pocket calculator was created by Texas Instruments in 1966.[5] This was followed in 1970/1971 by the first commercially available unit, a portable (hand-held) printing calculator called the Pocketronic. Created as a joint effort by Texas Instruments and Cannon, and priced at $150 (which was a *lot* of money at the time), this four-function device could add, subtract, multiply, and divide.

Today, of course, it's possible to purchase pocket calculators that boast enough computing power to guide a rocket to the moon, and cheap and cheerful versions are now so ubiquitous that it's not uncommon to find them as giveaways in boxes of breakfast cereal. However (believe it or not), we digress. . .

Déjà Vu (Didn't Someone Just Say That?)

Whether they are aware of it or not, the average person in the developed world comes into contact with dozens or hundreds of electronic calculating, computing, and controlling machines every day. These devices toil away, performing countless mathematical operations, but very few of us actually know what they are doing or, perhaps more importantly, how they are doing it.

Consider your own pocket calculator, for example. When you multiply two numbers like 36.984562 and 79.386431 together and you are presented with the result, you may well assume that the answer is correct, but just *how correct is it?* and *is it correct enough?* The accuracy and precision (as we will discover, these are two different things) required by an accountant and a rocket scientist may be poles apart.

As we noted earlier, in the not-so-distant past, anyone who was capable of performing relatively rudimentary mathematical operations could easily achieve a position of power and standing in the community. Well, we are in danger of finding ourselves in a déjà vu situation, because knowledge is power, and very few folks actually have any clue as to what is going on behind the scenes with regard to the way in which computers and calculators perform even simple operations.

---

[5]This machine is now preserved at the National Museum of American History (a part of the Smithsonian Institution) in Washington, DC, U.S.A.

But fear not, because there's nothing to fear but fear itself. You can turn that frown upside down into a smile, because we are going to explain just how computers and calculators perform their magic. The really "cool beans" part of all of this is the virtual microcomputer/calculator on the CD-ROM accompanying this book. When you first launch this DIY Calculator application on your home computer (PC) and click the buttons on the calculator interface nothing will happen. But wait, there's more! We are going to guide you through the process of creating your very own program to make the calculator function as required. On the way, we will discover all sorts of interesting things that will leave us grinning with delight and gasping for more. However, we digress. . .

Why Is this Chapter Numbered "0"?

Computer programmers and engineers typically start counting, indexing, and referencing things from zero.[6] Thus, in order to keep in the spirit of things, we decided to follow the same convention with our chapter numbers.

However, let us digress no more. You are poised on the brink of discovering all manner of weird and wonderful things, so proceed immediately to Chapter 1 and let the fun begin!

> **Note** Except where such interpretation is inconsistent with the context, the singular shall be deemed to include the plural, the masculine shall be deemed to include the feminine, and the spelling and punctuation shall be deemed to be correct!

---

[6]The reasons for this—and related conventions—will become apparent as we wend our way through the topics in this book.

CHAPTER 1

INTRODUCING BINARY AND HEXADECIMAL NUMBERS

I am ill at these numbers.

WILLIAM SHAKESPEARE
(1564–1616) in *Hamlet* (1601)

In this chapter we will learn about:

- Counting on fingers and toes
- Place-value number systems
- Using powers or exponents
- The binary number system
- The hexadecimal number system
- Counting in the binary and hexadecimal systems
- Using wires to represent numbers

Why Do We Need to Know this Stuff?

The number system with which we are most familiar is the *decimal system,* which is based on ten digits: 0, 1, 2, 3, 4, 5, 6, 7, 8, and 9. As we shall soon discover, however, it's easier for electronic systems to work with data that is represented using the *binary* number system, which comprises only two digits: 0 and 1.

Unfortunately, it's difficult for humans to visualize large values presented as strings of 0s and 1s. Thus, as an alternative, we often use the *hexadecimal* number system, which is based on sixteen digits that we represent by using the numbers 0 through 9 and the letters A through F.

Familiarity with the binary and hexadecimal number systems is necessary in order to truly understand how computers and calculators perform their magic. In this chapter, we will discover just enough to make us dangerous, and then we'll return to consider number systems and representations in more detail in Chapters 4, 5, and 6.

Counting on Fingers and Toes

The first tools used as aids to calculation were almost certainly man's own fingers. It is no coincidence, therefore, that the word "digit" is used to refer to a finger (or toe) as well as a numerical quantity. As the need grew to represent greater quantities, small stones or pebbles could be used to represent larger numbers than could fingers and toes. These had the added advantage of being able to store intermediate results for later use. Thus, it is also no coincidence that the word "calculate" is derived from the Latin word for pebble.

Throughout history, humans have experimented with a variety of different number systems. For example, you might use one of your thumbs to count the finger joints on the same hand (1, 2, 3 on the index finger; 4, 5, 6 on the next finger; up to 10, 11, 12 on the little finger). Based on this technique, some of our ancestors experimented with base-12 systems. This explains why we have special words like *dozen,* meaning "twelve," and *gross,* meaning "one hundred and forty-four" ($12 \times 12 = 144$). The fact that we have 24 hours in a day (2×12) is also related to these base-12 systems.

Similarly, some groups used their fingers *and* toes for counting, so they ended up with base-20 systems. This is why we still have special words like *score,* meaning "twenty." However, due to the fact that we have ten fingers, the number system with which we are most familiar is the decimal system, which is based on ten digits: 0, 1, 2, 3, 4, 5, 6, 7, 8, and 9 (see note). The word *decimal* is derived from the Latin *decam,* meaning "ten." As this system uses ten digits, it is said to be *base-10* or *radix-10;* the term *radix* comes from the Latin word meaning "root."

Place-Value Number Systems

Consider the concept of Roman numerals, in which I = 1, V = 5, X = 10, L = 50, C = 100, D = 500, M = 1,000, and so forth. Using this scheme, XXXV represents 35 (three tens and a five). One problem with this type of number system is that over time, as a civilization develops, it tends to become necessary to represent larger and larger quantities. This means that mathematicians either have to keep on inventing new symbols or start using lots and lots of their old ones. But the biggest disadvantage of this approach is that it's painfully difficult to work with (try multiplying CLXXX by DDCV and it won't take you long to discover what we mean).

An alternative technique is known as a *place-value system,* in which the value of a particular digit depends both on itself and on its position within the number. This is the way in which the decimal number system works. In this case, each column in the number has a "weight" associated with it, and the value of the number is determined by combining each digit with the weight of its column (Figure 1-1).

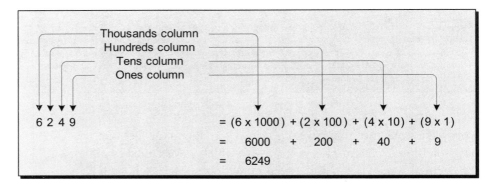

Figure 1-1. Combining digits and column weights in decimal.

Using Powers or Exponents

Another way of thinking about this is to use the concept of *powers;* for example, $100 = 10 \times 10$. This can be written as 10^2, meaning "ten to the power of two" or "ten multiplied by itself two times." Similarly, $1000 = 10 \times 10 \times 10 = 10^3$, $10,000 = 10 \times 10 \times 10 \times 10 = 10^4$, and so forth (Figure 1-2).

Rather than talking about using powers, some mathematicians prefer to refer to this type of representation as an *exponential form.* In the case of a number like 10^3, the number being multiplied (10) is known as the *base,* whereas the *exponent* (3) specifies how many times the base is to be multiplied by itself.

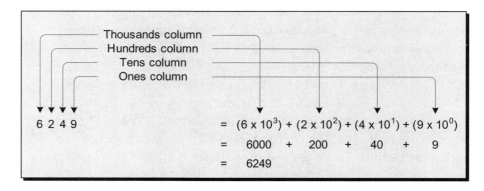

Figure 1-2. Using powers of ten.

There are a number of points associated with powers (or exponents) that are useful to remember:

- Any base raised to the power of 1 is the base itself, so $10^1 = 10$.

- Strictly speaking, a power of 0 is not really part of the series. By convention, however, any base to the power of 0 equals 1, so $10^0 = 1$.

- A value with an exponent of 2 is referred to as the *square* of the number; for example, $3^2 = 3 \times 3 = 9$, where 9 (or 3^2) is the square of 3.

- A value with an exponent of 3 is referred to as the *cube* of the number; for example, $3^3 = 3 \times 3 \times 3 = 27$, where 27 (or 3^3) is the cube of 3.

- The square of a *whole number* (0, 1, 2, 3, 4, etc.) is known as a *perfect square;* for example, $0^2 = 0$, $1^2 = 1$, $2^2 = 4$, $3^2 = 9$, $4^2 = 16$, $5^2 = 25$, and so on are perfect squares. (The concepts of whole numbers and their cousins are introduced in more detail in Chapter 4.)

- The cube of a *whole number* is known as a *perfect cube;* for example, $0^3 = 0$, $1^3 = 1$, $2^3 = 8$, $3^3 = 27$, $4^3 = 64$, $5^3 = 125$, and so on are perfect cubes.

But we digress. The key point here is that the column weights are actually powers of the number system's base. This will be of particular interest when we come to consider other systems.

Counting in Decimal

Counting in decimal is easy (mainly because we're so used to doing it). Commencing with 0, we increment the first column until we get to 9, at which point we've run out of available digits. Thus, on the next count we reset the first column to 0, increment the second column to 1, and continue on our way (Figure 1-3).

Similarly, once we've reached 99, the next count will set the first column to 0 and attempt to increment the second column. But the second column already contains a 9, so this will also be set to 0 and we'll increment the *third* column, resulting in 100, and so it goes.

Figure 1-3. Counting in decimal.

The Binary Number System

Unfortunately, the decimal number system is not well suited to the internal workings of computers. In fact, for a variety of reasons that will become apparent as we progress through this book, it is preferable to use the binary (base-2) number system, which employs only two digits: 0 and 1.

Binary is a place value system, so each column in a binary number has a weight, which is a power of the number system's base. In this case we're dealing with a base of two, so the column weights will be $2^0 = 1$, $2^1 = 2$, $2^2 = 4$, $2^3 = 8$, and so forth (Figure 1-4).

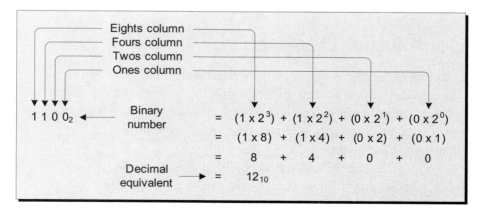

Figure 1-4. Combining digits and column weights in binary.

When working with number systems other than decimal, or working with a mixture of number systems as shown in Figure 1-4, it is common to use subscripts to indicate whichever base is in use at the time. For example, $1100_2 = 12_{10}$, which means $1100_{Binary} = 12_{Decimal}$.

Another common alternative is to prefix numbers with a character to indicate the base; for example, %1100, where the % indicates a binary value. (In fact there are a variety of such conventions, so you need to keep your eyes open and your wits about you as you plunge deeper into the mire.) Unless otherwise indicated, a number without a subscript or a prefix character is generally assumed to represent a decimal value.

Sometime in the late 1940s, the American chemist turned topologist turned statistician John Wilder Tukey realized that computers and the binary number system were destined to become increasingly important. In addition to coining the word "software," Tukey decided that saying "binary digit" was a bit of a mouthful, so he started to look for an alternative. He considered a variety of options, including *binit* and *bigit*, but eventually settled on *bit*, which is elegant in its simplicity and is used to this day.

Binary values of 1100_2 and 11001110_2 are said to be four and eight bits wide, respectively. Groupings of four bits are relatively common, so they are given the special name of *nybble* (or sometimes *nibble*). Similarly, groupings of eight bits are also common, so they are given the special name of *byte*. Thus, *"two nybbles make a byte,"* which goes to show that computer engineers do have a sense of humor (albeit not a tremendously sophisticated one).

Counting in binary

Counting in binary is even easier than counting in decimal; it just seems a little harder if you aren't familiar with it. As usual, we start counting from 0, and then we increment the first column to be 1, at which point we've run out of all the available digits. Thus, on the next count we reset the first column to 0, increment the second column to 1, and proceed on our merry way (Figure 1-5).

Similarly, once we've reached 11_2, the next count will set the first column to 0 and attempt to increment the second column. But the second column already contains a 1, so this will also be set to 0 and we'll increment the *third* column, resulting in 100_2. (Note that Figure 1-5 doesn't require us to use subscripts or special prefix symbols to indicate the binary values, because in this case they are obvious from the context.)

Binary	Decimal
0	0
1	1
1 0	2
1 1	3
1 0 0	4
1 0 1	5
1 1 0	6
1 1 1	7
1 0 0 0	8
:	:

Figure 1-5. Counting in binary.

Learning your binary "times tables"

Cast your mind back through the mists of time to those far-off days in elementary (junior) school. You probably remember learning your multiplication "tables" by rote, starting with the "two times table" ("one two is two, two twos are four, three twos are six, . . .") and wending your weary way onward and upward to the "twelve times table" ("ten twelves are one hundred and twenty, eleven twelves are one hundred and thirty-two, and—wait for it, wait for it—twelve twelves are one hundred and forty-four"). Phew!

The bad news is that we need to perform a similar exercise for binary: the good news is that, as illustrated in Figure 1-6, we can do the whole thing in about five seconds flat! That's all there is to it. This is the only binary multiplication table there is. This really is pretty much as complex as it gets!

Multiplication
0 x 0 = 0
0 x 1 = 0
1 x 0 = 0
1 x 1 = 1

Figure 1-6. The binary multiplication table.

Using Wires to Represent Numbers

Today's digital electronic computers are formed from large numbers of microscopic semiconductor switches called transistors, which can be turned on and off incredibly quickly (thousands of millions of times a second).

> **Note** Computers can be constructed using a variety of engineering disciplines, resulting in such beasts as electronic, hydraulic, pneumatic, and mechanical systems that process data using analog or digital techniques.
>
> For the purposes of this book however, we will consider only digital electronic implementations, because these account for the overwhelming majority of modern computer systems.

Transistors can be connected together to form a variety of primitive logical elements called *logic gates*. In turn, large numbers of logic gates can be connected together to form a computer.[1]

It's relatively easy for electronics engineers to create logic gates that can detect, process, and generate two distinct voltage levels.[2] For example, logic gates circa 1975 tended to use voltage levels of 0 volts and 5 volts. However, the actual voltage values used inside any particular computer are of interest only to the electronics engineers themselves. All we need know is that these two levels can be used to represent binary 0 and 1, respectively.

In the "counting in binary" example illustrated in Figure 1-5, there was an implication that our table could have continued forever. This is because numbers written with pencil and paper can be of any size, limited only by the length of your paper and your endurance. By comparison, the numbers inside a computer have to be mapped onto a physical system of logic gates and wires. For example, a single wire can be used to represent only $2^1 = 2$ binary values (0 and 1), two wires can be used to represent $2^2 = 2 \times 2 = 4$ different binary values (00, 01, 10, and 11), three wires can be used to represent $2^3 = 2 \times 2 \times 2 = 8$ differ-

---

[1]Transistors and logic gates are introduced in greater detail in the book *Bebop to the Boolean Boogie (An Unconventional Guide to Electronics)*, 2nd edition, by Clive "Max" Maxfield ISBN: 0-7506-7543-8.

[2]Voltage refers to electrical potential, which is measured in *volts*. For example, a 9-volt battery has more electrical potential than a 3-volt battery. Once again, the concept of voltage is introduced in greater detail in *Bebop to the Boolean Boogie (An Unconventional Guide to Electronics)*.

ent binary values (000_2, 001_2, 010_2, 011_2, 100_2, 101_2, 110_2, and 111_2), and so on.

The term *bus* is used to refer to a group of signals that carry similar information and perform a common function. In a computer, a bus called the *data bus* is, not surprisingly, used to convey data.

> **Note**   In this context, the term "data" refers to numerical, logical, or other information presented in a form suitable for processing by a computer. For the purposes of this book however, we will consider only digital electronic implementations, because these account for the overwhelming majority of modern computer systems.
>
> Actually, "data" is the plural of the Latin *datum*, meaning "something given." The plural usage is still common, especially among scientists, so it's not unusual to see expressions like "*These data are . . .*"
>
> However, it is becoming increasingly common to use "data" to refer to a singular group entity such as information. Thus, an expression like "*This data is . . .*" would also be acceptable to a modern audience.

For the purposes of these discussions, let's assume that we have a data bus comprising eight wires that we wish to use to convey binary data. In this case, we can use these wires to represent $2^8 = 2 \times 2 \times 2 \times 2 \times 2 \times 2 \times 2 \times 2 = 256$ different combinations of 0s and 1s. If we wished to use these different binary patterns to represent numbers, then we might decide to represent decimal values in the range 0 to 255 (Figure 1-7). For reasons that will be discussed in greater detail in Chapter 4, this form of representation is referred to as *unsigned binary numbers*.

In any numbering system, it is usual to write the most-significant digit on the left and the least-significant digit on the right. For example, when you see a decimal number such as 825, you immediately assume that the "8" on the left is the most-significant digit (representing eight hundred), whereas the "5" on the right is the least-significant digit (representing only five). Similarly, the most-significant binary digit, referred to as the *most-significant bit (MSB)*, is on the left-hand side of the binary number, whereas the *least-significant bit (LSB)* is on the right.

The Hexadecimal Number System

The problem with binary values is that, although we humans are generally good at making sense out of symbolic representations like words and numbers, we find it difficult to comprehend the meaning behind long strings of 0s and 1s. For example, the binary value 11001110_2 doesn't im-

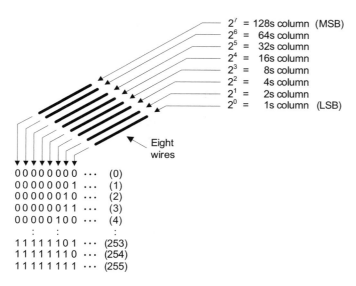

Figure 1-7. Using eight wires to represent unsigned binary numbers (values in parentheses are decimal equivalents).

mediately register with most of us, whereas its decimal equivalent of 206 is much easier to understand.

Unfortunately (as we shall come to see), translating values between binary and decimal can be a little awkward. However, any number system with a base that is a power of two (for example, 4, 8, 16, 32, etc.) can be easily mapped into its binary equivalent, and vice versa. In the early days of computing, it was common for computers to have data busses whose widths were wholly divisible by three (9 bits, 12 bits, 18 bits, and so forth). Thus, the octal (base-8) number system became very popular, because each octal digit can be directly mapped onto three binary digits (bits). More recently, computers have standardized on bus widths that are wholly divisible by eight (8 bits, 16 bits, 32 bits, and so forth). For this reason, the use of octal has declined and the hexadecimal (base-16) number system is now prevalent, because each hexadecimal digit can be directly mapped onto four bits.

As a base-16 system, hexadecimal requires 16 unique symbols, but inventing completely new symbols would not be a particularly efficacious solution, not the least that we would all have to learn them! Even worse, it would be completely impractical to modify every typewriter and computer keyboard on the face of the planet. As an alternative, we use a combination of the decimal numbers 0 through 9 and the alpha characters A through F (Figure 1-8).

Counting in hexadecimal

The rules for counting in hexadecimal are the same as for any place-value number system. We commence with 0 and count away until we've used up all of our symbols (that is, when we get to F); then, on the next count, we set the first column to 0 and increment the second column (Figure 1-9).

Observe that the binary values in Figure 1-9 have been split into two 4-bit groups. This form of representation is not typical, but it is used here to emphasize the fact that each hexadecimal digit maps directly onto four bits.

Also note that binary and hexadecimal values are often prefixed (padded) with leading zeros. This practice is used to illustrate the number of digits that will be used to represent these values inside the computer. For our purposes here, we're assuming a data bus that's eight bits wide, so the values shown in Figure 1-9 have been padded to reflect this width.

Combining digits and column weights in hexadecimal

As with any place-value system, each column in a hexadecimal number has a weight associated with it; the weights are derived from the base. In this case we're dealing with a base of sixteen, so the column weights will be $16^0 = 1$, $16^1 = 16$, $16^2 = 256$, $16^3 = 4,096$, and so forth (Figure 1-10).

As we discussed earlier in this chapter, it's common practice to use subscripts to indicate whichever base is in use at the time. For example, $4F0A_{16} = 20,234_{10}$, which means $4F0A_{Hexadecimal} = 20,234_{Decimal}$. As we

Decimal	0	1	2	3	4	5	6	7	8	9	10	11	12	13	14	15
Hexadecimal	0	1	2	3	4	5	6	7	8	9	A	B	C	D	E	F

Figure 1-8. The sixteen hexadecimal digits.

Value	Decimal	Binary	Hexadecimal
zero	0	0000 0000	00
one	1	0000 0001	01
two	2	0000 0010	02
three	3	0000 0011	03
four	4	0000 0100	04
five	5	0000 0101	05
six	6	0000 0110	06
seven	7	0000 0111	07
eight	8	0000 1000	08
nine	9	0000 1001	09
ten	10	0000 1010	0A
eleven	11	0000 1011	0B
twelve	12	0000 1100	0C
thirteen	13	0000 1101	0D
fourteen	14	0000 1110	0E
fifteen	15	0000 1111	0F
sixteen	16	0001 0000	10
seventeen	17	0001 0001	11
eighteen	18	0001 0010	12
:	:	:	:

Figure 1-9. Counting in hexadecimal.

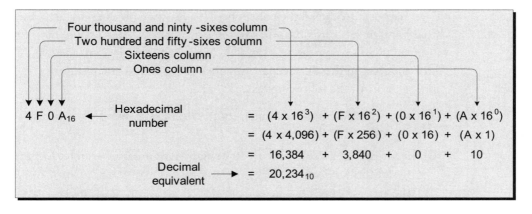

Four thousand and ninty -sixes column
Two hundred and fifty -sixes column
Sixteens column
Ones column

$4\ F\ 0\ A_{16}$ ← Hexadecimal number

$$= (4 \times 16^3) + (F \times 16^2) + (0 \times 16^1) + (A \times 16^0)$$
$$= (4 \times 4{,}096) + (F \times 256) + (0 \times 16) + (A \times 1)$$
$$= 16{,}384 + 3{,}840 + 0 + 10$$

Decimal equivalent → $= 20{,}234_{10}$

Figure 1-10. Combining digits and column weights in hexadecimal.

also noted, another common alternative is to prefix such numbers with a character to indicate the base. For example, $4F0A, where the $ indicates a hexadecimal value. (By now it may not surprise you to learn that, once again, there are a variety of such conventions.) As usual, any number without a subscript or a prefix character is generally assumed to represent a decimal value unless otherwise indicated.

Mapping between hexadecimal and binary

Mapping between hexadecimal and binary is extremely simple. For example, in order to convert the hexadecimal value $4F0A into binary, all we have to do is to replace each hexadecimal digit with its 4-bit binary equivalent. That is,

$$\$4 \rightarrow \%0100, \$F \rightarrow \%1111, \$0 \rightarrow \%0000, \text{ and } \$A \rightarrow \%1010$$

So $4F0A in hexadecimal equates to %0100111100001010 in binary.

Similarly, it's easy to convert a binary number such as %1100011010110010 into its hexadecimal equivalent. All we have to do is to split the binary value into 4-bit nybbles and to map each nybble onto its corresponding hexadecimal digit. That is,

$$\%1100 \rightarrow \$C, \%0110 \rightarrow \$6, \%1011 \rightarrow \$B, \text{ and } \%0010 \rightarrow \$2$$

So %1100011010110010 in binary equates to $C6B2 in hexadecimal.

Review

This chapter introduced the concepts of place-value number systems, the binary and hexadecimal number systems, using powers or exponents, and using wires to represent numbers. Now, just to make sure that we fully understand these concepts, let's answer the following questions and/or perform the following tasks:

1) What is the decimal equivalent of the Roman numeral MMMDCCLXXVIII?

2) In the case of a value such as 10^3, what terms are used to refer to the 10 and the 3, respectively?

3) What term is used to refer to a binary digit, and to what do the terms *nybble* and *byte* refer?

4) What are two different ways to indicate that the value 10100101 is a binary number?

5) What are two different ways to indicate that the value 10 is a hexadecimal number?

6) Convert the decimal value 125_{10} into an 8-bit unsigned binary equivalent.

7) Convert the binary value 10100101_2 into its hexadecimal equivalent.

8) Convert the hexadecimal value $E6 into its binary equivalent.

9) How many digits would there be in a quinary (base-5) place-value number system (list these digits); what would be the first five column weights in such a system; and how would one count in such a system?

10) Assuming positive numbers starting at 0, what range of decimal numbers could be represented using the different binary patterns that could be accommodated by a set of seven wires? How about nine wires?

Bonus Question   Cast your mind back to our discussions on counting with wires and Figure 1-7. There are two key problems associated with this form of binary representation. Can you spot what they are?

Answer: The first problem is that eight wires allow us to represent only 256 different combinations of 0s and 1s. For the purposes of these discussions, we decided to use these binary patterns to represent the decimal values 0 to 255, but this is obviously very restrictive. What happens if we wish to represent values greater than 255?

The second problem is that the scheme we demonstrated here allows us to represent only positive values. What happens if we wish to represent both positive and negative values?

Both of these issues are addressed in Chapter 4. However, it would be useful for you to start thinking about them now and see if you can come up with your own solutions. Later, when we reach Chapter 4, you can compare your ideas with the techniques that have been developed by computer scientists and see how well you did.

CHAPTER 2

COMPUTERS AND CALCULATORS

> *It is unworthy of excellent men to lose hours,*
> *like slaves, in the labors of calculation.*
>
> GOTTFRIED WILHELM LEIBNITZ (1646–1716)

> *Computers make it easier to do a lot of things,*
> *but most of the things they make it easier to*
> *do don't need to be done.*
>
> ANDY ROONEY (1919–)

000010 01101100 11100101010000101010110101001001001000110110000010100011

In this chapter we will learn about:

- The brain of a computer
- Memory devices
- Input and output ports
- The control, data, and address busses
- Using a computer to make a calculator
- Our DIY Calculator

Rampaging Around a Computer

In its broadest sense, a computer is a device that can accept information from the outside world, process that information using logical and/or mathematical operations, make decisions based on the results of this processing, and ultimately return the processed information to the outside world in its new form.

The main elements forming a computer system are its *central processing unit* (*CPU*)—the memory devices (ROM and RAM) that are used to store programs (sequences of instructions) and data—and the *input/output* (*I/O*) ports that are used to communicate with the outside world (Figure 2-1).

The "brain" of the computer is its CPU, which is where all of the number crunching and decision making is performed. Two of the CPU's input pins are driven by externally generated signals called *clock* and *reset*. The clock signal switches back and forth between two voltage levels millions of times a second (or billions of times a second in the case of today's computers). In much the same way that the measured beat of a drum can be used to keep a band marching in step, the clock signal is used to synchronize the internal and external actions of the CPU.

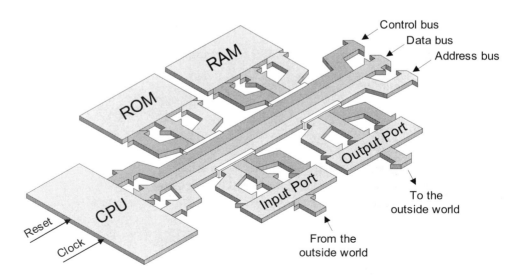

Figure 2-1. The main components forming a general-purpose digital computer.

Sometimes, things may end up "pear-shaped" (an English expression for something that has taken an unfortunate turn of events, such as a badly written program becoming locked up in an endless loop). In this case, the *reset* signal can be used to force the CPU into a known, well-behaved state. (This signal is automatically activated when power is first applied to the system, and the resulting *power-on reset* serves to initialize the CPU.)

> **Note** Prior to the availability of ROM devices, computer users had to manually install the initialization instructions each time the computer was powered on. This was usually accomplished by setting a bank of switches for each instruction and then pushing a button to input that instruction. This process, referred to as "booting up the computer," could be long and tedious.
>
> "Bootstraps" are small loops of leather sewn into the tops of high boots as an aid to pulling them on. Since the 18th century, the phrase "pulling oneself up by one's own bootstraps" has meant "to succeed by one's own efforts." Although "bootstrap" has been used in computing circles since the 1950s, the now-common shortened form "boot" only became popular with the "personal computer explosion" of the 1980s.

Memory Devices and I/O Ports

With regard to the other components forming a computer system, the ones in which we are interested here are the memory devices (each of which may contain thousands or millions of pieces of data) and the input and output ports. (In this context, the term "data" may be used to encompass the raw data upon which programs operate along with the sequences of instructions forming the programs themselves.)

As their name might suggest, the data contained in *read only memories* (*ROMs*) is hard-coded during their construction. The CPU can read (extract) data from ROM devices, but it cannot write (insert) new data into them. Thus, ROMs may be used to store such things as the low-level initialization and control routines that are required when the computer is first powered up.

By comparison, data can be read out of *random access memories* (*RAMs*) and new data can be written back into them (the act of reading data from a RAM does not affect the master copy of the data stored inside the device). When power is first applied to the system, the RAMs end up containing random values; this means that any meaningful data stored inside a RAM must be written into it by the CPU (or via some other mechanism) *after* the system has powered up. RAMs can be used to store programs, data, and intermediate results.

Last but not least, the computer uses its input and output ports to communicate with the outside world. (Figure 2-1 shows only individual input and output ports, but a computer can effectively have as many of each type as its designers wish to use.)

> **Note** In addition to being somewhat esoteric, early attempts to create computer memories were typically sequential in nature. By this we mean that it was only possible to retrieve data in the order in which it had been stored (consider a magnetic tape, for example).
>
> Thus, when it was invented, the term *random access memory (RAM)* was coined to emphasize the fact that this new type of memory allowed data to be directly read from or written to any location in the device.
>
> Some of us would prefer to use the more meaningful appellation *read–write memory (RWM)*, but there is little chance of this nomenclature becoming widely adopted.

The Control, Data, and Address Busses

As we previously noted, the term *bus* is used to refer to a group of signals that carry similar information and perform a common function. A computer actually makes use of three buses called the *control bus, address bus,* and *data bus.* The CPU uses its address bus to "point" to other components in the system, it uses the control bus to indicate whether it wishes to "talk" (output/write/transmit data) or "listen" (input/read/receive data), and it uses the data bus to convey information back and forth between itself and the other components. (To be honest, the *clock* and *reset* signals are also considered to be part of the control bus, but these signals are often treated separately as is illustrated in our diagrams.)

The data bus

For the purposes of these discussions, we'll assume that we're working with a simple computer that has an 8-bit data bus. Note that the illustration shown in Figure 2-1 is rather abstract and makes the busses look like solid entities. In reality, they are formed from groups of wires; for example, let's take a closer look at the data bus (Figure 2-2).

The address bus

Now try to visualize a series of thousands upon thousands of boxes standing side-by-side and stretching away into the dim and distant beyond. Each box is numbered sequentially, commencing with zero (Figure 2-3).

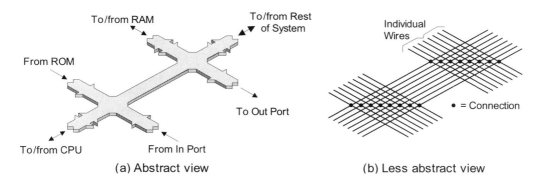

(a) Abstract view   (b) Less abstract view

Figure 2-2. The data bus.

Let's imagine that some of these boxes have pieces of transparent plastic sealing their ends, whereas the others are open to the elements. Let's further suppose that each of the sealed boxes contains a slip of paper with a number written on it. We can read the numbers (data) on the slips inside these boxes but we can't alter them in any way. This is similar to the way in which our ROM devices function.

In the case of an open box, we can write a number on a new slip of paper and insert it into that box. If an open box already contains a slip of paper, we might simply read the number on the slip but leave the slip where it is, we might copy the slip and place the copy in another box, or we might erase the number on the slip and write a new number in its place. This is similar to the way in which our RAM devices operate.

Strange as it may seem, this is the way that the CPU views its world—as a series of boxes that can be used to store instructions and data (although we prefer to use the term *memory locations* rather than "boxes"). Each location in the memory has a unique identification num-

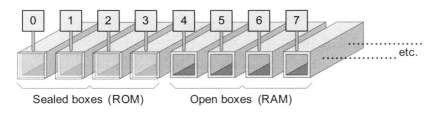

Sealed boxes (ROM)   Open boxes (RAM)

Figure 2-3. A series of boxes.

ber, referred to as its *address,* and the CPU uses its address bus to "point" to any memory location in which it is currently interested (Figure 2-4).

Each location in the memory is referred to as a *word,* and each word has the same width as the data bus. Thus, as our CPU's data bus is 8 bits wide, each word in the memory must also be 8 bits wide. These bits are numbered from 0 to 7, as shown in Figure 2-4 (the reason for this numbering scheme is revealed in Chapter 4). Each bit in a memory word can be used to store a 0 or a 1, and all of the bits forming a word are typically written to or read from simultaneously (the 0s and 1s shown in the memory word at address $000A in Figure 2-4 have no meaning beyond providing some example values).

The totality of memory locations that can be addressed by the computer are referred to as its *address space.* In the case of our CPU, we're going to assume that the address bus is 16 bits (two bytes) wide. This means that our address bus can carry 2^{16} unique combinations of 0s and 1s, which can therefore be used to point to 65,536 different memory locations numbered from 0 to 65,535. For a variety of reasons, however, it is not convenient to refer to the addresses of memory locations using a decimal notation. Thus, rather than referring to a location such as 48,259 using its decimal value (or worse, its binary equivalent of %1011110010000011), it is common practice to refer to addresses using a hexadecimal notation such as $BC83 (remember that we're using the % and $ characters to indicate binary and hexadecimal values, respectively).

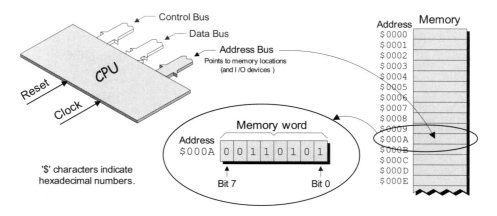

Figure 2-4. The address bus.

Actually, we should note that the computer architecture we're describing here reflects a very simple system that one might find in a low-end application like a pocket calculator. This type of architecture was also common in personal computers circa 1975. The busses and bus protocols used in modern computers like PCs are much more sophisticated; but, as was noted in Chapter 0, once you fully comprehend how a simple computer functions, you can easily extrapolate to more complex machines.

Careening Around a Calculator

Once we've conceived the idea of a general-purpose computer, the next step is to think of something to do with it. In fact, there are millions of tasks to which computers can be assigned, but the application we're interested in here is that of a simple calculator. So what does it take to coerce a computer to adopt the role of a calculator? Well, one thing we require is some form of user interface, which will allow us to present data to, and view results from, the computer (Figure 2-5).

The calculator's user interface primarily consists of buttons and some type of display. Each button has a unique binary code associated with it, and this code will be presented to the computer's input port whenever that button is pressed. Meanwhile, one or more of the com-

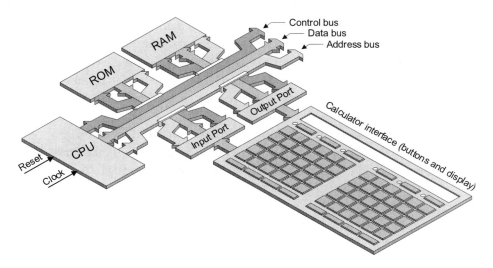

Figure 2-5. A calculator requires some form of user interface.

puter's output ports can be used to drive the display portion(s) of the interface.

So this interface is really a "dumb device," because all of the actual number crunching is performed by the general-purpose computer. However, in order for the computer to perform its cunning machinations, it requires a program, and creating this program requires *us* to decide on the ways in which we wish to represent and manipulate numbers. Ultimately, the purpose of this book is to introduce methods of representing numerical data inside a computer and techniques for manipulating this data so as to perform common arithmetic operations.

The DIY Calculator

As we mentioned in Chapter 0, one of the best ways to learn something and remember it afterward is by means of hands-on ("fumble and stumble") experience, which you are about to gain in huge dollops. As a key element in this "ordeal by fire," you are going to create a program to implement a simple calculator by performing a series of interactive laboratories featuring the internationally acclaimed DIY Calculator.

What is a DIY Calculator and where might one be found? As fate would have it, this little rapscallion, which was conceived and designed by the authors, comprises a calculator front panel (the user interface discussed in the previous section) connected to a simple general-purpose computer with a 16-bit address bus, an 8-bit data bus, some RAM and ROM, and a number of input and output ports. But the really cunning thing about the DIY Calculator is that, as opposed to being constructed from physical components, it has been implemented as a virtual machine (a simulator) that runs on your PC and is delivered on the CD-ROM accompanying this book.

If you haven't already done so, install your DIY Calculator now (see Appendix A). Once you've performed this installation, use the **Start** > **Programs** > **DIY Calculator** > **DIY Calculator** command (or double-click the DIY Calculator icon on your desktop) to invoke the little scamp (Figure 2-6).

In the middle of the main DIY Calculator window is the *calculator front panel* (sometimes we'll simply refer to this as the "calculator" or "front panel" for short). This is the graphical interface that can be used to feed data to, display results from, and control our calculator (Figure 2-7).

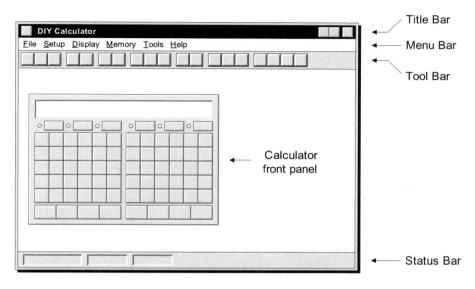

Figure 2-6. The main DIY Calculator window.

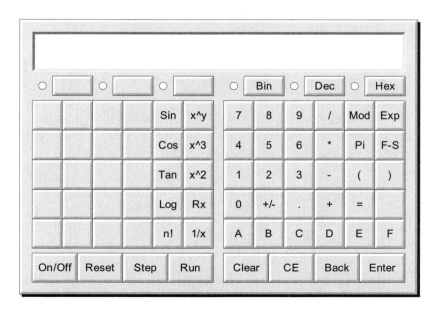

Figure 2-7. The calculator front panel.

As you can see, the calculator front panel is festooned with buttons with which you are soon to become intimately acquainted. Observe that some of these buttons have been left blank for you to assign your own operations. Also note that you can reconfigure the front panel to modify the legends on the buttons, change the colors of the legends, and so forth. These aspects of the front panel are discussed in Chapter 6.

At the bottom left-hand side of the front panel are a group of four special control buttons: **On/Off**, **Reset**, **Step**, and **Run**. Just to get a feel for what's going on, click the **On/Off** button to power-up the calculator (that is, the front panel and the computer, our simulator, "behind" the front panel). Note that the main display area fills up with "-" characters, indicating that it's in an uninitialized state. Similarly, the six little lights next to the **Bin**, **Dec**, and **Hex** (and related) buttons, located just beneath the main display area, all glow red to indicate that they too are in an uninitialized state.

Try clicking the calculator's 0 through 9 button, and note that nothing happens whatsoever. This is because, thus far, you haven't loaded a program into the calculator's memory.

Next, use the main DIY Calculator window's **Memory > Load RAM** pull-down menu to invoke a dialog offering a list of the programs that are currently available. By default, this dialog should be looking in the *C:\DIY Calculator\Work* folder; if not, set the context to this folder. Select (click on) the *test.ram* file you'll find in the *Work* folder and then click the **Load** button to load the contents of this file into the calculator's memory.

Now click the front panel's **Run** button to execute the program and note that the main display is cleared and the six little lights go out. Next, try clicking the 0 through 9 buttons again and observe what happens.

Note that these are the only buttons that do anything; the other function buttons have no effect (don't play with the **On/Off**, **Reset**, **Step**, or **Run** control buttons until instructed to do so). The reason for this is quite simple. After clearing the display, all this particular program does is to loop around waiting for you to click a button. If this button corresponds to 0 through 9, the program copies that value back to the display; otherwise, it ignores you and returns to looping around waiting for you to click another button.

Interactive Laboratories

When you've finished playing, click the **On/Off** button to power-down the calculator. Now you're really ready to rock and roll, so take a deep breath and perform the following interactive laboratories that you'll find toward the back of this book:

Lab 2a: Creating a Simple Program
Lab 2b Constant Labels and .EQU Statements
Lab 2c: Driving the Front Panel's Main Display
Lab 2d: Reading from the Front Panel's Keypad
Lab 2e: Writing to the Front Panel's Six LEDs
Lab 2f: Using the Memory Walker and Other Diagnostic Displays

Review

This chapter introduced the main elements of a computer system and described how a computer could be used in the role of a calculator.

1) What does the abbreviation CPU stand for and what is its role in a computer system?

2) What do the abbreviations ROM and RAM stand for?

3) Summarize the key features of ROM and RAM devices.

4) Name the three busses used by a computer and describe their various roles.

5) Summarize the actions of the *clock* and *reset* control signals.

6) What is the purpose of the *power-on reset* circuit?

7) What elements does the computer use to "talk" to the outside world?

8) To what do the terms *word* and *address space* refer?

9) What key services are provided by a calculator's front panel?

10) What are the names of the CPU's status flags?[1]

[1]You will need to have performed the interactive laboratories associated with this chapter in order to be able to answer this question.

CHAPTER 3

SUBROUTINES AND OTHER STUFF

Though this be madness, yet there be method in it.

WILLIAM SHAKESPEARE (1564–1616)
in *Hamlet* (1601)

In this chapter we will learn about:

- Logical instructions (AND, OR, and XOR)
- Shift instructions (SHL and SHR)
- Rotate instructions (ROLC and RORC)
- The program counter (PC)
- The index register (X)
- The stack and stack pointer (SP)
- Subroutines
- Recursion
- Self-modifying code

Logical Instructions (AND, OR, and XOR)

As we're sure you noticed, this chapter is entitled "Subroutines and Other Stuff." Well, by some strange quirk of fate, we're actually going to start with the "other stuff."

Primitive logic gates

In Chapter 1, we noted that computers are formed from large numbers of primitive logical elements called *logic gates*. We also noted that it's relatively easy for electronics engineers to create logic gates that can detect, process, and generate two distinct voltage levels, and that these two levels can be used to represent binary 0 and 1, respectively.

In addition to considering 0 and 1 as representing numerical information in the form of binary digits, we sometimes prefer to think of them in terms of the logical values *False* and *True*, respectively. As an example of this way of looking at things, consider three different types of logic gates called AND, OR, and XOR (Figure 3-1).

Each symbol has an associated truth table, which provides a useful way of representing the different possible input combinations and their corresponding output values. In the case of the AND gate, the output Y is only *True* (logic 1) if both inputs A *and* B are *True*; for any other combination of inputs, the output is *False* (logic 0). By comparison, in the case of the OR gate, the output Y is *True* if either input A *or* B is *True*. In fact, this form of OR should more properly be called an "Inclusive OR,"

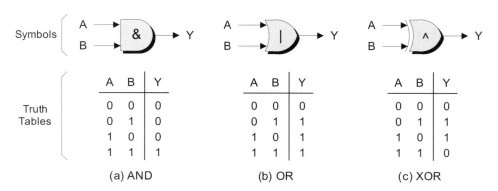

Figure 3-1. AND, OR, and XOR logic gates.

because the cases where the output is *True* include the one where both of the inputs are *True*. Contrast this with the XOR ("Exclusive OR"), where the *True* output cases exclude the case where both inputs are *True*.

Logical instructions

The point behind all of this is that, in addition to performing arithmetic operations such as incrementing (adding 1 to) the accumulator using an INCA[1] instruction, or adding two numbers together using an ADD[2] instruction, we often desire to perform logical operations on the contents of the accumulator. Thus, our virtual computer is equipped with the logical instructions AND, OR, and XOR, which cause a byte of data stored in the memory to be AND-ed, OR-ed, or XOR-ed with the current contents of the accumulator, respectively. The results from these operations are stored in the accumulator, thereby overwriting its original contents (the contents of the memory remain undisturbed).

In order to illustrate the way in which these instructions perform their magic, let's assume that the accumulator originally contains 00001111_2 and that we wish to either AND, OR, or XOR this value with a memory location containing 01010101_2 (Figure 3-2).

The key point to note about these logical instructions is that they operate in a bit-wise fashion (that is, on a bit-by-bit basis). In the case of the AND, for example, bit 0 in the memory is AND-ed with bit 0 in the accumulator and the result is stored in bit 0 in the accumulator, bit 1 in the memory is AND-ed with bit 1 in the accumulator and the result is stored in bit 1 in the accumulator, and so forth.

As we shall come to discover, these logical operators can be extremely useful for a wide variety of tasks. For example, suppose the accumulator contains a value of 11100011_2 and we wish to *set* bit 3 (the middle 0 in the group of three 0s) to a 1. We can achieve this by OR-ing the accumulator with 00001000_2, resulting in 11101011_2. (Remember that the eight bits forming the accumulator are numbered left to right from 0 to 7, as shown in Figure 3-2.)

Alternatively, consider the case where the accumulator contains a value of 11100011_2 and we wish to *clear* bit 6 (the middle 1 in the group of three 1s) to a 0. We can achieve this by AND-ing the accumulator with 10111111_2, resulting in 10100011_2.

[1]The INCA instruction was introduced in Lab 2c.
[2]The ADD instruction is presented in Chapter 4.

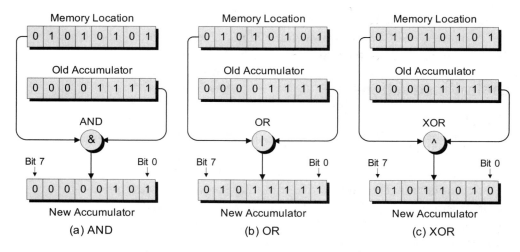

Figure 3-2. AND, OR, and XOR instructions.

Another useful logical operation is the NOT. Sad to relate, our virtual computer does not actually support a NOT instruction, although, if it did, this instruction would invert all of the bits in the accumulator (swapping the 0s for 1s, and vice versa). But turn that frown upside down into a smile, because we can achieve the same effect by XOR-ing the contents of the accumulator with a value comprising all 1s. For example, if the accumulator originally contains a value of 00110101_2, then XOR-ing it with 11111111_2 will return a result of 11001010_2.

Shift Instructions (SHL and SHR)

Some computers offer two different "flavors" of shift instructions known as *arithmetic shifts* and *logical shifts*. In such a case, each of these flavors would have "shift left" and "shift right" variants, which means we would end up with four instructions as follows:

Mnemonic	Function	Note
ASHL	Arithmetic shift left	Called SHL in our virtual computer
ASHR	Arithmetic shift right	Called SHR in our virtual computer
LSHL	Logical shift left	Not supported by our virtual computer
LSHR	Logical shift right	Not supported by our virtual computer

Arithmetic shifts

The easiest approach to visualize the way in which these work, and to compare the commonalities and differences between them, is by means of diagrams. Let's begin by considering the arithmetic shifts that our virtual computer *does* support (Figure 3-3).

Note that the question mark "?" shown as the original value in the C (carry) status flag indicates that, for the purposes of this example, we don't know (or care) whether this flag starts off containing a 0 or a 1. (The status flags were introduced in Lab 2c.)

In the case of the arithmetic shift left (for which our mnemonic is SHL), a logic 0 is shifted into the least-significant (right-hand) bit, all of the other bits shift one place to the left, and when the original value in the most-significant (left-hand) bit "falls off the end" it is copied into the carry flag. By comparison, in the case of the arithmetic shift right (for which our mnemonic is SHR), the most-significant (left-hand) bit is copied back into itself. At the same time, the most-significant bit, along with all of the other bits, shifts one place to the right, and when the original value in the least-significant (right-hand) bit "falls off the end" it is copied into the carry flag.

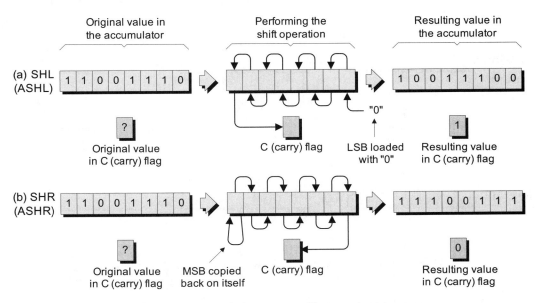

Figure 3-3. Arithmetic shifts supported by our virtual computer.

Logical shifts

Now let's take a look at the logical shifts that our virtual computer *does not* support (Figure 3-4).

As we can see, the logical shift left instruction functions in an identical manner to its arithmetic counterpart. In fact, even in a system that does purport to offer both flavors, this would, in reality, be an illusion. In such a case, the assembler would accept both ASHL and LSHL mnemonics, but each of these would map onto the same opcode as seen by the CPU. So why would an assembler offer both mnemonics in the first place? For aesthetic reasons mostly (in order to provide one-for-one matches with the different flavors of shift right), but also, to a lesser extent, because it helps to signify what the programmer had in mind when creating the code.

The real difference is between the arithmetic and logical versions of a shift right. Remember that the arithmetic version copies the most-significant bit in the accumulator back on itself. By comparison, in the case of a logical shift right instruction, a logic 0 is shifted into the most-significant bit.

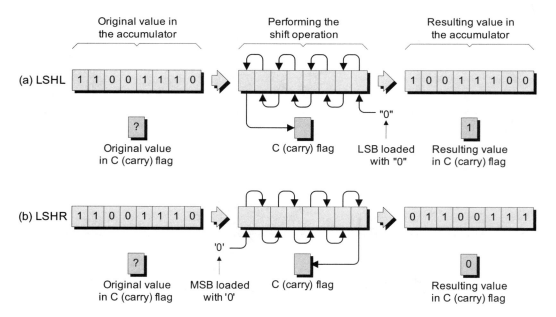

Figure 3-4. Logical shifts *not* supported by our virtual computer.

Coercing the results from arithmetic shift rights into their logical counterparts

The reasons why we are interested in arithmetic versus logical shift right instructions are presented in more detail in Chapter 4. For our purposes here, we need only reiterate that our virtual computer supports only arithmetic shifts, which we decided to call SHL ("shift left") and SHR ("shift right") for simplicity.

But will the lack of logical shift instructions be detrimental to our ability to write useful programs and our general standing in the community? Fear not, because as we already noted, arithmetic and logical shift lefts do the same thing anyway. Meanwhile, there's a simple trick we can play to coerce the results from arithmetic shift rights into their logical shift right equivalents. Once we've performed an arithmetic shift right, we can force it into its logical counterpart by means of an AND instruction as discussed earlier in this chapter; for example:

```
:
LDA     %11001110     # Load accumulator some value
SHR                   # Perform an arithmetic shift right
AND     %01111111     # Coerce result into logical version
:
```

First, we load the accumulator with the same "seed value" (11001110_2) we used in the examples shown in Figures 3-3(b) and 3-4(b). Next, we use our SHR instruction to perform an arithmetic shift right, which returns the same result we saw in Figure 3-3(b) (11100111_2). Finally, we use our AND instruction to mask out the most-significant bit and clear it to a 0, which leaves us with the same logical shift right result we saw in Figure 3-4(b) (01100111_2). (A good example of an actually usage of this technique is to be found in Lab 3a.)

Rotate Instructions (ROLC and RORC)

Our virtual calculator supports two rotate instructions: ROLC ("rotate left through the carry flag") and RORC ("rotate right through the carry flag"), as illustrated in Figure 3-5.

These are very similar to the logical shift instructions we discussed in the previous section. The big difference is that, instead of shifting a

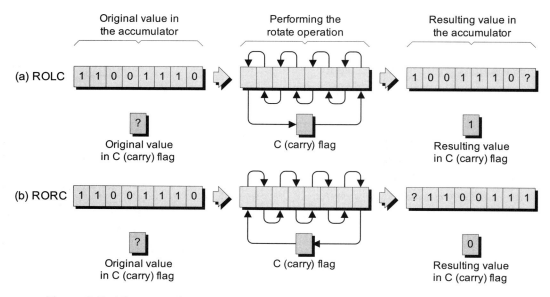

Figure 3-5. The rotate through carry instructions supported by our virtual computer.

logic "0" into the "input end" of the accumulator, we actually feed in the original value from the carry flag. Once again, the question mark "?" shown as the original value in the carry flag indicates that, for the purposes of this example, we don't know whether this flag starts off containing a 0 or a 1. The point is that, whatever this value is, it ends up in the least-significant bit of the accumulator following the ROLC instruction, or the most-significant bit following the RORC instruction.

Rotates that bypass the carry bit

Some computers also support ROL ("rotate left") and ROR ("rotate right") instructions, which we might class as "rotate bypassing carry." In this case, whichever bit is conceptually "falling off the end" of the accumulator is fed directly back into the other end of accumulator (Figure 3-6).

Coercing rotates through the carry bit into their bypass counterparts

All four rotate instructions (ROLC, RORC, ROL, and ROR) are useful depending on the occasion, so why did we decide that our virtual computer

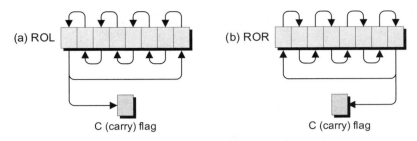

Figure 3-6. The rotate bypassing carry instructions *not* supported by our virtual computer.

would support only the rotate through carry versions? The answer is that we wanted to keep our computer's instruction set as simple and minimalist as possible; also, it's easier to convert the results from a rotate through carry instruction into its rotate bypassing carry counterpart than to do things the other way around. For example, consider the following technique for coercing the result from a ROLC into the result from a ROL:

```
           :
           LDA    %11001110    # Load accumulator with some value
           ROLC                # Perform a rotate-left-through-carry
           JC     [FORCE1]     # If carry = 1, jump to label FORCE1
FORCE0:    AND    %11111110    # ... otherwise clear LS bit to 0
           JMP    [CONTINUE]   # Unconditional jump to label CONTINUE
FORCE1:    OR     %00000001    # Set LS bit to 1
CONTINUE:  :         :         # Continue with rest of program
```

The way this works is that, following the ROLC instruction, we know that whatever value is in the carry flag is the value we actually want to see in the least-significant bit (LSB) of the accumulator. If the value in the carry flag is 0, the JC instruction will fail, and the AND %11111110 instruction will be used to clear the LSB to 0. Alternatively, if the value in the carry flag is 1, the JC instruction will pass, and the OR %00000001 instruction will be used to set the LSB to 1.

Question Suppose that we had opted to implement the ROL and ROR instructions in our virtual computer and not to implement the ROLC and RORC versions. In this case, how would you coerce the result from a ROL into the result from a ROLC (or the result from a ROR into the result from a RORC)?

Note   This would be a really good time to perform Lab 3a, which provides some "hands-on" use of the logical and shift instructions.

The Program Counter (PC)

The CPU contains a number of registers that it uses to point to instructions and data in the memory. First and foremost is the *program counter* (*PC*), which keeps track of where the CPU is in a program.

The program counter is a 16-bit register whose bits are numbered from 0 to 15, but it may help us to visualize this register as being formed from a *most-significant byte* (*MSB*) and a *least-significant byte* (*LSB*)[3] (Figure 3-7).

Most of the time, the program counter is used to drive the CPU's address bus; that is, the contents of the program counter are presented to the outside world via the address bus. As the program counter and address bus are both 16 bits (two bytes) wide, they can be used to point to $2^{16} = 65,536$ unique memory locations numbered from $0000 to $FFFF (or 0 to 65,535 in decimal).

It's interesting to note that there really isn't any way for us to control the program counter per se. When the CPU is first powered up, the program counter potentially initializes with an unknown value. What saves it from this fate is the system's power-on reset circuit, which applies a pulse to the *reset* input driving the CPU. This initializes a lot of things, including loading the program counter with some known good value; for example, $0000, which is the first instruction in the virtual monitor program stored in our virtual computer's ROM. After this point, the program counter follows the dictates of the program itself.

Note   Now would be the perfect time to perform Lab 3b, in which the actions of the program counter are described in more detail.

[3]Yes, we know that the abbreviations for *most-significant bit* (*MSB*) and most-significant byte (*MSB*) are the same, and similarly for *least-significant bit* (*LSB*) and *least-significant byte* (*LSB*). This is how engineers get their fun!

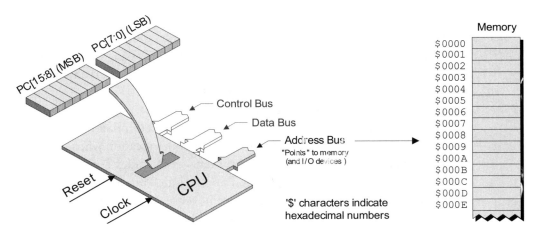

Figure 3-7. The 16-bit program counter drives the CPU's address bus.

The Index Register (X)

Another CPU register that often proves to be very useful is the *index register (X)* (Figure 3-8). Like the program counter, the index register is also a 16-bit entity whose bits are numbered from 0 to 15 but, once again, it may help us to visualize this register as being formed from a most-significant byte (MSB) and a least-significant byte (LSB).

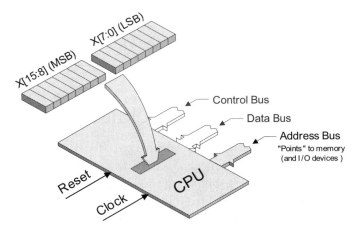

Figure 3-8. The 16-bit index register can be used to modify the value on the address bus.

Of particular interest is the fact that the index register can be used to modify the value being driven out on the address bus. For example, consider an LDA instruction using the absolute addressing mode, such as LDA [$5000]. This loads the accumulator with the byte of data stored in the memory location whose address is $5000.

By comparison, consider an LDA using the indexed addressing mode, such as LDA [$5000,X]. In this case, the accumulator will be loaded with the byte of data stored in the memory location whose address is calculated by adding $5000 to the contents of the index register (X).

When the CPU is first powered up, the index register initializes with an unknown value. Thus, if we want to use this register, we first have to load it with something useful. Purely for the sake of an example, let's assume that we want to load it with a value of $0006 (that's just 6 in decimal). In this case, we'd use a BLDX ("big load index register") instruction: BLDX $0006. The reason this is referred to as a "big load" is that the index register is 16 bits wide, as compared to the accumulator, which is only 8 bits wide.

Once we've loaded the index register, we can modify its contents by means of the INCX ("increment index register") and DECX ("decrement index register") instructions, which add or subtract 1 from the contents of the register, respectively.

If we wish, we can also use a BSTX ("big store index register") instruction, which copies the current contents of the index register to a specified location (2 bytes long) in the main memory.

Note   This would be a jolly good time to perform Lab 3c, in which different ways of using the index register are presented and discussed.

The Stack and Stack Pointer (SP)

Most of us have been to a cafeteria in which a pile of serving plates is stacked on top of a spring-like mechanism. Now let's suppose that we are in a warehouse and that there is something similar in the middle of the floor, involving a flat metal plate mounted on top of a big spring (Figure 3-9).

Initially, the mechanism is empty [Figure 3-9(a)]. Now suppose someone hands you a box with a big number "1" painted on its side. After standing around for a few moments looking rather foolish, you de-

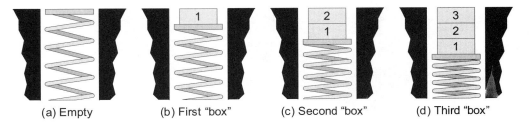

| (a) Empty | (b) First "box" | (c) Second "box" | (d) Third "box" |

Figure 3-9. A spring-like mechanism in the middle of the floor.

cide to place the box on the plate [Figure 3-9(b)]. A little later, someone else thrusts another package into your hands; this one has the number "2" painted on its side and you decide to put it on top of the first box [Figure 3-9(c)]. Still later, you are presented with a third box carrying the number "3" and you set this on top of the pile [Figure 3-9(d)].

Time passes. You hang around twiddling your thumbs until someone new rushes in and requests that you give them a package. Which one are you going to select? The obvious choice is the parcel on the top of the pile, so this is the one you retrieve and hand over. In computing terms, this form of storage and retrieval would be classed as a *last in, first out* (*LIFO*) process.

Introducing the stack and stack pointer

The point of these ramblings is that a CPU can be persuaded to make use of a concept known as the *stack,* which, if we really stretch our imaginations, we can visualize as working something like the spring-like mechanism discussed above. In order to point to the stack, the CPU employs yet another 16-bit register called the *stack pointer* (*SP*) (Figure 3-10).

When our CPU is first powered up, the stack pointer initializes with random 0s and 1s, which means that we have to load it with some value if we wish to use it. For example, let's assume that we want to load it with a value of $EFFF (we'll return to this point in a moment). In this case, we'd use a BLDSP ("big load stack pointer") instruction of the form BLDSP $EFFF.[4] (As usual, the reason this is referred to as a "big load" is that the stack pointer is 16 bits wide.)

[4]There is also a corresponding BSTSP ("big store stack pointer") instruction, which copies the current contents of the stack pointer to a specified location (2 bytes long) in the main memory.

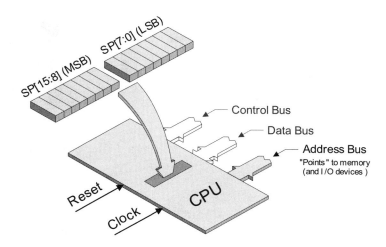

Figure 3-10. The 16-bit stack pointer is used to point to the stack.

Pushing the accumulator onto the stack

Let's assume that we have indeed just loaded the stack pointer with a value of $EFFF. We would say that the stack pointer is currently pointing to the top of the stack, which, at this particular moment in time, happens to be at memory location $EFFF [Figure 3-11(a)].

Note that the $XX values shown in this figure indicate that these memory locations are uninitialized and currently contain random, unknown values.

Now suppose that the accumulator contains a value (say, $01) that we wish to save for some future purpose. In this case, we could store it in some memory location using the absolute addressing mode; for example, STA [$5000]. As we know, this will copy the accumulator into memory location $5000. Alternatively, we could store the accumulator using the indexed addressing mode as discussed in the previous section, for example, STA [$5000,X], which will cause a copy of the accumulator to be stored in the memory location whose address is calculated by adding $5000 to the contents of the index register (X).

But now we have a third possibility, which is to use a PUSHA ("push accumulator") instruction. This causes a copy of the accumulator to be placed on the top of the stack at the location being referenced by the stack pointer [Figure 3-11(b)], and then the stack pointer is auto-

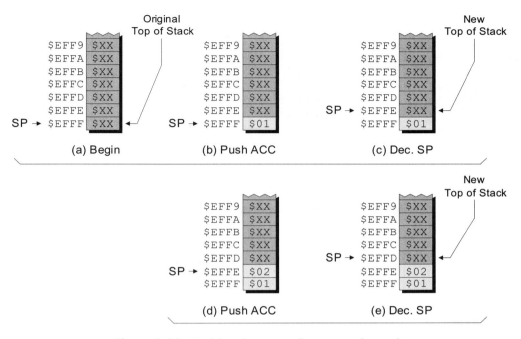

Figure 3-11. Pushing the accumulator onto the stack.

matically decremented to point to the next free location, the *new* top of the stack [Figure 3-11(c)].

Similarly, if we now loaded the accumulator with a new value (say $02) and performed another PUSHA instruction, this copy of the accumulator would be pushed onto the top of the stack [Figure 11(d)], and the stack pointer would then be automatically decremented to point to the next free location [Figure 3-11(e)]. Thus, the reason we call this the "stack" is that the bytes of data are conceptually stacked on top of one another.

Popping the accumulator back off the stack

Purely for the purposes of these discussions, let's assume that we originally loaded our accumulator with a value of $01 as discussed above; then we cycled five times around a loop, pushing the accumulator onto the stack and then incrementing it each time, which would leave things as shown in Figure 3-12(a).

Now suppose that we execute a POPA ("pop accumulator") instruction. Remember that the stack pointer is currently pointing at the first

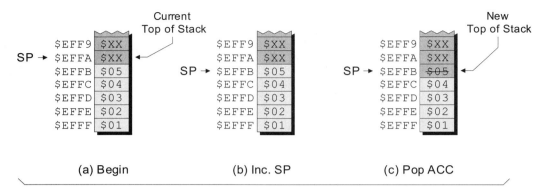

Figure 3-12. Popping the accumulator off the stack.

free location on the top of the stack. Thus, the CPU first increments the stack pointer to point to the last piece of data we pushed onto the stack [Figure 3-12(b)], and then it copies ("pops") that data off the stack into the accumulator [Figure 3-12(c)].

Note that the act of popping data off the stack doesn't actually change the contents of that memory location ($EFFB in this example). However, as far as we are concerned, we cannot re-access that the data currently stored in that location by means of standard stack pointer-based operations. For example, if we were to execute another POPA, the stack pointer would move further away from location $EFFB. Alternatively, if we were to execute another PUSHA, this would overwrite the contents of location $EFFB.

Pushing and popping the status register

In addition to the PUSHA ("push accumulator") and POPA ("pop accumulator") instructions, we also have their PUSHSR ("push status register") and POPSR ("pop status register") counterparts. This opens the door to a raft of possibilities. For example, we could use a PUSHSR followed by a POPA to first push a copy of the contents of the status register onto the stack and then pop a copy into the accumulator. We could then use the logical AND, OR, and XOR instructions to manipulate individual bits [for example, setting or clearing the Z (zero) flag]. Finally, we could use a PUSHA followed by a POPSR to transfer this new value back into the status register.

Why use the stack at all?

There are several reasons for using the stack (not the least being the use of subroutines, which are introduced in the next section). For example, if we are writing a program in which we know that we are never going to need to store more than ten bytes of data, then we might decide to simply use something like a `.BYTE *10` directive to reserve these memory locations for future use.

But what happens if we are performing some task such as reading (and storing) codes from buttons pushed on a keypad and we don't know exactly how many values we are going to have to deal with. One solution would be to reserve the largest amount of memory possible, but we might need that memory for other tasks later on. The alternative would be to use the stack, which grows to whatever size is required and then shrinks again once its data is no longer required.

Another point that is well worth considering is that `PUSHA` and `POPA` instructions take significantly fewer CPU clock cycles than do their `STA` and `LDA` counterparts, which means that using the stack can help speed up your programs (at least, this would be the case when working with a real, physical computer).

Why does the stack pointer "appear" to work the wrong way round?

One point you may have speculated about is "Why does a push instruction cause the stack pointer to be *decremented* while a pop instruction causes it to be *incremented*?" After all, it would seem to be far more intuitive to make it work the other way around. Well, the rationale behind this is really rather cunning. Let's consider our virtual computer's memory map (Figure 3-13).

The program counter (PC) starts off pointing to the first instruction in our program, which is always located in the first location in the RAM at address $4000. As we add more instructions into our program, we can visualize it as "growing" toward the higher memory locations.

As we know, we can initialize the stack pointer (SP) with any value. The reason we typically set it to the highest possible location in the RAM (address $EFFF in the case of our virtual computer) and the reason the stack "grows" in the opposite direction to the program is to save us a lot of pain.

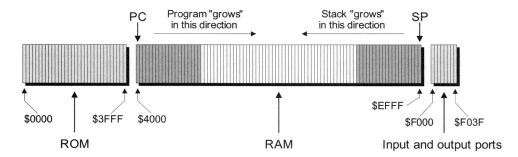

Figure 3-13. Our virtual computer's memory map.

The point is, when we start running a program, we often don't know exactly how much data we are going to want to push onto the stack. However, if we commence the program at the lowest possible RAM address and we initialize the stack pointer to the highest possible address, then we can use all of the available memory to its maximum advantage.

Of course, if we push too much data onto the stack, at some point the top of the stack will "collide" with the top of the program and start to overwrite it. This condition, which is known as a *stack overflow,* is generally not considered to be a good thing to happen. Similarly, if we attempt to "pop" more data off the stack than we have "pushed" onto it, the result will be a *stack underflow* (once again, we try not to let this occur).

> **Note**   Now would be an appropriate time to perform Lab 3d, in which the stack and stack pointer are used to implement a new technique for displaying button values in reverse order.

Subroutines

As we discovered in the previous section and in Lab 3d, the stack and stack pointer, when used in conjunction with the PUSHA, POPA, PUSHSR, and POPSR instructions, are useful in their own right. But they really come into their own when combined with the concept of subroutines.

A simple example

Before we start, it's important to note that the stack pointer must be initialized in order for our subroutines to work (the reason for this will become clear as we progress). As a very simple example, let's assume that we have a program in which we have already used a BLDSP ("big load stack pointer") instruction to initialize the stack pointer to point to memory location $EFFF. Let's further assume that somewhere in the body of our program we wish to clear the calculator's main display, which we can do by loading the accumulator with our special clear code and writing it to the display (these lines of code are highlighted in gray).

```
BEFORE:    NOP                     # A "no operation" instruction
           LDA      CLRCODE        # Load accumulator with clear code
           STA      [MAINDISP]     # Write clear code to main display
AFTER:     NOP                     # Another "no operation" instruction
           :
```

Note that the NOP ("no operation") instruction was introduced in Lab 3d. The two NOP instructions shown above are used only to provide points of reference for our future discussions.

Now let's suppose we wish to perform this "clearing the display" task multiple times throughout the course of our program. We could, of course, replicate the LDA and STA instructions wherever we wished to perform this chore. Alternatively, we could create a subroutine called CLEAR (or any other label we choose to use), and then call this routine from the body of the program as required (the call to the subroutine and the lines of code comprising the routine are highlighted in gray).

```
BEFORE:    NOP                     # A "no operation" instruction
           JSR      [CLEAR]        # JSR = "Jump to subroutine"
AFTER:     NOP                     # Another "no operation" instruction
           :
           :                       # Body of the program
           :
########## Start of "clear display" subroutine
CLEAR:     LDA      CLRCODE        # Load accumulator with clear code
           STA      [MAINDISP]     # Write clear code to main display
RETURN:    RTS                     # RTS = "return from subroutine"
########## End of "clear display" subroutine
```

The way in which this works is actually quite simple. When the CPU "sees" the JSR ("jump to subroutine") instruction, it automatically pushes the address of the following instruction (the NOP associated with the AFTER label in this example) onto the top of the stack (this is known as the "return address"); then it loads the program counter with the address of the first instruction in the subroutine (the LDA associated with the CLEAR label in this example).

The CPU then continues to execute the subroutine until it reaches a RTS ("return from subroutine") instruction, at which point it automatically pops the return address off the top of the stack and loads this value back into the program counter.

A more detailed breakdown

In order to truly wrap our brains around this, it would be useful to know where the various portions of our program end up in the calculator's memory. Purely for the sake of discussion, let's assume that the NOP associated with the BEFORE label occurs at memory location $5000, the first instruction in our subroutine (the LDA associated with the CLEAR label) resides in memory location $6000, and, as was previously discussed, the stack pointer has been initialized to point to memory location $EFFF (Figure 3-14).

Now, keeping this illustration in mind, let's consider how the subroutine is executed in more detail. Let's assume that we've just executed the NOP associated with the label BEFORE, which means that the program

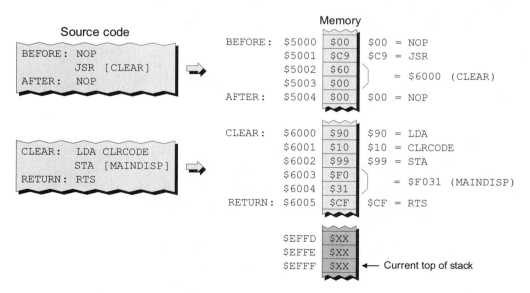

Figure 3-14. Visualizing the program in the computer's memory.

counter (PC) is currently pointing to the JSR ("jump to subroutine") instruction at address $5001 [Figure 3-15(a)].

When the CPU executes the JSR instruction, it first pushes the return address (the address of the NOP at label AFTER, which is $5004) onto the top of the stack, and then it loads the target subroutine address into the program counter. This leaves the stack pointer (SP) pointing to the first free location on top of the stack at address $EFFD and the program counter pointing at the first instruction in the subroutine, the LDA associated with label CLEAR at address $6000 [Figure 3-15(b)].

Now let's assume that the CPU has executed the instructions forming the subroutine and that the program counter is currently pointing to the RTS ("return from subroutine") instruction at address $6005 [Figure 3-15(c)].

When the CPU executes the RTS instruction, it pops the return address (the address of the NOP at label AFTER, which is $5004) off the top of the stack and loads this address into the program counter. This leaves the stack pointer pointing at the old top of the stack at address $EFFF and the program counter pointing at the instruction following the JSR [Figure 3-15(d)].

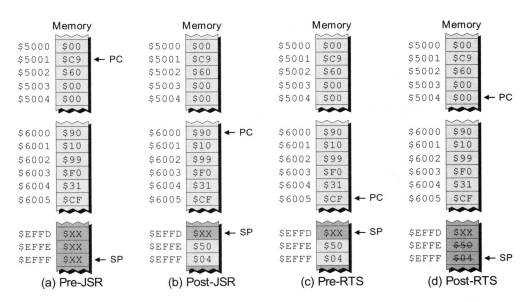

Figure 3-15. The actions of the program counter (PC) and stack pointer (SP).

Multiple calls to the same subroutine

There are a few additional points relating to subroutines that are worth noting before we actually start to play with them. First of all, a subroutine may be called multiple times from the main body of the program, for example:

```
######    This is the main body of the program
   :
"Inst"                # Some instruction or other
 JSR     [CLEAR]      # Call CLEAR subroutine
"Inst"                # Some instruction or other
   :
   :                  # More instructions
   :
"Inst"                # Some instruction or other
 JSR     [CLEAR]      # Call CLEAR subroutine
"Inst"                # Some instruction or other
   :
```

Of course, our "clear display" subroutine, comprising just three instructions (including the RTS), is a trivial example that is hardly worth making into a subroutine in the first place. However, let's suppose that this subroutine actually contained, say, 100 lines of code. In this case, the ability to call the routine multiple times from different places in the main body of the program, as opposed to recreating this code over and over again, is going to save us a lot of time (writing the code) and memory (storing it).

Multiple return points from within a subroutine

A subroutine may contain multiple RTS ("return from subroutine") instructions. For example, let's assume we have a subroutine called TEST that compares the contents of the accumulator to some value that has been assigned to a constant label called TSTVAL ("test value"). If the value in the accumulator is less than the value associated with TSTVAL, we will want to write the letter "S" ("smaller") to the main display; if the two values are equal, we will want to write the letter "E" to the main display; and if the value in the accumulator is greater than the value associated with TSTVAL, we will want to write the letter "G" to the main display:

```
##########  Start of "test" subroutine
TEST:       CMPA    TSTVAL      # Compare ACC with TSTVAL
            JC      [GREATER]   # The value in ACC is greater
            JZ      [EQUAL]     # The values are equal
SMALLER:    LDA     $53         # ASCII code for 'S'
            STA     [MAINDISP]  # Write code to main display
RETURN1:    RTS                 # Return from subroutine
EQUAL:      LDA     $45         # ASCII code for 'E'
            STA     [MAINDISP]  # Write code to main display
RETURN2:    RTS                 # Return from subroutine
GREATER:    LDA     $47         # ASCII code for 'G'
            STA     [MAINDISP]  # Write code to main display
RETURN3:    RTS                 # Return from subroutine
##########  End of "test" subroutine
```

Observe the use of three different RTS instructions. Each of these has exactly the same effect, which is to pop the return address off the top of the stack, load it into the program counter, and return to the point in the program that called this subroutine in the first place.

Pushing the ACC *before* calling a subroutine

One of the things of which we have to be aware with regard to the CLEAR subroutine we created earlier is that it modifies the contents of the accumulator (in reality, it's hard to think of a subroutine that wouldn't).

In some cases, this may be a problem. For example, let's assume that we wish to write a question mark character ("?") to the main display, but first we wish to clear the display. Suppose that a program to perform this task was created as follows:

```
            :
BEFORE:     LDA     $3F         # Load ACC with ASCII code for "?"
            JSR     [CLEAR]     # JSR = "Jump to subroutine"
AFTER:      STA     [MAINDISP]  # Copy ACC to display
            :
            :                   # Body of the program
            :
##########  Start of "clear display" subroutine
CLEAR:      LDA     CLRCODE     # Load accumulator with clear code
            STA     [MAINDISP]  # Write clear code to main display
RETURN:     RTS                 # RTS = "return from subroutine"
##########  End of "clear display" subroutine
```

Observe the three lines highlighted in gray. In this case, we have a problem, because immediately after we've loaded the accumulator with $3F (the ASCII code for a question mark) we call the CLEAR subroutine, which overwrites the contents of the accumulator with the code used to clear the display.

This is obviously a contrived example, and we could easily resolve this particular situation by relocating the LDA $3F instruction such that it appeared after the JSR. However, the point is that we often find ourselves in a similar situation, which is that at some stage in the future we are going to need the current contents in the accumulator, but we are poised to call a subroutine that will overwrite these contents. So let's assume that we have to maintain the relative locations of the LDA, JSR, and STA instructions in the main body of the program.

One solution is to push the contents of the accumulator onto the stack *before* we call the subroutine, and to subsequently retrieve them by popping the accumulator off the stack *after* we return from the subroutine as follows (note the lines highlighted in gray):

```
                :
            LDA     $3F         # Load ACC with ASCII code for "?"
BEFORE:     PUSHA               # Push the ACC onto the stack
            JSR     [CLEAR]     # JSR = "Jump to subroutine"
AFTER:      POPA                # Pop the ACC off the stack
            STA     [MAINDISP]  # Copy ACC to display
                :
                :               # Body of the program
                :
##########  Start of "clear display" subroutine
CLEAR:      LDA     CLRCODE     # Load accumulator with clear code
            STA     [MAINDISP]  # Write clear code to main display
RETURN:     RTS                 # RTS = "return from subroutine"
##########  End of "clear display" subroutine
```

Pushing the ACC *after* calling a subroutine

One problem with our previous solution is that we will have to push and pop the accumulator every time we call this particular subroutine. As an alternative, we can remove the push and pop instructions from the main body of the program and relocate them into the subroutine itself as follows (note the lines highlighted in gray):

```
          :
BEFORE:   LDA      $3F          # Load ACC with ASCII code for "?"
          JSR      [CLEAR]      # JSR = "Jump to subroutine"
AFTER:    STA      [MAINDISP]   # Copy ACC to display
          :
          :                     # Body of the program
          :
########## Start of "clear display" subroutine
CLEAR:    PUSHA                 # Push the ACC onto the stack
          LDA      CLRCODE      # Load accumulator with clear code
          STA      [MAINDISP]   # Write clear code to main display
          POPA                  # Pop the ACC off the stack
RETURN:   RTS                   # RTS = "return from subroutine"
########## End of "clear display" subroutine
```

Passing parameters into a subroutine

Let's assume that we wish to create a subroutine called GETBIG that will compare two 8-bit numbers and return the one with the largest value. Let's further assume that we want to be able to call this subroutine from different points in the main program and that the source of the two numbers to be compared may vary from one call to another.

One way to address this is—you guessed it—based on the stack. First consider some preparations we might make in the body of the program prior to calling our subroutine:

```
          :
          LDA      $32          # Load ACC with a number
FIRST:    PUSHA                 # Push it onto the stack
          LDA      $45          # Load ACC with another number
SECOND:   PUSHA                 # Push it onto the stack
          JSR      [GETBIG]     # Jump to the subroutine
RESULT:   POPA                  # Retrieve the result
          :
```

As usual, this is a very contrived case because we already know which is the bigger of the two numbers without having to call the subroutine. In the real world, however, the two numbers could have come from a variety of sources; for example, we could have read two buttons from the calculator's keypad and wish to compare their codes.

One key point here is that the two numbers are pushed onto the stack before we call the subroutine, so we can regard them as parameters being passed into the subroutine. Another key point is that our subroutine is going to leave the result (the biggest number) on the top of the stack such that we can use a POPA to pop it off the stack into the accumulator as soon as we return from the subroutine.

In order to understand how our subroutine performs its magic, we first need to have a clear picture as to the state of the stack when we enter the subroutine. Let's assume that the stack pointer was initialized with $EFFF [Figure 3-16(a)], that the POPA instruction associated with the label RESULT happens to occur at address $5000, and that the program snippet shown above reflects the first time the stack has been used thus far.

In this case, as soon as the CPU has executed the PUSHA instruction associated with the FIRST label, the stack will appear as shown in Figure 3-16(b). Similarly, once the CPU has executed the PUSHA associated with the SECOND label, the stack will appear as shown in Figure 3-16(c). And after the CPU has executed the JSR instruction, which causes a copy of the return address ($5000) associated with the POPA instruction at label RESULT to be pushed onto the stack, the situation will appear as shown in Figure 3-16(d).

Last but not least, once the subroutine has finished and used an RTS ("return from subroutine") to return to the main body of the program (just prior to actually executing the POPA at label RESULT), the stack will appear as shown in Figure 3-16(e).

Now let's focus on the state of the stack when we actually enter the subroutine [Figure 3-16(d)]. It's immediately obvious that we have a slight problem in that the numbers we wish to compare are located at the

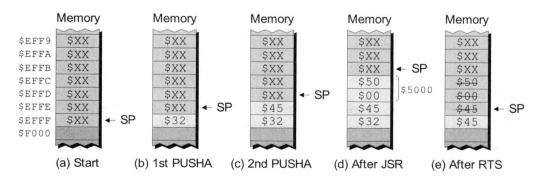

Figure 3-16. What's happening on the stack.

bottom of the stack underneath the return address of $5000. Let's bear this in mind as we consider how we might go about implementing our GETBIG subroutine itself:

```
########## Start of "find biggest number" subroutine
## Pop and save the return address
GETBIG:    POPA                  # Pop MS byte of return address
           STA     [GB_RAD]      # Store it away
GB_GLSA:   POPA                  # Pop LS byte of return address
           STA     [GB_RAD+1]    # Store it away
## Pop and compare the two numbers
GB_GET2:   POPA                  # Pop the 2nd number off the stack
           STA     [GB_TEMP8]    # Store it away
GB_GET1:   POPA                  # Pop the 1st number off the stack
           CMPA    [GB_TEMP8]    # Compare it to the 2nd number
GB_WHICH:  JC      [GB_TIDY]     # The 1st num (in the ACC)is bigger
           LDA     [GB_TEMP8]    # The 2nd num (in temp) is bigger
## Push the larger number and return address back onto the stack
GB_TIDY:   PUSHA                 # Push bigger number onto stack
           LDA     [GB_RAD+1]    # Retrieve LS byte of return address
           PUSHA                 # Push it onto the stack
GB_PMSA:   LDA     [GB_RAD]      # Retrieve MS byte of return address
           PUSHA                 # Push it onto the stack
GB_RET:    RTS                   # RTS = "return from subroutine"
########## Reserve temporary storage for this subroutine
GB_RAD:    .2BYTE                # Two bytes for the return address
GB_TEMP8:  .BYTE                 # A one byte temp location
########## End of "find biggest number" subroutine
```

Note that all of the internal labels associated with this routine start off with the same two letters. This is a simple technique to help ensure that we don't try to reuse the same labels in multiple subroutines.

Also note that we reserve several bytes for temporary storage at the end of this subroutine (shown highlighted in gray). The first, GB_RAD, is a two-byte location that we are going to use to store the return address when we pop it off the stack. The second, GB_TEMP8, is just a one-byte temporary location.

Let's take a brief stroll through this subroutine. The first thing we do is to pop the most-significant byte of the return address off the top of the stack and store it in the most-significant byte of our two-byte GB_RAD field. Next (at label GB_GLSA), we pop the least-significant byte of the return address off the top of the stack and store it in the least-significant byte of our GB_RAD field.

Now (at label GB_GET2), we pop the second number (that we wish to use in our comparison) off the top of the stack and squirrel it away in our GB_TEMP8 location, and then (at label GB_GET1) we pop the first number off the top of the stack into the accumulator, at which point we're really ready to rock and roll.

The CMPA ("compare accumulator") is used to compare the first number (that's currently stored in the accumulator) with the second number (that's currently stored in our GB_TEMP8 location). If the number in the accumulator is the larger, the carry flag will be set to 1, in which case the JC ("jump if carry") instruction (at label GB_WHICH) will cause us to jump to label GB_TIDY. Otherwise, the JC instruction will fail and we will load the accumulator with the larger value that is stored in our GB_TEMP8 location.[5]

The point is, that by the time we get to label GB_TIDY, we know that the accumulator contains the larger number, so we push this value onto the stack. Next, we load the accumulator with the least-significant byte of the return address (from the least-significant byte of our GB_RAD field) and push this onto the stack. Then we load the accumulator with the most-significant byte of the return address (from the most-significant byte of our GB_RAD field) and push this onto the stack.

Thus, when we reach the GB_RET label, just before we execute the RTS ("return from subroutine") instruction, the stack contains the largest of the two numbers ($45) and the return address ($5000) as shown in Figure 3-17(a).

This means that, when we execute the RTS, the return address is popped off the stack and copied into the program counter, leaving the result (the largest number) on the top of the stack from whence it can be accessed by the body of the program [Figure 3-17(b)].

It now becomes obvious that we can play all sorts of cunning games. For example, when a subroutine is called, we know its return address is stored on the top of the stack. If we wished, our subroutine could pop this address off the top of the stack and replace it with a completely different address. This new return address would subsequently be used by an RTS instruction that doesn't know any better. Of course this sort of thing is not recommended practice per se, and it makes debugging your

---

[5]Of course, the two values may be equal, but in this case it doesn't matter which one is returned!

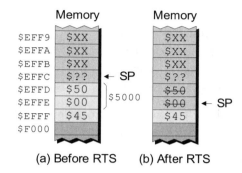

(a) Before RTS   (b) After RTS

Figure 3-17. The stack before and after the RTS instruction.

program extremely difficult, but a trick like this can occasionally come in really handy.

Nested subroutines

It's important to note that there is nothing to prevent a subroutine from calling one or more sub-subroutines. Similarly, these sub-subroutines might themselves call sub-sub-subroutines ad infinitum. This situation, which is very common, is referred to as *nesting* or *nested* subroutines.

> **Note** This would be a rather good time to perform Lab 3e, in which a subroutine is used to display the hex codes associated with a number of bytes in memory.

Recursion

The term recursion comes from the Latin *recursi*, meaning "running back." In a mathematical context, recursion refers to an expression in which each term is generated by repeating a particular mathematical operation.

By comparison, in the context of a computer program, when a subroutine (or function or procedure) calls itself, that subroutine is said to be "recursive." For example, if the body of the program calls a subroutine named SUBA, and SUBA subsequently calls itself, then this would be a classic example of recursion.

In the case where the recursion involves a series of subroutines—for example, subroutine SUBA calls subroutine SUBB, and then SUBB calls SUBA, then SUBA calls SUBB again, and so on—this group of sub-routines are said to be "mutually recursive."

The canonical example of a recursive function is a factorial. For example, 5! (which indicates "the factorial of 5") is evaluated as $5 \times 4 \times 3 \times 2 \times 1 = 120$. For a positive integer n, this can be represented as:

If $n = 1$, then factorial $n = 1$, otherwise factorial $n = n \times$ factorial $(n - 1)$

That is, the factorial of some integer n (where n is greater than 1) is equal to n multiplied by the factorial of $(n - 1)$, where the factorial of $(n - 1)$ is itself equal to $(n - 1)$ multiplied by the factorial $(n - 1 - 1)$, and so forth.

In practice, recursion can be mind-bogglingly complicated, and it's usually a lot easier to use a more standard alternative such as creating a simple loop that uses a variable as a counter. On the other hand, every now and then one runs across a task for which a recursive implementa-tion provides the most elegant solution.

The most important point to remember if you are creating a recur-sive function is to fully define some "end condition" that will terminate things; otherwise your subroutine may well end up calling itself forever.

> **Note** This would be a good time to perform Lab 3f, in which we use a recursive subroutine to display a string of characters in reverse order.

Self-Modifying Code

It should not be surprising to see a program modifying data values; for example, consider the following program snippet:

```
            :
            LDA     [TEMP8]    # Load ACC from temp location
SHIFT:      SHL                # Shift the value left by one bit
            STA     [TEMP8]    # Store ACC to the temp location
            :

            :
TEMP8:      .BYTE   $04        # Reserve temp loc and initialize it
            :
```

When we reserve the 1-byte location called TEMP8, we initialize it with a value of $04 in hexadecimal (%00000100) in binary. But, as we see from our program, at some stage we load this data value into the accumulator, shift it left by one bit, and store the result ($08 in hexadecimal or %00001000 in binary) back into the temporary location.[6]

What isn't quite so obvious is that a program can also modify its own instructions. As a simple example, consider a slight modification to our previous program (modifications are shown highlighted in gray):

```
           :
           LDA     $71        # Load ACC with opcode for SHR
           STA     [SHIFT]    # Store it to location SHIFT
           :
           LDA     [TEMP8]    # Load ACC with value in temp location
SHIFT:     SHL                # Shift the value left by one bit
           STA     [TEMP8]    # Store it back into the temp location
           :
           :
TEMP8:     .BYTE   $04        # Reserve temp loc and initialize it
           :
```

When the program is first assembled, the memory location associated with the SHIFT label will be assigned a value of $70, which is the op-code for a SHL ("shift left") instruction. But when the program is run (before it reaches our SHL instruction), we load the accumulator with a value of $71, and then we use the STA [SHIFT] instruction to store this value in the memory location associated with the SHIFT label.

The point is that $71 is the opcode for a SHR ("shift right") instruction. Thus, when the program reaches the SHIFT label, it will actually shift the contents of the accumulator *right* by one bit, and store the result ($02 in hexadecimal or %00000010 in binary) back into the temporary location.[7]

Don't do it, but if you must . . .

Of course, we are in no way recommending this as an everyday practice. Debugging programs that feature this type of self-modification can

---

[6]Shifting a value left by one bit is the equivalent of multiplying it by 2 (we shall consider this in more detail in Chapter 4).

[7]Shifting a value right by one bit is the equivalent of dividing it by 2 (once again, we shall consider this in more detail in Chapter 4).

bring the strongest among us to our knees. Generally speaking, this sort of coding would be considered to be very, very naughty, and programmers who use these techniques usually deserve to be soundly chastised.

However, now we've opened this Pandora's box, we may as well mention that there are several different flavors of self-modifying code. One class is a program that "tweaks" existing instructions as discussed above. Every now and then, one runs across a task for which this type of self-modifying implementation provides a surprisingly elegant and efficient solution.

Another flavor of self-modifying code is "on-the-fly" code generation, whereby a program may generate a whole batch of new instructions (and possibly data) in some free area of the computer's memory and then transfer control to these new instructions.

Although this latter case may seem a little weird, it is actually not uncommon in such things as graphical applications, in which the programmer is striving for extreme speed. In this case, it may be possible to generate a special-purpose routine to perform a specific set of operations and then execute this routine faster than would be possible in the case of a preexisting, general-purpose routine. This is because the general-purpose routine may be carrying a lot of "baggage" around to make it "general-purpose" in the first place.

Interactive Laboratories

If you haven't already done so, you really need to take the time to perform the following interactive laboratories that you'll find towards the back of this book:

Lab 3a: Using Logical Instructions, Shifts, and Rotates

Lab 3b: Understanding the Program Counter (PC)

Lab 3c: Using the Index Register (X)

Lab 3d: Using the Stack and Stack Pointer (SP)

Lab 3e: Using Subroutines

Lab 3f: Using Recursion

Review

This chapter introduced the logical instructions (AND, OR, and XOR), the shift instructions (SHL and SHR), and the rotate instructions (ROLC and RORC); the program counter (PC), the index register (X), and the stack and stack pointer (SP); and subroutines, recursion, and self-modifying code.

1) What would be the result of AND-ing the binary values 10011001 and 01101101 together? How about OR-ing and XOR-ing these values together?

2) Summarize any similarities and/or differences between (a) logical and arithmetic shift left instructions and (b) logical and arithmetic shift right instructions.

3) If the accumulator originally contained 00111100_2, what would it contain following a SHL instruction and what value would be in the carry flag? Assuming the same initial value, what would be the contents of the accumulator and carry flag following a SHR instruction? Now, assuming an initial value of 11000011_2, what would be the contents of the accumulator and carry flag following (a) a SHL instruction or (b) a SHR instruction?

4) If the accumulator originally contained 00111100_2 and the carry flag originally contained 1, what would be the contents of the accumulator and carry flag following (a) a ROLC instruction or (b) a RORC instruction?

5) Summarize the actions of the program counter (PC), index register (X), and stack pointer (SP).

6) What is the name of the instruction used to copy the value in the accumulator onto the stack? When this operation is performed, is the stack pointer incremented or decremented?

7) What is the name of the instruction used to take a value off the top of the stack and copy it into the status register? When this action is performed, is the stack pointer incremented or decremented?

8) Summarize the actions that occur when a JSR ("jump to subroutine") instruction is performed.

9) Summarize the actions that occur when an RTS ("return from subroutine") instruction is performed.

10) Define recursion in the context of subroutines. Try to think of an example in which recursion might be used (different to the examples presented in the interactive laboratories associated with this chapter).

CHAPTER
4 INTEGER ARITHMETIC

> *The four branches of arithmetic—ambition, distraciton, uglification and derision.*
>
> LEWIS CARROLL (1832–1898) in
> *Alice in Wonderland* (1865)

> *Arithmetic is being able to count up to twenty without taking off your shoes.*
>
> MICKEY MOUSE (1828–)

In this chapter we will learn about:

- Signed and unsigned binary numbers
- Adding and subtracting binary numbers
- Multiplying and dividing binary numbers
- Multibyte operations
- Building binary adders and subtractors
- The use of the carry, borrow, and overflow flags

Unsigned Binary Numbers

In order for computers to work with numbers, we first have to decide which numbers we wish to represent, and then we need to determine exactly how we are going to represent them. As we will come to discover, almost every aspect of this is more complicated than you might at first suppose. As a simple example, do we wish to represent natural numbers, whole numbers, or integers?

Natural (counting) numbers: $1, 2, 3, 4, 5, \ldots, \infty$

Whole numbers: $0, 1, 2, 3, 4, 5, \ldots, \infty$

Integers: $-\infty, \ldots, -5, -4, -3, -2, -1, 0, 1, 2, 3, 4, 5, \ldots, \infty$

Note that the ∞ symbol represents infinity, meaning "boundless," "limitless," or "without end." Also, by default, any number without a minus sign is assumed to be positive. This latter point is relevant, because we shall begin by considering only the positive integers (which correspond to the whole numbers). Since we know that we are going to consider only positive quantities, we don't need to worry about representing the sign of these values.

Another point to consider is that numbers written by hand with pencil and paper can be as large as we wish (limited only by the size of the paper and our endurance). By comparison, the size of numbers stored and mapiulated inside a computer is dictated by the physical elements used to represent them.

As we discussed in Chapters 1 and 2, our virtual computer's memory is only 8 bits wide (similarly for its data bus). One bit can represent two distinct binary values: 0 and 1. Two bits can represent $2^2 = 4$ distinct binary values: 00, 01, 10, and 11. Three bits can represent $2^3 = 8$ distinct binary values: 000, 001, 010, 011, 100, 101, 110, and 111, and so forth. Thus, 8 bits can be used to represent $2^8 = 256$ binary values, which can, in turn, be used to represent positive integers in the range 0_{10} to 255_{10} (Figure 4-1).

> **Note**   Just in case anyone should happen to ask, with respect to referencing numerical values in figures and words, there are two types of numbers, *carinal* and *ordinal*.
>
> A *cardinal number* such as 1, 2, 3, . . . ("one," "two," "three," . . .) is used in counting to indicate quantity but not order.
>
> By comparison, an *ordinal number* such as 1st, 2nd, 3rd, . . . ("first," "second," "third," . . .) indicates its position in a series or order.

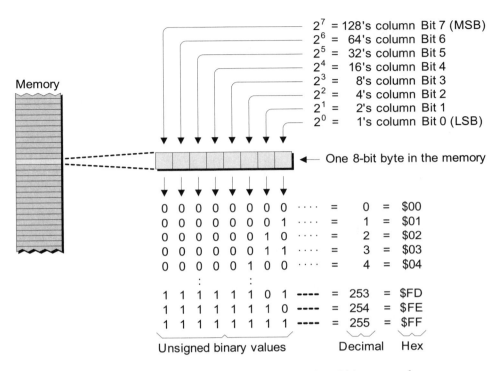

Figure 4-1. Using eight bits to represent unsigned binary numbers.

Whenever we're dealing with computers, we typically start counting things from zero; thus, the bits forming an 8-bit byte are numbered from bit 0 to bit 7. This may seem a little weird the first time you see it, but it actually makes a lot of sense, because bits 0 through bit 7 represent the 2^0 through 2^7 columns of a binary number, respectively. Bit 0 is also known as the *least-significant bit* (*LSB*), because it represents the least significant value. Similarly, bit 7 is known as the *most-significant bit* (*MSB*), because it represents the most significant value.

Remember that we're only considering positive quantities here, which means we don't need any way to represent the sign of these numbers. For this reason, these values are referred to as *unsigned binary numbers*.

Adding Unsigned Binary Numbers

The process of adding binary numbers is identical in concept to working in decimal, but it can be a little tricky the first time you do it so we'll take

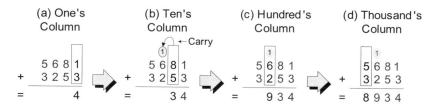

Figure 4-2. Adding decimal numbers.

things gradually. Let's start by considering a standard decimal addition (Figure 4-2).

We commence with the least-significant digits, which are in the one's column [Figure 4-2(a)]. As we see, the one's column contains $1 + 3 = 4$, so there are no surprises here. When we move to the ten's column [Figure 4-2(b)], we find $8 + 5 = 13$, which gives us a result of 3 in the ten's column and a carry of 1 into the hundred's column. Thus, when we reach the hundred's column [Figure 4-2(c)], we see $6 + 2 + (1) = 9$ [where the (1) in parentheses was carried over from the ten's column]. Finally, we're back on home ground in the thousand's column with $5 + 3 = 8$ [Figure 4-2(d)].

The point about adding in decimal is that we do it so often we don't even think about it. Even the process of generating a carry in one column and using it in the next doesn't give us any pause. Now let's ponder the process of adding two unsigned binary numbers together, 01010101_2 and 01110100_2, which represent 85_{10} and 116_{10} in decimal, respectively (Figure 4-3).

As usual, we commence with the least-significant digits (the bit 0 column), which are $1 + 0 = 1$ in this example [Figure 4-3(a)]. Moving on to the bit 1 column, we see $0 + 0 = 0$ [Figure 4-3(b)].

This isn't too difficult, is it? But *wait*! What's this? The bit 2 column contains a brace of ones [Figure 4-3(c)]. Don't Panic! This is just like

Note The number to which another number is to be added is called the *augend*, from the Latin *augendum*, meaning "a thing to be increased" (this would be the 5,681 value in the example shown in Figure 4.2).

By comparison, a number that is added to another number is called the *addend*, from the Latin *addere*, meaning "to add" (this would be the 3,253 value in the example shown in Figure 4-2).

The result from the addition of two or more numbers is called the *sum*, from the Latin *summa*, from the feminine of *summus*, meaning "highest."

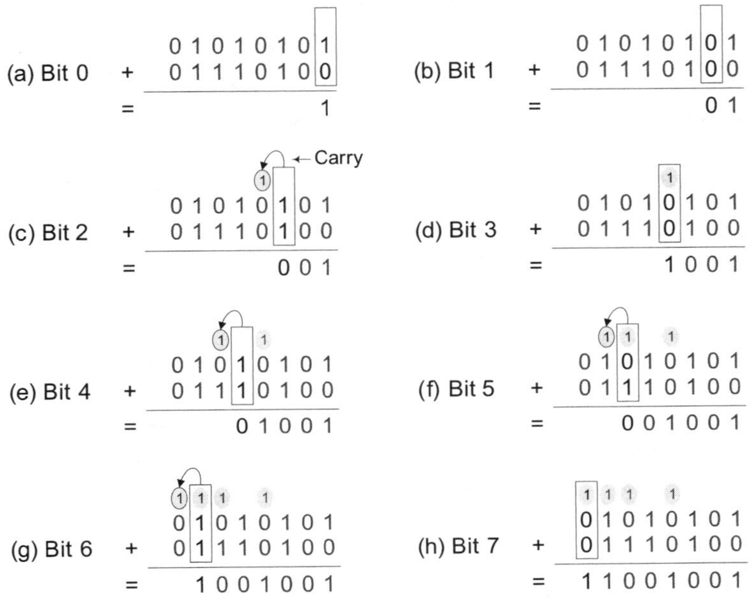

Figure 4-3. Adding unsigned binary numbers.

generating a carry in decimal. That is, $1 + 1 = 10_2$ in binary (2 in decimal), which gives us a result of 0 in the bit 2 column and a carry of 1 into the bit 3 column. Thus, when we reach the bit 3 column [Figure 4-3(d)] we see $0 + 0 + (1) = 1$ [where the (1) in parentheses was carried over from the bit 2 column].

Based on our newfound knowledge, the bit 4 column holds no dread for heroes such as we [Figure 4-3(e)]. In fact, we sneer contemptuously as we once again calculate that $1 + 1 = 10_2$ (in binary), which gives us a result of 0 in the bit 4 column and a carry of 1 into the bit 5 column. Similarly, we scoff when we reach the bit 5 column itself [Figure 4-3(f)], because $0 + 1 + (1) = 10_2$ (in binary), which gives us a result of 0 in the bit 5 column and a carry of 1 into the bit 6 column.

There may, however, be much gnashing of teeth and rending of clothing when we reach the bit 6 column [Figure 4-3(g)], because now we have to add three 1s together! But hold hard for just a moment. Let's take a step back and put things in perspective. All we're talking about is adding three 1s together, and how difficult can this be? In fact it's not difficult at all, because $1 + 1 + (1) = 11_2$ in binary (3 in decimal), which

> **Note** In the Middle Ages,[1] it was common practice to use the word *più* (from the Latin *plus*, meaning "more") to indicate an addition. It was also common to use the Latin *et*, meaning "and." For example, "*6 più 2*" or "*6 et 2*" were often used to *indicate* "*6 plus 2*" or "*6 and 2*" or "*6 add 2*" in modern terms.
>
> Some manuscripts from around the 1350s show a symbol that looks like a modern "+" that is believed to be an abbreviation of the word *et*.
>
> In 1489, the "+" symbol appeared for the first time in print in a book written by the German mathematician Johann Widmann. In this case, it was associated with a numerical quantity and used to indicate a surplus in business problems. Over time, however, the "+" symbol was increasingly used to indicate an addition operation or a positive number.

gives us a result of 1 in the bit 6 column and a carry of 1 into the bit 7 column.

Last, but not least, we can race to the finish line that is the bit 7 column, because $0 + 0 + (1) = 1$. Thus, $01010101_2 + 01110100_2 = 11001001_2$ in binary (or $85 + 116 = 201$ in decimal).

Building a Binary Adder

As we discussed in Chapters 1 and 3, computers are built out of simple logic gates such as ANDs, ORs, and XORs, which are, in turn, constructed out of transistors. If we were designing a computer, one of the things we would have to build would be an adder. In this case, our first step might be to consider constructing a functional block that could add just two bits together (Figure 4-4).

Even though we're adding two bits together, this would actually be referred to as a "1-bit adder" because it generates a 1-bit result. More correctly, it would be classed as a "1-bit full adder" because it includes the `cin` ("carry-in") and `cout` ("carry-out") signals as discussed below.

The symbol provides a pictorial view of the adder, whereas the truth table describes the relationships between its inputs (on the left) and outputs (on the right).[2] In the case of the inputs, `a` and `b` are the two bits to be added together, while `cin` is the carry-in from the previous stage. Meanwhile, the `s` (for "sum") output represents the result of the addition, while `cout` is the carry-out to the next stage.

---

[1]The period in European history between Antiquity and the Renaissance, often dated from 476 AD to 1453 AD.

[2]Transistors, logic gates, and truth tables are introduced in excruciating detail in the book *Bebop to the Boolean Boogie (An Unconventional Guide to Electronics)*, 2nd Edition, ISBN: 0-7506-7543-8.

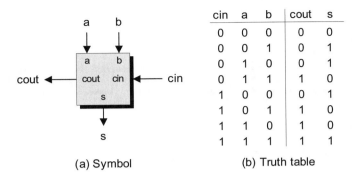

cin	a	b	cout	s
0	0	0	0	0
0	0	1	0	1
0	1	0	0	1
0	1	1	1	0
1	0	0	0	1
1	0	1	1	0
1	1	0	1	0
1	1	1	1	1

(a) Symbol   (b) Truth table

Figure 4-4. Adding two bits together (along with a carry-in bit).

Purely for the sake of interest (we don't really need to know this but, generally speaking, the more we know, the happier we are), we could construct such an adder from two AND gates, two XOR gates, and a single OR gate as shown in Figure 4-5.

But we digress. What do we mean when we talk about "carry-in from the previous stage" and "carry-out to the next stage"? Well, in the case of a computer with an 8-bit data bus (like our virtual computer) we really want an 8-bit adder that can add two 8-bit numbers, and one way to achieve this is to connect eight of our simple adders together (Figure 4-6).

Observe that we've named our signals $a[7]$, $a[6]$, $a[5]$, and so on (using what is referred to as a *vector notation*), as opposed to calling them $a7$, $a6$, $a5$, and so on (where this latter form would be referred to

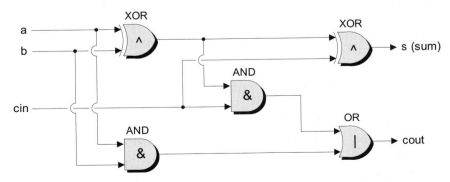

Figure 4-5. Constructing an adder from primitive logic gates.

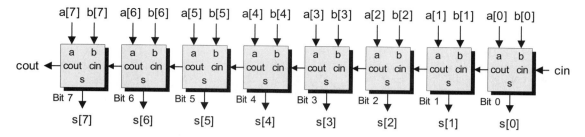

Figure 4-6. Creating an 8-bit adder.

as a *scalar notation*). To all intents and purposes, the two notations are functionally equivalent. The advantage of the vector form is that it allows us to easily and efficiently reference ranges (groups) of signals like a[7:0] (meaning a[7], a[6], a[5],..., all the way to a[0]).

In the case of the simple unsigned binary additions we're considering here, the cin ("carry-in") signal to the bit 0 stage would be connected to a logic 0. Bearing this in mind, if you return to the two binary numbers we added together in Figure 4-3 and applied one of these numbers to the a[7:0] inputs and the other to the b[7:0] inputs, the correct result would appear on the s[7:0] outputs.

The advantage of this type of adder is that it's relatively simple to understand and construct. However, there is a disadvantage in that each stage has to wait for the output from the previous stage to become valid. That is, the bit 1 adder can't complete its calculation until the carry-out signal from the bit 0 adder becomes valid, the bit 2 adder can't complete *its* calculation until the carry-out from the bit 1 adder is valid, and so forth. Thus, this type of adder is known as a *ripple adder* (or a *ripple carry adder*) because the intermediate results "ripple" through it. (It is possible to create more sophisticated adders that can add all of the bits together simultaneously, but these are outside the scope of this book.)

The ADD Instruction

Before we proceed further, we should note that our virtual computer supports an ADD instruction, which adds a byte of data to the current contents of the accumulator. For example, consider the addition illustrated in Figure 4-3; one possible assembly code version of this addition would be as follows:

```
            :
        LDA      [NUMA]        # Load ACC with value in NUMA
        ADD      [NUMB]        # Add value in NUMB
        STA      [RESULT]      # Store the result in RESULT
            :
            :
NUMA:       .BYTE    %01010101     #  85 in decimal, $55 in hex
NUMB:       .BYTE    %01110100     # 116 in decimal, $74 in hex
RESULT:     .BYTE                  # Store result here
            :
```

Inside the CPU, the ADD instruction causes two 8-bit values to be presented to an adder function similar (in concept) to the one shown in Figure 4-6. The way in which the above program snippet works can be visualized as illustrated in Figure 4-7.

At point (1), the LDA [NUMA] instruction loads the accumulator with the data byte from the memory location associated with the NUMA label. At point (2), the ADD [NUMB] instruction causes the current value in the accumulator to be added to the data byte from the memory location associated with the NUMB label. Note that the ADD instruction causes the cin ("carry-in") signal to the adder to be forced to logic 0. Also observe that the cout ("carry-out") signal from the adder is loaded into the

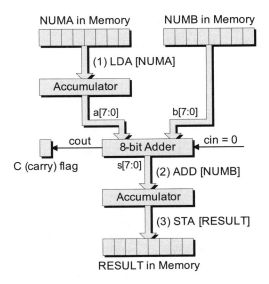

Figure 4-7. Adding two 8-bit numbers together.

C (carry) status flag and the result from the addition is stored back in the accumulator (thereby overwriting its previous contents). Finally, at point (3), the STA [RESULT] instruction stores the current contents of the accumulator (the result from the addition) into the byte in memory associated with the RESULT label.

The Carry Flag

Now let's consider the cout ("carry-out") signal in a little more detail. This signal is used to indicate whether or not the result from the addition was too big to be held in our 8-bit accumulator.

The key point to remember here is that, assuming we're working with unsigned binary numbers, our 8-bit fields can be used to represent positive integers in the range 0 to 255. When we add two 8-bit values such as 01010101_2 and 01110100_2 together, this returns an 8-bit result: 11001001_2. In this case (which equates to $85 + 116 = 201$ in decimal), the cout signal from our adder would be 0, thereby indicating that the result *will* fit in our 8-bit accumulator.

By comparison, when we add two 8-bit values such as 10100100_2 and 01110100_2 together, this returns a 9-bit result: 100011000_2. In this case (which equates to $164 + 116 = 280$ in decimal), the cout signal from our adder would be 1, thereby indicating that the result will not fit in our 8-bit accumulator. (The cout signal will be 1 for any addition that results in a value greater than 255.)

The point is that, when performing an addition, the value on the cout signal is stored in the carry flag. We can subsequently test this flag and respond accordingly—by displaying an error message, for example—using one of our JC ("jump if carry") or JNC ("jump if not carry") instructions. Alternatively, in the case of the multibyte additions discussed below, we can visualize the C (carry) flag as forming the ninth bit in the result.

Note   Prior to the invention of the "=" character, equality was usually indicated by a word such as *aequales* or *aequantur*, from the Latin *aequlis*, meaning "even" or "level" (for example, "*6 + 2 aequales 8*").

The first use of the modern equal symbol appeared in a 1557 AD book called *The Whetstone of Witte*, which was authored by the English mathematician Robert Recorde (1510–1558). In this tome, Recorde noted: "I will sette as I doe often in woorke use, a paire of paralles, or Gemowe lines of one lengthe, thus "=" bicause noe 2 thynges can be moare equalle."

Multibyte Additions and the ADDC Instruction

It would obviously be somewhat limiting to work only with positive integers in the range 0 to 255. The solution is to use multiple bytes to represent each number. For example, let's suppose that we decide to use two bytes (16 bits) to represent an unsigned binary number. In reality, these bytes would be located in adjacent locations in memory [Figure 4-8(a)], but we can visualize them as being rearranged to form a single 16-bit value comprising a *most-significant byte* (*MSB*) containing bits 15 to 8 and a *least-significant byte* (*LSB*) containing bits 7 to 0 [Figure 4-8(b) and Figure 4-8(c)].

In this case, our two bytes can assume any one of $2^{16} = 65,536$ different binary patterns from 0000000000000000_2 to 1111111111111111_2, which can be used to represent positive integers in the range 0 to 65,535 in decimal (or \$0000 to \$FFFF in hexadecimal).

Now let's consider some assembly code that can add two of these 16-bit numbers together and store the result in a 16-bit field:

```
            :
        LDA     [NUMA+1]      # Load ACC with LSB of NUMA
        ADD     [NUMB+1]      # Add LSB of NUMB
        STA     [RESULT+1]    # Store LSB of result in RESULT
        LDA     [NUMA]        # Load ACC with MSB of NUMA
        ADDC    [NUMB]        # Add MSB of NUMB
        STA     [RESULT]      # Store MSB of result in RESULT
            :
            :
NUMA:    .2BYTE  $1234         # 4660 in decimal
NUMB:    .2BYTE  $2468         # 9320 in decimal
RESULT:  .2BYTE                # Store result here
            :
```

In particular, observe the use of the ADDC ("add with carry") instruction for the second portion of the addition (the line highlighted in gray). We can visualize the way in which the above program snippet works as illustrated in Figure 4-9.

At point (1), the LDA [NUMA+1] instruction loads the accumulator with the LSB of the number associated with the NUMA label. At point (2), the ADD [NUMB+1] instruction causes the current value in the accumulator to be added to the LSB of the number associated with the NUMB label.

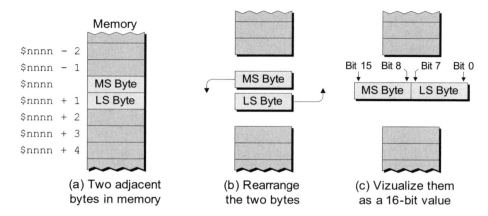

Figure 4-8. Visualizing two bytes in memory forming a single value.

Note once again that the ADD instruction causes the cin ("carry-in") signal to the adder to be forced to logic 0. As before, the cout ("carry-out") signal from the adder is loaded into the C (carry) status flag and the result from the addition is stored back in the accumulator (thereby overwriting its previous contents). Next, at point (3), the STA [RESULT+1]

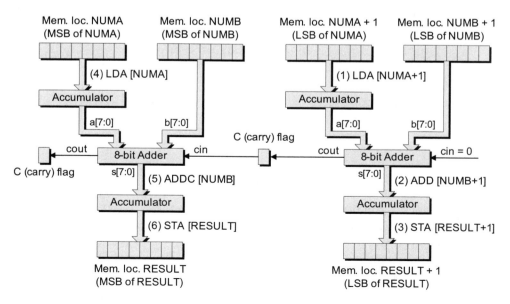

Figure 4-9. Adding two 16-bit numbers together.

instruction stores the current contents of the accumulator (the LSB portion of the addition) to be stored in the LSB of the field associated with the RESULT label.

Now things start to get interesting. At point (4), the LDA [NUMA] instruction loads the accumulator with the MSB of the number associated with the NUMA label. At point (5), the ADDC [NUMB] instruction causes the current value in the accumulator to be added to the MSB of the number associated with the NUMB label. In this case, however, the ADDC ("add with carry") instruction causes the cin ("carry-in") signal to be presented with the current value in the C (carry) flag. As usual, the cout ("carry-out") signal from the adder is loaded into the C (carry) flag and the result from the addition is stored back in the accumulator (once again overwriting its previous contents). Finally, at point (6), the STA [RESULT] instruction stores the current contents of the accumulator (the MSB portion of the addition) to be stored in the MSB of the field associated with the RESULT label.

The key to all of this is the combination of the ADD and ADDC instructions. When we use the ADD to sum the least-significant bytes of our two numbers, it forces the adder's cin ("carry-in") input to logic 0. Furthermore, in this case we are treating the carry flag as being the ninth bit in the result. Even if the carry flag contains a 1, this doesn't mean that the result is too big, because we still have the most-significant bytes to consider. When we subsequently use the ADDC to sum the most-significant bytes of our two numbers, it causes the current value in the carry flag (the carry-out from the previous addition) to be used as the carry-in value for this addition. It may help to think of the contents of the carry flag acting like the baton in a relay race and being passed from one runner to another.

Of course, we could use more bytes to represent our numbers if required. For example, 3-byte (24-bit) values can assume any one of $2^{24} = 16,777,216$ different binary patterns, which can be used to represent positive integers in the range 0 to 16,777,215 in decimal (or \$000000 to \$FFFFFF in hexadecimal). Similarly, 4-byte values can assume any one of $2^{32} = 4,294,967,296$ different binary patterns, which can be used to represent positive integers in the range 0 to 4,294,967,295 in decimal (or \$00000000 to \$FFFFFFFF in hexadecimal).[3] The only thing to remember

[3]See also the discussions in Chapter 6 with regard to using values of arbitrary (and dynamically variable) precision.

is to use an ADD instruction to sum the least-significant bytes followed by ADDC instructions for the remaining bytes.

> **Note**   This would be a good time to perform Lab 4a, in which we construct a testbench program that we'll use to verify the integer math subroutines we create throughout the course of this chapter. Next, work your way through Lab 4b, in which we create a general-purpose subroutine to perform 16-bit additions (we will be using this subroutine as part of the integer-based calculator program we are going to build in Chapter 5).

Subtracting Unsigned Binary Numbers

Unsigned binary numbers can be subtracted from each other using a process identical to that used for decimal subtraction. For reasons of efficiency, however, computers rarely perform subtractions in this way; instead, they perform these operations using *complement* techniques.

This does require a little thought, but once you've wrapped your brain around it you'll find it to be a lot simpler than you might suppose. And, as usual, we'll take things step by step. Let's start by considering a standard decimal subtraction so as to ensure we're all tap-dancing to the same drumbeat (Figure 4-10).

Don't worry; we won't overdo this. The important point (at least in this example) occurs when we reach the ten's column [Figure 4-10(b)]. We start by wanting to subtract 3 from 2, but 3 is bigger than 2, so we borrow 1 from the hundreds column. This means that we subtract 1 from the 6 in the hundreds column of the minuend (see note), leaving 5; and then we use our borrowed 1 to augment the 2 in the tens column of the minuend to form 12. Thus, our ten's column now requires us to subtract 3 from 12, leaving 9; and when we reach the hundred's column, instead of subtracting 2 from 6, we now subtract 2 from 5, leaving 3.

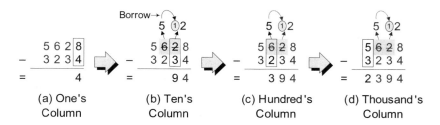

Figure 4-10. Subtracting decimal numbers (the "American borrow").

> **Note** The number from which another number is to be subtracted is called the *minuend*, from the Latin *minuendum,* meaning "thing to be diminished" (this would be the 5,628 value in the example shown in Figure 4-10).
>
> By comparison, a number that is to be subtracted from another number is called the *subtrahend,* from the Latin *subtrahendum,* meaning "to subtract" (this would be the 3,234 value in the example shown in Figure 4-10).
>
> Last but not least, the result obtained by subtracting one number from another is called the *remainder* or the *difference.*

Actually, the above reflects the American method for performing a borrow operation. Purely for the sake of interest, the way in which this would have been performed in England when the authors were young lads (and still is for all we know) is illustrated in Figure 4-11.

In this case, when we come to performing our borrow operation, we again augment the 2 in the ten's column of the minuend with a 1 borrowed from the hundred's column to form 12. However, rather than *subtracting* 1 from the 6 in the hundred's column of the minuend to leave 5, we instead *add* 1 to the 2 in the hundreds column of the subtrahend to give 3. Thus, when we come to the hundred's column, we now perform the operation $6 - 3 = 3$ (as opposed to $5 - 2 = 3$ using the American technique).

The end result is the same, of course, because we'd be in something of a pickle if performing a simple math operation such as an integer subtraction gave conflicting results on the opposite sides of the Atlantic Ocean. The advantage of the American scheme is that it's more intuitive when it comes to visualizing where the "borrow" comes from; the disadvantage comes in the form of the special case that occurs should you have to borrow (subtract 1) from the next column when that column contains a 0. By comparison, the English approach is slightly less intuitive, but there are no special cases.

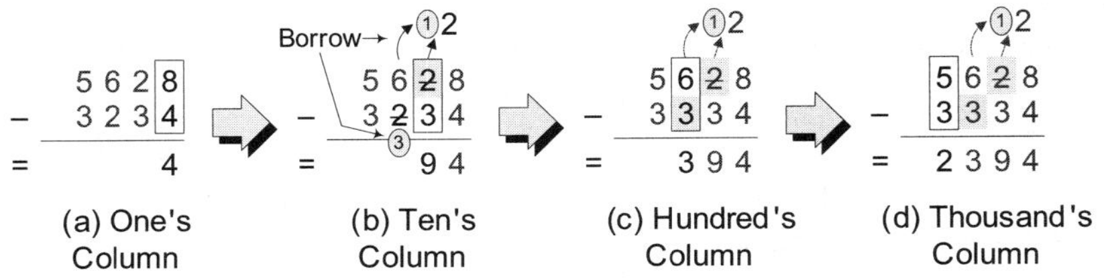

Figure 4-11. Subtracting decimal numbers (the "English borrow").

Just to humor us, quickly perform the following subtraction on a piece of paper. If you are used to working with the American scheme, then try using the English technique, and vice versa:

$$74,654,008$$
$$- 13,995,623$$
$$??,???,???$$

The answer, of course, is 60,658,385, but it wouldn't surprise us if performing this calculation took a little longer than you expected because you were using an approach that is unfamiliar. The point is that, in the not-so-distant past, the average person was much less familiar with performing even simple math operations, so the folks who did know what they were doing came up with tricks to make things easier.

Nines complement decimal subtraction

Every number system has something called a *radix complement* and a *diminished radix complement* associated with it, where the term "radix" refers to the base of that number system. With regard to the decimal (base-10) system, its radix complement is also known as the *tens complement,* whereas its diminished radix complement is referred to as the *nines complement.*

First, let's consider our original decimal subtraction (5,628 − 3,234) performed using the nines complement technique. The first step is to generate our nines complement value, which we achieve by subtracting our original subtrahend (3,234) from 9,999 [Figure 4-12(a)].

Next we add our nines complement value (6,765) to our original minuend (5,628) to generate an intermediate result (12,393) as shown in Figure 4-12(b). Finally, we perform an *end-around-carry* operation, which involves taking the most significant "1" from our intermediate result, moving it into the units (ones) column under that result, and adding it to what remains of the intermediate result to generate the final result (2,394) as illustrated in Figure 4-12(c).

The advantage of the nines complement technique is that it's never necessary to perform a borrow operation. The disadvantages are that there are more steps than in our regular way of doing things and also that we have to perform the end-around-carry operation.

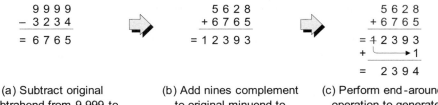

```
   9 9 9 9              5 6 2 8              5 6 2 8
 - 3 2 3 4            + 6 7 6 5            + 6 7 6 5
 = 6 7 6 5            = 1 2 3 9 3          = 1 2 3 9 3
                                          +        1
                                          =   2 3 9 4
```

(a) Subtract original   (b) Add nines complement   (c) Perform end-around-carry
subtrahend from 9,999 to   to original minuend to   operation to generate the
form nines complement   generate intermediate result   final result

Figure 4-12. Performing a decimal subtraction using the nines complement technique.

Tens complement decimal subtraction

Now consider the same subtraction performed using the tens complement technique. In this case, we first have to generate a tens complement
value by subtracting our original subtrahend (3,234) from 10,000
[Figure 4-13(a)].

Next, we add our tens complement value (6,766) to our original
minuend (5,628) to generate an intermediate result (12,394) as shown
in Figure 4-13(b). And then we simply drop the carry digit from the
intermediate result to leave the final result (2,394) as illustrated in
Figure 4-13(c).

The main advantage of the tens complement approach is that it is
unnecessary to perform an end-around-carry, because the carry-out resulting from the addition of the most significant digits is simply dropped
from the final result. The disadvantage is that, during the process of creating the tens complement, it is necessary to perform a borrow operation
for every nonzero digit in the subtrahend (this problem can be overcome
by first generating the nines complement, adding "1" to the result, and
then performing the remaining operations as for the tens complement).

```
  1 0 0 0 0            5 6 2 8              5 6 2 8
 -   3 2 3 4          + 6 7 6 6            + 6 7 6 5
 =   6 7 6 6          = 1 2 3 9 4          = 1 2 3 9 4
```

(a) Subtract original   (b) Add tens complement   (c) Drop the carry
subtrahend from 10,000 to   to original minuend to   digit to generate
form tens complement   generate intermediate result   the final result

Figure 4-13. Performing a decimal subtraction using the tens complement technique.

Ones complement binary subtraction

The point of all of the above is that similar techniques may be employed with binary (base-2) numbers, where the radix complement is known as the *twos complement* and the diminished radix complement is known as the *ones complement*.

Let's assume that we wish to subtract 00011110_2 from 00111001_2 in binary ($57 - 30 = 27$ in decimal). There's no point in beating our heads against the wall trying to understand a standard binary subtraction—that is, subtracting the subtrahend (00011110_2) directly from the minuend (00111001_2)—because we simply never do things this way in the real world. Instead, we use one of the complement techniques.

First, we'll consider a binary subtraction performed using the ones complement procedure. In this case, we generate our ones complement value by subtracting the subtrahend (00011110_2) from an all-ones value (11111111_2) as shown in Figure 4-14(a).

The resulting ones complement value (11100001_2) is now added to our original minuend (00111001_2) to generate a 9-bit intermediate result (100011010_2), as illustrated in Figure 4-14(b). Finally, an end-around-carry operation is performed to generate the final result (00011011_2) (which equates to 27 in decimal), as shown in Figure 4-14(c).

The advantage of the ones complement technique is that it is never necessary to perform a borrow operation. In fact, *it isn't even necessary to perform a subtraction operation,* because the ones complement of a value can be generated by simply inverting all of its bits, that is, by exchanging all of its 0s with 1s, and vice versa. This means that even if we were to stop here, you already know how to perform a simple binary subtraction using only inversion and addition, without any actual nitty-gritty sub-

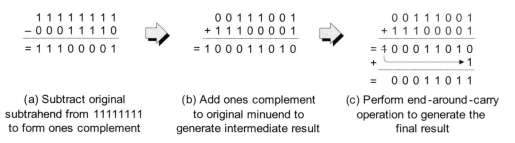

(a) Subtract original subtrahend from 11111111 to form ones complement

(b) Add ones complement to original minuend to generate intermediate result

(c) Perform end-around-carry operation to generate the final result

Figure 4-14. Performing a binary subtraction using the ones complement technique.

traction being involved at all. But watch in wonder and awe, because things are going to get better and better.

Twos complement binary subtraction

In the case of a twos complement binary subtraction, we first generate a twos complement value by subtracting our original subtrahend (00011110_2) from 100000000_2, as shown in Figure 4-15(a).

Next, we add the resulting twos complement value (11100010_2) to our original minuend (00111001_2) to generate a 9-bit intermediate result of 100011011_2, as illustrated in Figure 4-15(b). And then we simply drop the carry digit to leave the final result (00011011_2), which, once again, equates to 27 in decimal, as shown in Figure 4-15(c).

As is usual when using a full complement technique, the advantage of the twos complement approach is that it isn't necessary to perform an end-around-carry operation, because the carry digit from the intermediate result is simply dropped to leave the final value. The disadvantage is that, during the process of creating the twos complement value, it is necessary to perform a borrow operation for every nonzero digit in the subtrahend. Of course this problem can be overcome by first generating the ones complement, adding "1" to the result, and then performing the remaining operations as for the twos complement (this point is *very* important when it comes to building the arithmetic unit in a computer, as we'll see in the next section).

However, before we proceed, it's worth noting that a very useful short-cut method is available to us when it comes to generating twos complement values by hand. For example, let's assume that we wish to generate the twos complement of our original subtrahend (00011110_2). Commencing with the least-significant bit of the value to be complemented, we copy each bit up to and including the first 1 as illustrated in

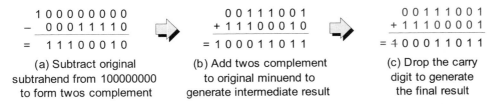

(a) Subtract original subtrahend from 100000000 to form twos complement

(b) Add twos complement to original minuend to generate intermediate result

(c) Drop the carry digit to generate the final result

Figure 4-15. Performing a binary subtraction using the twos complement technique.

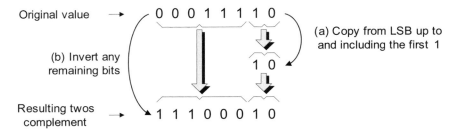

Figure 4-16. Shortcut for generating a twos complement by hand.

Figure 4-16(a); then the remaining bits are simply inverted (the 0s are replaced with 1s, and vice versa) as shown in Figure 4-16(b).

Building a Binary Subtractor

If we wished, we could construct a binary subtractor from first principles in a similar manner to the way in which we approached the binary adder earlier in this chapter. Such a unit could directly subtract one value from another, but now our CPU would have to contain two relatively complex processing units: the adder and the subtractor.

As an alternative, computer designers employ some cunning tricks derived from the way in which ones and twos complement values work. In order to understand this, let's assume that we have two 8-bit binary numbers called $A[7:0]$ and $B[7:0]$. If we wish to subtract the latter value from the former, we can represent this operation as follows:

$$S[7:0] = A[7:0] - B[7:0]$$

> **Note**　In early manuscripts, it was common practice to use the word *minus* (from the Latin *minor*, meaning "less") to indicate a subtraction. For example, "6 *minus* 2" would be used to indicate "6 *take away 2*" in modern terms.
>
> People then started abbreviating the word minus by simply using the letter "m" with a horizontal line drawn over the top ($\overline{m}$). Still later, they omitted the letter "m," leaving only the horizontal line.
>
> This "–" symbol appeared for the first time in print in a 1489 book by the German mathematician Johann Widmann. In this case, it was associated with a numerical quantity and used to indicate a deficit in business problems. Over time, however, the "–" symbol was increasingly used to indicate a subtraction operation or a negative number.

Where S[7:0] represents the 8-bit result. Alternatively, as we now know, we can represent the same operation as follows:

$$S[7:0] = A[7:0] + (\text{twos complement of } B[7:0])$$

Note that we will have to "throw away" the carry-out from this operation, but that's a "no-brainer" (in fact, as we shall soon discover, this value appears as the cout ("carry-out") signal from our adder, and it is stored in the C (carry) status flag as usual). And finally—and this is the really important point—the twos complement of a number is equal to its ones complement plus 1, so we can actually represent our original subtraction as follows:

$$S[7:0] = A[7:0] + (\text{ones complement of } B[7:0]) + 1$$

Now consider the diagram shown in Figure 4-17, which represents the portion of the *arithmetic logic unit* (*ALU*) inside a computer that actually performs additions and subtractions. Don't panic! This is much less complicated than it may at first appear.

Observe the inverter block, which simply inverts each bit of an 8-bit value; that is, it exchanges 0s for 1s, and vice versa. For example, if the inputs to the inverter block are 00011110_2, then the outputs from that

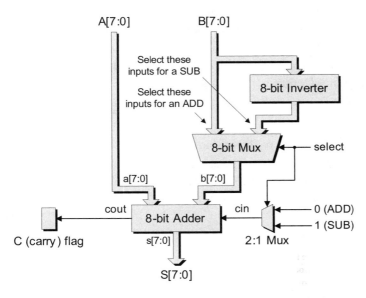

Figure 4-17. Block diagram of combined adder/subtractor function.

block will be 11100001_2. Thus, the outputs from the inverter block are the ones complement of the main B[7:0] inputs.

Next, observe the 8-bit multiplexer ("mux" for short). Depending on the value presented to its control ("select") input, the multiplexer will either select the unmodified version of the B[7:0] inputs and convey this value to its outputs, or it will select the inverted (ones complement) version of the B[7:0] inputs and convey *this* value to its outputs.

Finally, observe the 2:1 ("2-to-1") multiplexer whose output drives the adder's cin ("carry-in") input. This is controlled by the same select signal as the 8-bit multiplexer.

Now, consider what happens if we wish to perform an addition. In this case, the A[7:0] inputs (which feed the adder's a[7:0] inputs) represent the *augend*. Meanwhile, the select signal causes the 8-bit multiplexer to choose the unmodified version of the B[7:0] inputs, which represent the *addend* in this case. The multiplexer conveys this value to its outputs, which are used to feed the adder's b[7:0] inputs. The select signal also causes the 2:1 multiplexer to present a 0 to the adder's cin input. Thus, the addition works in exactly the same way as was discussed earlier in this chapter.

By comparison, if we wish to perform a subtraction, the A[7:0] inputs (which feed the adder's a[7:0] inputs) represent the *minuend*. This time, the select signal causes the 8-bit multiplexer to choose the value from the inverter; that is, the ones complement of the B[7:0] inputs, which are now assuming the role of the *subtrahend*. This ones complement value is fed to the adder's b[7:0] inputs. Thus, the adder is now presented with the following:

$$A[7:0] + (\text{ones complement of } B[7:0]) + cin$$

The clever part of all of this is that, in the case of the subtraction, the select signal also causes the 2:1 multiplexer to present a 1 to the adder's cin input, which means the adder is actually performing the following operation:

$$A[7:0] + (\text{ones complement of } B[7:0]) + 1$$

And, of course, this is exactly what we wish to do in order to perform the complement version of our subtraction.

The SUB Instruction

It should not surprise anyone to discover that our virtual computer supports a SUB instruction, which subtracts a byte of data from the current contents of the accumulator. For example, if we wish to perform the operation $00111001_2 - 00011110_2$, one possible assembly code version of this subtraction would be as follows:

```
            :
            LDA     [NUMA]       # Load ACC with NUMA value
            SUB     [NUMB]       # Subtract value in NUMB
            STA     [RESULT]     # Store the result in RESULT
            :
            :
NUMA:       .BYTE   %00111001    # 57 in decimal, $39 in hex
NUMB:       .BYTE   %00011110    # 30 in decimal, $1E in hex
RESULT:     .BYTE                # Store result here
            :
```

Inside the CPU, the SUB instruction causes the unmodified version of NUMA and the inverted (ones complement) version of NUMB to be presented to an adder function, and the adder's cin ("carry-in") input to be presented with a 1 value. The result is automatically stored in the accumulator, while the cout ("carry-out") signal from the adder is automatically stored in the C (carry) status flag.

The Borrow Flag

Just to add to the confusion, when we subtract one binary number from another, we don't actually generate *carries* but, instead, we generate *borrows*. Thus, although we typically continue to refer to this status bit as the carry flag, it actually ends up containing the status pertaining to any requirements for a borrow in the case of a SUB instruction.

To further muddy the waters, *borrows* act in an opposite manner to *carries*. By this we mean that if the carry flag contains a 1 following a SUB instruction, this indicates a "no borrow" condition; but if the carry flag ends up containing a 0, then this reflects the fact that a borrow was generated. Due to this inversion, the contents of the carry flag following a SUB instruction (or it's SUBC cousin introduced in the next section) are sometimes referred to as representing a *borrow-out-not* or a *borrow-not* condition.

In order to fully grasp this concept, let's remind ourselves that we're still dealing with unsigned binary numbers, in which case an 8-bit field can be used to represent values in the range 00000000_2 to 11111111_2 (or 0 to 255 in decimal). First, let's assume that we wish to perform two additions as illustrated in Figure 4-18.

In the case of Figure 4-18(a), we're adding 184_{10} and 27_{10}, which means that the result of 211_{10} will fit in an 8-bit field; thus, no carry is generated and the carry flag ends up containing a 0. By comparison, if we add 184_{10} and 86_{10} as shown in Figure 4-18(b), the result of 270_{10} is too big to store in an 8-bit field; thus, a carry is generated and the carry flag ends up containing a 1.

With regard to this second addition, if we consider only the value stored in the 8-bit result field (00001110_2 in binary), this equates to 14_{10}, which is somewhat meaningless in this context. In this case, the carry flag simply serves to indicate that the result was too big to fit in an 8-bit field. On the other hand, if we consider the carry flag to also represent the 256s

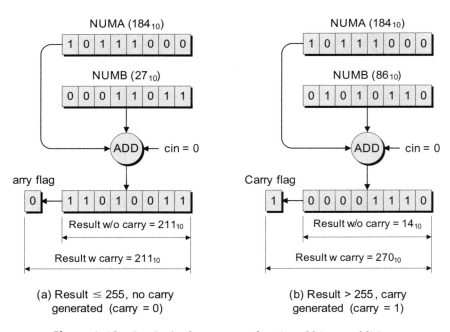

Figure 4-18. Carries in the context of unsigned binary additions.

column of a 9-bit field, then the result is 100001110_2 in binary, which equates to the correct answer of 270_{10} in decimal. This is, of course, the way in which our multibyte additions function when we use the ADDC instruction to feed the carry from a previous addition forward into the current addition.

Now let's consider two subtractions as illustrated in Figure 4-19. In the first example [Figure 4-19(a)], we're subtracting 86_{10} from 184_{10}. In this case, the result of 98_{10} is greater than 0, so the carry flag ends up containing a 1, indicating that no borrow was generated.

By comparison, in the second example [Figure 4-19(b)], we're attempting to subtract 184_{10} from 86_{10}. The problem is that this results in -98_{10}, which is a negative value. However, we are currently dealing with unsigned binary numbers, which can be used to represent only positive values. Thus, the value stored in the 8-bit result field (10011110_2 in binary) equates to 158_{10}, which has little meaning in this context. In this case, the carry flag ends up containing a 0, thereby indicating that a borrow *was* generated because the result was too *small* to fit in our 8-bit field.

We will see how this borrow is used in the next section, after which we will start to consider signed binary numbers that do allow us to represent both positive and negative values.

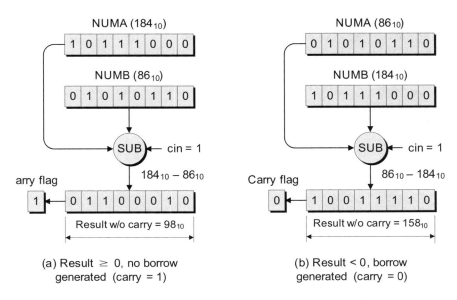

Figure 4-19. Borrows in the context of unsigned binary subtractions.

> **Note**   The symbols < and > are used to represent the concepts *less than* and *greater than*, respectively. These first appeared in a book called *Artis Analyticae Praxis ad Aequationes Algebraicas Resolvendas* ("The Analytical Arts Applied to Solving Algebraic Equations"), which was written by the English mathematician Thomas Harriot (1560–1621) and published posthumously in 1631.
>
> Later mathematicians started to combine these symbols with the equality character to give two new symbols ≤ and ≥, meaning *less than or equal to* and *greater than or equal to*, respectively. Similarly, the symbol ≠ is used to indicate inequality or *not equal*.
>
> Due to that fact that these special symbols are not always available in word-processing programs and text editors, and also the fact that programming languages (and their associated parsers, compilers, and/or assemblers) tend to employ only ASCII characters, the character combinations <=, >=, and != are often used to represent the concepts of *less than or equal, greater than or equal,* and *not equal*, respectively.

Multibyte Subtractions and the SUBC Instruction

Earlier in this chapter, we discussed the concept of using multiple bytes to represent larger numbers than could be held in a single 8-bit field. For example, a 2-byte (16-bit) field can assume any one of $2^{16} = 65,536$ different binary patterns, which can be used to represent positive integers in the range 0 to 65,535 in decimal (or $0000 to $FFFF in hexadecimal).

As we previously noted, in the case of multibyte additions, our virtual computer supports two instructions called ADD ("add") and ADDC ("add with carry"). Similarly, in the case of multibyte subtractions, we have the corresponding SUB ("subtract") and SUBC ("subtract with carry") instructions.[4]

For example, consider some assembly code that can subtract one 16-bit number called NUMB from another called NUMA and store the result in a 16-bit field:

```
        :
    LDA     [NUMA+1]    # Load ACC with LSB of NUMA
    SUB     [NUMB+1]    # Subtract LSB of NUMB
    STA     [RESULT+1]  # Store LSB of result in RESULT
    LDA     [NUMA]      # Load ACC with MSB of NUMA
    SUBC    [NUMB]      # Subtract MSB of NUMB
    STA     [RESULT]    # Store MSB of result in RESULT
```

[4]As opposed to a SUBC ("subtract with carry") instruction, some assembly languages use the mnemonic SUBB ("subtract with borrow"). Irrespective of the mnemonic, these would both end up using the carry flag and the same machine code instruction as seen by the computer.

```
                :
                :
NUMA:       .2BYTE   $2056        # 8278 in decimal
NUMB:       .2BYTE   $10B8        # 4280 in decimal
RESULT:     .2BYTE                # Store result here
                :
```

In particular, observe the use of the SUBC ("subtract with carry") instruction for the second portion of the addition (the line highlighted in gray). Unlike the SUB instruction, in which the cin ("carry-in") input to the adder is forced to 1, the SUBC instruction causes the current contents of the C (carry) flag, representing the borrow from the subtraction of the least-significant bytes, to be presented to the adder's cin input.

Now consider the least-significant bytes of NUMA and NUMB, which are $56 (or 86_{10} in decimal) and $B8 (or 184_{10} in decimal). By some strange quirk of fate, these are the same values we used when performing the 1-byte subtraction in Figure 4-19(b). In that case, the result was meaningless because it was smaller than zero, but the same result does have meaning now that we're dealing with multibyte values.

Let's work this problem through by hand in the same way in which our virtual computer will execute the assembly code above. The operation we wish to perform is as follows:

$$\$2056 - \$10B8$$

First, we use a SUB instruction to subtract the least-significant byte of NUMB from the least-significant byte of NUMA. If we were performing this subtraction directly, it would appear as follows:

$$\$56 - \$B8$$

But we know that the way in which the computer will actually execute this task is to take the ones complement of $B8 (or %10111000 in binary), which will return $47 (or %01000111 in binary), to force the adder's cin input to 1, and to perform the following operation:

$$\$56 + \$47 + 1$$

The result is $9E with a carry-out of 0, which will be stored in the carry flag. As we now know, this carry-out of 0 reflects a borrow. In turn, this indicates that the result is too small (that is, negative) to fit in our 8-bit field, but this isn't an issue in this case because we have still the most-significant bytes to consider. Now, we use a SUBC instruction to subtract the most-significant byte of NUMB ($10) from the most-significant byte of NUMA ($20) This time, we know that the way the computer will execute this task is to take the ones complement of $10 (or %00010000 in binary), which will return $EF (or %11101111 in binary), and to perform the following operation:

$$\$20 + \$EF + cin$$

In this case, we know that cin will be fed from the current contents in the C (carry) flag, and that this value will be 0, representing the borrow from the previous subtraction, so the result will be $0F with a carry-out of 1. This carry-out of 1 reflects a no-borrow situation, which, in turn, indicates that the output from this subtraction will fit in the 8-bit field that represents the most-significant byte of our result. In summary, we just performed the following multibyte subtraction:

$$\$2056 - \$10B8 = \$0F9E$$

which equates to $8,278_{10} - 4,280_{10} = 3,998_{10}$ in decimal.

> **Note** This would be a good time to work your way through Lab 4c, in which we create a general-purpose subroutine to perform 16-bit subtractions (we will be using this subroutine as part of the integer-based calculator program we are going to build in Chapter 5).

Sign-Magnitude Representations

As we've come to appreciate, the problem with using unsigned binary numbers is that, by definition, they cannot be used to represent negative quantities. In the real world, this would obviously be somewhat limiting, because we would only ever be able to subtract smaller numbers from bigger ones.

In the case of standard decimal representations, negative values are indicated using a minus sign. For example, negative twenty-seven would be shown as –27. This is referred to as a *sign-magnitude* format (that is, a sign followed by an associated magnitude), and it means that the only difference between a positive quantity like +27 and its negative counterpart –27 is the sign.

If we wished, we could apply the same sign-magnitude concept to binary numbers. For example, assuming we're dealing with 8-bit values, we could designate the most-significant bit as being the *sign bit* (where 0 and 1 in this sign bit indicate positive and negative values, respectively). Meanwhile, the remaining seven bits could be used to represent the magnitude of the number. This would allow us to represent negative and positive integers in the range –127 to +127 (Figure 4-20).

There are several problems with this format, not the least that we now have both positive and negative versions of zero (+0 and –0). What should happen if we were to compare these values (is +0 conceptually bigger than –0)?

In fact, if a computer were forced to use sign-magnitude representations, it would be obliged to perform a whole slew of operations. Even in the case of a simple addition of two 8-bit values such as A[7:0] +

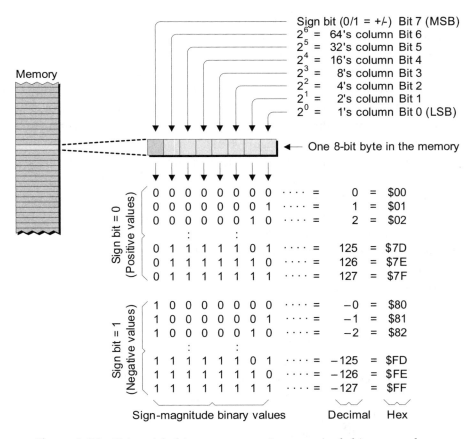

Figure 4-20. Using eight bits to represent sign-magnitude binary numbers.

B[7:0], for example, the computer would first have to check the signs of the numbers. If the signs were the same (both positive or both negative), the computer could simply add the two values together (excluding the sign bits), because the result would always have the same sign as the original numbers. If the signs were different, however, the computer would have to compare the two magnitudes to determine which was the smaller, subtract this from the larger value, and then ensure that the correct sign was appended to the result.

This would be terribly tedious and time-consuming to say the least. If only there were some other way...

Signed Binary Numbers

The cunning ruse underlying the concept of signed binary numbers is that the sign bit is used to signify a negative *quantity* (not just the sign). As we originally showed in Figure 4-1, in the case of an *unsigned* binary number, the most significant bit in an 8-bit field is used to represent 2^7 (the +128s column). By comparison, in the case of a *signed* binary number, this bit (the sign bit) is now used to represent -2^7 (the −128s column). Meanwhile, the remaining seven bits continue to represent their original positive quantities (Figure 4-21).

This means that we start off with 00000000_2 representing 0_{10} as usual, and work our way up to 01111111_2, which represents $+127_{10}$. However, instead of representing $+128_{10}$ as is the case with an unsigned binary number, the value 10000000_2 now represents -128_{10}.

As we already noted, even when the sign bit is 1, the remaining bits continue to represent positive quantities. Thus, 10000001_2 equates to $-128_{10} + 1_{10} = -127_{10}$; 10000010_2 equates to $-128_{10} + 2_{10} = -126_{10}$; 10000011_2 equates to $-128_{10} + 2_{10} + 1_{10} = -125_{10}$, and so on to 11111111_2, which equates to $-128_{10} + 64_{10} + 32_{10} + 16_{10} + 8_{10} + 4_{10} + 2_{10} + 1_{10} = -1_{10}$.

All of this means that an 8-bit signed binary number can be used to represent values in the range -128_{10} to $+127_{10}$. To more fully illustrate the differences between the sign-magnitude and signed binary formats, consider a positive sign-magnitude decimal number and its negative equivalent; for example, +27 and −27. As usual, the numerical digits are identical in both cases and only the sign changes. Now consider the same

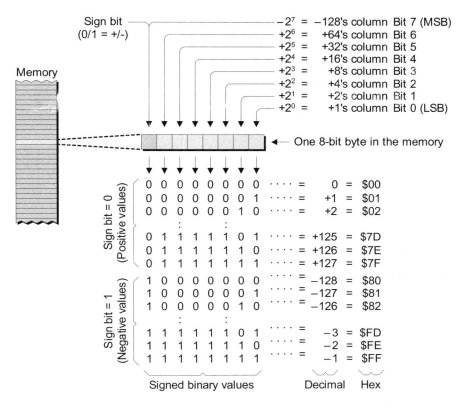

Figure 4-21. Using eight bits to represent signed binary numbers.

values represented as sign-magnitude binary numbers and signed binary numbers (Figure 4-22).

In the case of the positive and negative sign-magnitude binary numbers, only the sign bit changes, whereas the bit patterns for the magnitude components are identical. By comparison, the bit patterns associated with the positive and negative signed binary numbers are very different. As we know, this is because the sign bit represents an actual quantity (-128_{10}) rather than a simple plus or minus; thus, the signed binary equivalent of -27_{10} is formed by combining -128_{10} with $+101_{10}$.

Good grief Charlie Brown!

At a first glance, the signed binary concept may appear to be an outrageously complex solution to a fairly simple problem. In addition to rep-

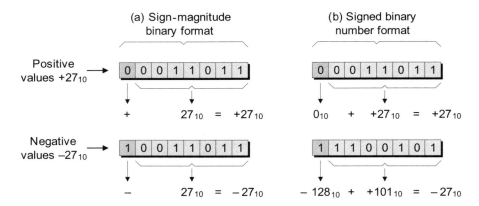

Figure 4-22. Comparison of sign-magnitude and signed binary formats.

resenting an asymmetrical range of negative and positive numbers (-128_{10} to $+127_{10}$ in the case of an 8-bit value), the way in which these representations are formed is, to put it mildly, alien to the way we're used to thinking about numbers.

As you might expect, however, there's reason behind the (apparent) madness, not the least that signed binary numbers only have one version of 0 (as opposed to the +0 and –0 inherent in the sign-magnitude format). Now pay close attention, because this is the really clever part; closer investigation of the two signed binary values shown in Figure 4-22(b) reveals that each bit pattern is the *twos complement* of the other! To put this another way, taking the twos complement of a positive signed binary value returns its negative equivalent, and vice versa.

As we shall soon discover, using this format makes our lives much easier. In fact, surprising as it may seem, signed binary numbers actually make sense at an intuitive level. Consider what happens if we start with 00000000_2 in the accumulator and then loop around incrementing (adding one to) the accumulator until we reach 11111111_2. The next increment will cause the accumulator to overflow and return it to containing 00000000_2. "So what," you may ask. Well, now consider what happens if we go the other way; that is, start with 00000000_2 in the accumulator and then decrement (subtract one from) the accumulator, which will leave the accumulator containing 11111111_2. This exactly matches what happens in decimal, in that subtracting 1 from 0 results in –1; similarly, subtracting 1 from 00000000_2 (or \$00 in hexadecimal) results in 11111111_2, (or \$FF in hexadecimal), which equates to -1 in the signed binary format.

Adding signed binary numbers

The end result of using signed binary numbers (which may also be referred to as twos complement numbers) is to greatly reduce the complexity of operations inside a computer. To illustrate why this is so, consider one of the simpler operations: addition. Compare the additions of positive and negative decimal values in sign-magnitude form with their signed binary counterparts as shown in Figure 4-23.

Let's start by examining the standard decimal calculations on the left-hand side of Figure 4-23. The one in Figure 4-23(a) is easy to understand because it's a straightforward addition of two positive values. However, even though we are extremely familiar with decimal addition, the three other problems in Figures 4-23(b), 4-23(c), and 4-23(d) typically require a little pause for thought while we decide exactly how to handle the negative values.

By comparison, the signed binary calculations on the right-hand side of Figure 4-23 are all simple additions (note that we have omitted

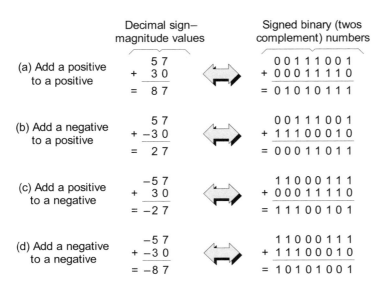

Figure 4-23. Comparison of sign-magnitude decimal and signed binary additions.

any carry-out values from these additions for simplicity; such carry-outs would be automatically stored in the carry flag). Thus, the advantage of the signed binary format for addition operations is apparent: signed binary values can always be directly added together to give the correct result, irrespective of whether they represent positive or negative quantities. That is, when using signed binary numbers, the operations $(+a) + (+b)$, $(+a) + (-b)$, $(-a) + (+b)$, and $(-a) + (-b)$ are all performed in exactly the same way, which is by simply adding the two values together. This results in computers that are simpler to design, use fewer logic gates, and perform calculations faster.

Subtracting signed binary numbers

Now let's consider the case of subtraction. We all know that $10 - 7 = 3$ when using decimal arithmetic. We also know that we can obtain the same result by negating the right-hand value and inverting the operation; that is, $10 + (-7) = 3$.

The same technique is also true for signed binary numbers. In this case, however, the negation of the right-hand value is performed by taking its twos complement rather than changing its sign. For example, consider a generic signed binary subtraction represented by $a - b$. Generating the twos complement of b results in $(-b)$, thereby allowing the operation to be performed as an addition: $a + (-b)$.

This means that computers don't require an adder and a subtractor; instead, they require only an adder and some way to generate the twos complement of a number. And, of course, this is exactly what we were doing all along when we introduced the concept of binary subtraction in the context of unsigned binary numbers.

One final point to ponder before we continue is that it's up to us to decide what the binary values stored in a computer represent at any particular time. For example, depending on the task we are trying to perform, we may consider the 8-bit value 11111111_2 to be an *unsigned* binary number, in which case it represents 255_{10} in decimal. But a little later we may decide to consider the same 11111111_2 value to be a *signed* binary number, in which case it represents -1_{10} in decimal. The really cool thing is that the computer doesn't care: it performs additions and subtractions exactly the same way, irrespective of whether we are considering the numbers to represent signed or unsigned quantities [see also the discussions on the O (overflow) flag in the next section].

The only problem

The only problem with the signed binary format is that, due to its asymmetrical range, the largest negative number cannot be negated. In the case of an 8-bit (1-byte) signed binary number, for example, we can't negate -128_{10} to form $+128_{10}$ because the largest positive value that's supported is $+127_{10}$. We typically get around this issue by restricting ourselves to working only with the range -127_{10} to $+127_{10}$ in the case of an 8-bit field. Alternatively, if we are using 16-bit (2-byte) signed binary numbers, which can be used to represent values from $-32,768_{10}$ to $+32,767_{10}$, we will restrict ourselves to working only with the range $-32,767_{10}$ to $+32,767_{10}$.

> **Note** This would be a good time to perform Lab 4d, in which we create a 16-bit negation subroutine; that is, one that converts a 16-bit positive signed binary number into its negative counterpart, and vice versa.

The Overflow Flag

As we know, the C (carry flag) is used to store any carries generated by the ADD and ADDC instructions and any borrows generated by the SUB and SUBC instructions. In the case of *unsigned* binary numbers, the carry flag is also used to indicate whether or not the result from the operation will fit into the 8-bit accumulator. Sad to relate, however, there's a slight "gotcha" when it comes to working with *signed* binary representations.

Let's consider a simple ADD instruction. When we were discussing this instruction in the context of *unsigned* binary numbers, we stated that, following the addition, a 0 in the carry flag would indicate that the operation was successful; that is, the result fit in the 8-bit accumulator. By comparison, a 1 in the C (carry) flag would indicate that the result was too large to fit in the accumulator.

Consider the addition illustrated in Figure 4-24. If we assume that we're working with *unsigned* binary numbers, then our 8-bit accumulator can represent values in the range 0_{10} to 255_{10}. In this case, adding 01010100_2 (84_{10} in decimal) to 00111001_2 (57_{10} in decimal) will result in 10001101_2 (141_{10} in decimal). As this value is less than 255_{10}, it fits in our accumulator and the carry flag is loaded with 0.

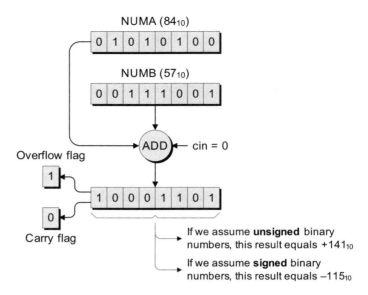

Figure 4-24. Introducing the O (overflow) flag.

Now consider what happens if we add exactly the same numbers together, but this time we regard them, and the result, as representing *signed* binary values. In this case, our 8-bit accumulator can represent values in the range -128_{10} to $+127_{10}$, which means that the result of $+141_{10}$ is too large. Due to the fact that we're assuming signed binary numbers, we would regard the ensuing 10001101_2 value as equating to -115_{10}, which is meaningless in the context of adding 84_{10} to 57_{10}.

The problem is that the C (carry) flag doesn't help us here. Thus, the computer also features an O (overflow) status flag, which is generated as an XOR (exclusive OR)[5] of the carry signals going in to and out of bit 7 of the result. (The overflow flag works exactly the same way with additions and subtractions of positive and negative signed binary values.)

What all of this means is that, if we decide to treat our numbers as representing *unsigned* binary values, then we ignore the O (overflow) flag and we use the C (carry) flag to determine whether or not the result fits in the accumulator. Alternatively, if we decide to treat our numbers as representing *signed* binary values, then we rely on the O (overflow) flag

[5]The concept of the logical XOR function was introduced in Chapter 3.

to determine whether or not the result is OK (overflow = 0), or whether we have an overflow or underflow condition (overflow = 1).

In order to make use of the overflow flag, the computer supports two special jump instructions: JO ("jump if overflow"), which will jump if the overflow flag contains a 1, and JNO ("jump if not overflow") which will jump if the overflow flag contains a 0.

Note that in the case of multibyte additions and subtractions on both *signed* and *unsigned* binary values, we still use the contents of the carry flag to propagate carries and borrows up the line. Following the final addition or subtraction operation on the most significant bytes of *unsigned* values, we ignore the overflow flag and we test the value in the carry flag (in the case of an addition, a 0 in the carry flag tells us that the result is OK, whereas a 1 indicates an overflow condition; in the case of a subtraction, a 1 in the carry flag tells us that the result is OK, whereas a 0 indicates an underflow condition).

By comparison, following the final addition or subtraction operation on the most significant bytes of *signed* values, we ignore the carry flag and instead test the overflow flag (irrespective of whether we are performing an addition or a subtraction, a 0 in the overflow flag tells us that the result is OK, whereas a 1 indicates either an overflow or an underflow condition).

Note   This would be a good time to perform Lab 4e, in which we modify our 16-bit add and subtract routines to check the overflow flag and generate an error message if an overflow (or underflow) condition occurs.

Multiplying Signed Binary Numbers
Multiplying by shifting left

Before we plunge headfirst into the fray, it's worth noting that there's a shortcut to multiplying binary numbers by powers of two (that is, $2^1 = 2$, $2^2 = 4$, $2^3 = 8$, etc.), which is to shift the value to the left by the same number of bits as the exponent.

For example, if we take the binary value 00011011_2 (or $+27_{10}$ in decimal) and shift it one bit left, we end up with 00110110_2 (or $+54_{10}$ in decimal), which is the same as multiplying our original number by

> **Note**   A number that is to be multiplied by another is called the *multiplicand,* from the Latin *multiplicandum,* meaning "to multiply." By comparison, the number with which another number is to be multiplied is called the *multiplier.*
>
> The result obtained from the multiplication of two or more numbers is called the *product,* from the Latin *producere,* meaning "to bring forth."
>
> Thus, in the equation $3 \times 6 = 18$, the multiplier is 3, the multiplicand is 6, and the product is 18. (If it seems to you that the terms multiplier and multiplicand have been applied the wrong way round, try thinking of the above equation in terms of "three sixes are."

$2^1 = 2$. Similarly, if we had taken our original number and shifted it two bits to the left, we would have ended up with 01101100_2 (or $+108_{10}$ in decimal), which is the same as multiplying our original number by $2^2 = 4$.

This also works for signed (twos complement) binary numbers. For example, if we assume the binary value 11100010_2 is in the signed binary format (equating to -30_{10} in decimal), then shifting this value one bit to the left results in 11000100_2 (or -60_{10} in decimal).

High-level computer languages such as the C programming language use software tools called *compilers* to translate their human-readable source code representations into the machine code that is to be executed by the computer.[6] Depending on the capabilities of the target machine, these compilers may be instructed to look for special cases like

```
fred = bert * 2;
```

where `fred` and `bert` are integer variables and the asterisk character "*" indicates a multiplication operation. In such a case, the compiler might replace this multiplication by a simple shift left. Alternatively, the original programmer may decide to do this by hand using

```
fred = bert << 1;
```

which means the variable `bert` is to be shifted left by one bit and the result is to be assigned to the variable `fred`. If fact, when computers were less powerful than they are now, programmers used to employ all sorts of tricks like this; for example, suppose we wanted `fred` to be made equal to `bert` multiplied by nine:

```
fred = bert * 9;
```

---

[6]In some cases, an intermediate step may be used whereby the compiler translates the source code into assembly language, which is subsequently assembled into machine code.

then we could achieve the same effect by means of the following statement:

```
fred = (bert << 3) + bert;
```

where "(bert << 3)" means "shift bert three bits to the left," which has the same effect as multiplying bert by eight. So the whole expression equates to "(bert * 8)+ bert." This gives the same result as our original "bert * 9" statement, but the "shift and add" approach is a lot faster on microcomputers or microcontrollers with limited computational resources.[7]

Beware! It's easy to run into trouble here if you aren't careful. For example, if we were to take the binary value 10110110 and shift it one bit to the left, we'd end up with 01101100. This is, of course, perfectly acceptable if we're considering the accumulator to contain a simple pattern of 0s and 1s, because the result is exactly what we'd expect. However, if we're visualizing these values to represent signed (twos complement) binary numbers then we have a problem, because we started with

---

Note   The now-familiar × symbol was first used to indicate multiplication in the 1631 tome *Clavis Mathematicae* ("Key to Mathematics"), which was written by the English mathematician and clergyman William Oughtred (1574–1660).[8]

Unfortunately, although the meaning of the × symbol is clear when the operands are numbers (for example, 3 × 6 = 18), it can easily be confused in algebraic formulas that use letters to represent variables (for example, x × y = z). For this reason, sometime during the 1690s, the German mathematician Gottfried von Leibniz (1646–1716) started to use a raised dot to indicate multiplication (for example, x·y = z).[9, 10]

In his 1659 book *Teutsche Algebra* ("German Algebra"), the Swiss mathematician Johann Rahn (also known as Johann Rhonius, 1622–1676) used the asterisk character "*" to indicate a multiplication (for example, 6 * 3 = 18), and it is this symbol that is now commonly used in computer programming languages and on calculator front panels.

---

[7]Depending on the application and the capabilities of the target machine, programmers who need as much speed as possible may still resort to using tricks like this.

[8]In 1621, Oughtred invented the slide rule, which was based on the concept of logarithms. In turn, logarithms were invented by the Scottish mathematician and Laird of Merchiston, John Napier (1550–1617). See Appendix D for more details on additional resources, including *The History of Calculators, Computers, and Other Stuff* on the CD accompanying this book.

[9]Others may have used the raised dot before Leibnitz, but this convention didn't really catch on until he started to champion it.

[10]Leibniz also invented a mechanical calculator called the Step Reckoner in the 1670s. Once again, see Appendix D for more details on additional resources, including *The History of Calculators, Computers, and Other Stuff* on the CD accompanying this book.

10110110_2 (or -74_{10} in decimal), multiplied this value by two by shifting it one bit to the left, and ended up with 01101100_2 (or $+108_{10}$ in decimal), which is obviously incorrect. The problem is that, as per the discussions associated with Figure 4-21, our 8-bit accumulator can only represent signed binary numbers in the range -128_{10} to $+127_{10}$, which means that it isn't large enough to store the result generated by multiplying -74_{10} by two.

Decimal multiplication using a "shift-and-add" technique

Modern computers typically have special multiplication units that they use to multiply two binary values together (we'll discuss this in a little more detail below). However, early microprocessors didn't have this capability, and neither does our virtual computer, which means that we will have to perform multiplication operations using only the simple instructions (add, subtract, shift, rotate, etc.) that are available to us.

One technique for performing multiplication in any number base is by means of repeated addition; in decimal, for example: $6 \times 4 = 4 + 4 + 4 + 4 + 4 + 4 = 24$. Although simple, this technique can be very time-consuming if the numbers being multiplied together are large.

Another technique is based on the concept of "shift-and-add." As usual, we'll begin by considering a standard decimal multiplication in order to ensure that we're all marching along in step. Suppose we wish to perform the operation $3,120_{10} \times 33,201_{10}$. We can visualize this as illustrated in Figure 4-25.

Note that the values chosen for the multiplier and multiplicand have no significance beyond the purposes of this example.

Using this technique, a *partial product* is generated for each digit in the multiplier. Each partial product is a copy of the multiplicand that has been multiplied by its associated multiplier digit. Also, each partial product is left-shifted by the number of digits equal to the value of the exponent in the multiplier column with which it is associated. For example, consider the partial product associated with the units column in the multiplier ($33,201 \times 0 = 00,000$). As this is the 10^0 column, the exponent is 0 (zero), so this partial product is left-shifted by zero digits, which means that it isn't left-shifted at all. In the case of the multiplier's tens (10^1) column, the exponent is 1, so this partial product ($33,201 \times 2 = 66,402$) is left-shifted by one digit, which corresponds to an additional multipli-

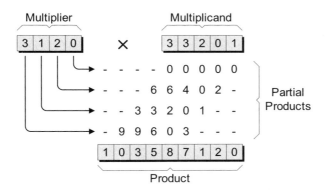

Figure 4-25. Decimal multiplication using a shift-and-add technique.

cation by 10. Similarly, the partial product associated with the hundreds (10^2) column in the multiplier ($33,201 \times 1 = 33,201$) is left-shifted by two digits, which corresponds to an additional multiplication by 100. And the partial product associated with the thousands (10^3) column in the multiplier ($33,201 \times 3 = 99,603$) is left-shifted by three digits, which corresponds to an additional multiplication by 1000.

Once all of the partial products have been generated, they are added together to give the final product (result). Observe that the width of the result (nine decimal digits) equals the sum of the widths of the multiplier (five digits) and the multiplicand (four digits).

Binary multiplication using a "shift-and-add" technique

The same shift-and-add technique as discussed above can also be used to multiply binary numbers together. For example, suppose we wish to perform the operation $01110011_2 \times 00010110_2$. We can visualize this as illustrated in Figure 4-26.

Once again, a *partial product* is generated for each digit in the multiplier. Each partial product is a copy of the multiplicand that has been multiplied by its associated multiplier digit. Also, each partial product is left-shifted by the number of digits equal to the value of the exponent in the multiplier column with which it is associated. Thus, the partial product associated with bit 0 in the multiplier, which represents the 2^0 column, is left-shifted by 0 bits (that is, it isn't shifted at all); the partial product associated with bit 1 in the multiplier, which represents the

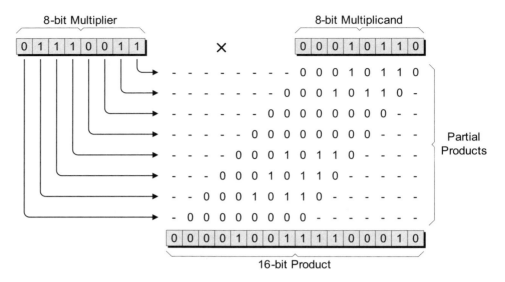

Figure 4-26. Binary multiplication using a shift-and-add technique.

2^1column, is left-shifted by 1 bit; the partial product associated with bit 2 in the multiplier, which represents the 2^2 column, is left-shifted by 2 bits; and so forth.

As fate would have it, the fact that we're working in binary actually makes things a whole lot easier. In the case of a multiplier bit that's 0, the corresponding partial product is all 0s; whereas in the case of a multiplier bit that's 1, the corresponding partial product is simply a copy of the multiplicand.

Once all of the partial products have been generated, they are added together to give the final product (result). In this case, the width of the result equals the sum of the widths of the multiplier and the multiplicand; that is, 8 + 8 = 16 binary digits.

Implementing a hardware multiplier for signed (twos complement) binary numbers

One minor "glitch" with binary multiplication using the shift-and-add technique presented in the previous section is that, assuming we're working with signed binary numbers, it only works if both the multiplier and multiplicand have positive values. We can easily get around this, of course, by converting any negative values into their positive counterparts, performing the multiplication using positive values, and then correcting the sign of the product if necessary.

For some reason, it's generally easier to visualize the way in which this sort of thing works in terms of a hardware implementation using a special block of logic whose sole task is to multiply two binary values together. For example, consider the block diagram for a hardware multiplier intended to multiply two 8-bit values together (Figure 4-27).

As usual, this is much less complicated than it might at first appear, so don't be unduly alarmed. Let's start by considering the 8-bit multiplier in the upper-left corner of this illustration. A direct copy of the multiplier is fed to one side of a multiplexer. The multiplier is also presented to a block of logic that will generate the twos complement of this value (that is, it will turn a positive value into its negative counterpart, and vice versa).

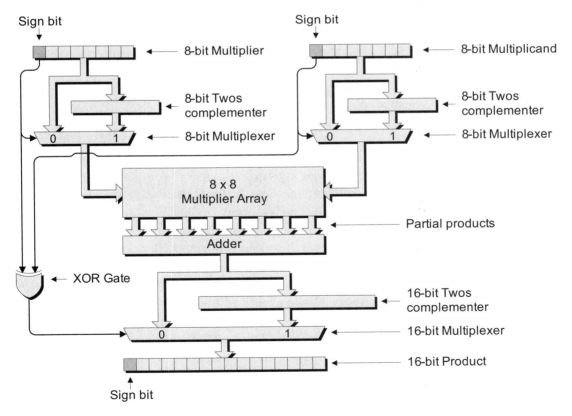

Figure 4-27. Block diagram of an 8-bit × 8-bit hardware multiplier.

The multiplier's sign bit is used to control a multiplexer. If the sign bit is 0, the multiplier represents a positive value, so the sign bit causes the multiplexer to select the direct copy of the multiplier and convey this value to its outputs. Alternatively, if the sign bit is 1, the multiplier represents a negative value, in which case the sign bit causes the multiplexer to select the output from the twos complement block (where this value is the positive equivalent of the multiplier).

Similar actions are performed on the 8-bit multiplicand in the upper-right corner of Figure 4-27. Thus, the multiplier and multiplicand values fed into the multiplier array are always positive. The multiplier array simultaneously generates all eight partial products, which are fed into an adder block that sums them all together.

At this point, we have to decide whether or not to correct the sign of the result (that is, whether we have to convert the output from the adder into a negative value). The way in which we do this is rather elegant: consider the rules associated with multiplying positive and negative values together:

+ multiplied by + equals +
+ multiplied by − equals −
− multiplied by + equals −
− multiplied by − equals +

Now observe that the two sign bits from the original multiplier and multiplicand are fed into an XOR gate, and that the output from this XOR gate is used to control a multiplexer. From Chapter 3, we know that the truth table for an XOR gate is as shown in Figure 4-28.

If we consider the 0 and 1 values in this truth table to represent + and −, respectively, then the output from the XOR gate tells us whether the result from our multiplication should be positive or negative.

The 16-bit output from the adder block is fed directly to one side of the multiplexer and also to a block of logic that will generate the twos complement of this value (the output from this twos complement block is presented to the other side of the multiplexer). In turn, the output from the multiplexer represents the final product (result) from the multiplication. If the output from the XOR gate is 0, the result from the multiplication should be positive, so the XOR gate causes the multiplexer to select the output from the adder block and convey this value to its outputs. Alternatively, if the output from the XOR gate is 1, the result from

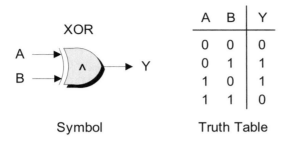

Figure 4-28. Truth table for an XOR gate.

the multiplication should be negative, so the multiplexer is directed to select the output from the twos complement block.

Some CPUs are equipped with this type of hardware multiplier (although in the real world it would feature a few more "bells and whistles"), in which case their associated assembly language will contain a MULT ("multiply") instruction. Sad to relate, simple CPUs, such as the one featured in our virtual DIY Calculator, do not contain such a unit. This means that they are obliged to implement the multiplication using only the basic add, subtract, shift, rotate, and other instructions that are available to them.

> **Note** This would be a good time to work your way through Lab 4f, in which we create a general-purpose subroutine to perform 16-bit multiplications (we will be using this subroutine as part of the integer-based calculator program we are going to build in Chapter 5).

Dividing Signed Binary Numbers
Dividing by shifting right

And so, finally, we turn our attention to division, which is one of the trickiest of the basic math operations we have to perform.

> **Note** A number that is to be divided by another is called the *dividend*, whereas a number that is to be divided into another number is called the *divisor* (this may also be known as the *factor* if it divides a given quantity without leaving any remainder).
>
> The result obtained from dividing one number by another is called the *quotient*, from the Latin *quotiens*, meaning "how many times." And the term *remainder* refers to anything that is left over when one integer is divided by another.

Before we tackle this topic, however, it's worth noting that there's a shortcut to dividing binary numbers by powers of two (that is, $2^1 = 2$, $2^2 = 4$, $2^3 = 8$, etc.), which is to shift the value to the right by the same number of bits as the exponent.

For example, if we take the binary value 01101100_2 (or $+108_{10}$ in decimal) and shift it one bit right, we end up with 00110110_2 (or $+54_{10}$ in decimal), which is the same as dividing our original number by $2^1 = 2$. Similarly, if we had taken our original number and shifted it two bits to the right, we would have ended up with 00011011_2 (or $+27_{10}$ in decimal), which is the same as dividing our original number by $2^2 = 4$.

Note that, if we divide an odd number like 00010001_2 (or $+17_{10}$ in decimal) by two using this shift right technique, we will end up with 00001000_2 (or $+8_{10}$ in decimal), and the 1 that ends up in the carry flag following the shift operation represents the remainder from the division (that is, $17 \div 2 = 8$ with a remainder of 1).

Of course, this also works for signed (twos complement) binary numbers. For example, if we assume the binary value 11000100_2 is in the signed binary format (equating to -60_{10} in decimal), then shifting this value one bit to the right results in 11100010_2 (or -30_{10} in decimal).

This is very important! With regard to the previous point, it's important to note that the reason this works is because our virtual computer's SHR ("shift right") instruction is an *arithmetic* shift right, which means that the most-significant bit (the sign bit) is copied back on itself (you may wish to review the discussions on arithmetic versus logical shifts in Chapter 3; in particular, take a glance at Figure 3-3).

Performing a division in decimal

In order to set the scene, let's first consider the way in which we might perform a division operation in decimal; for example, dividing 14_{10} into 662_{10} (Figure 4-29).

In order to perform this calculation, our thought processes may go something like this. We know that we can't divide 14 into 6, so we take the next option and try to divide 14 into 66, but how should we set about doing this? The majority of us would probably take a stab at it and say: "Hmmm, let's try $3 \times 14 = 42$. No, that's too small. How about $5 \times 14 = 70$? No, that's too big. Let's try $4 \times 14 = 56$. Yes, that's about as close as we can get."

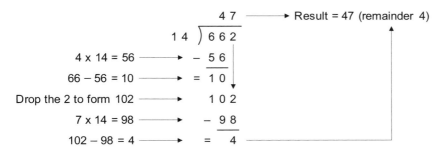

Figure 4-29. Performing a long division in decimal.

Now, we know that the first digit in our quotient (result) is 4 (from the 4 × 14), so we write this down. Next, we subtract 56 from 66 leaving 10; drop the 2 to form 102; and go through our "Goldilocks and the three bears" process again,[11] saying: "6 × 14 = 84, but that's too small; 8 × 14 = 112, but that's too big; 7 × 14 = 98, and that's just right (or, at least, as close as we can get)."

So we now know that the second digit in our quotient is 7, and we write it down. Finally, we subtract 98 from 102 leaving 4, which (remembering that we're performing an integer division) is too small to be divided by 14. Thus, our final result is 47 with a remainder of 4.

Note   As we previously noted, division is conceptually the most complex of the fundamental arithmetic operations (+, −, *, /), which may serve to explain why there are so many symbols associated with it!

One common symbol is a *vinculum*, from the Latin *vincre*, meaning "to bind" or "to tie together." In its general sense, this refers to a horizontal line drawn over two or more algebraic terms to indicate that they are to be treated as a single entity. Figure 4-30(a) shows a vinculum used to bind the values "2 + 1" together. Over time, however, it also became common to use a vinculum to represent divisions involving only single terms, as illustrated in Figure 4-30(b).

In the early 1600s, a colon ":" was used to indicate fractions and ratios, such as 3:5 (meaning "three fifths"). Sometime later, around the early to mid-1680s, the German mathematician Gottfried von Leibniz started to use colons for both fractions and to indicate division operations, as shown in Figure 4-30(c).

In his 1659 book, *Teutsche Algebra* ("German Algebra"), the Swiss mathematician Johann Rahn was the first to use a symbol called the *obelus* as illustrated in Figure 4-30(d). More recently, it became common to use the forward slash symbol "/"—called a *solidus* or *virgule*, as illustrated in Figure 4-30(d), and it is this symbol that is now commonly used in computer programming languages and on calculator front panels.

Last but not least, the symbol used in Figures 4-29 and 4-30(e) (a combination of a vinculum attached to the top of a closing parenthesis) first appeared in U.S. textbooks around the early 1880s. Although this symbol is now very commonly used in the case of long division problems, it does not have an official name.

---

[11]Or the "suck it and see" approach in the vernacular used by engineers.

$$\frac{6}{2+1}$$

a) Vinculum
(multiple terms)

$$\frac{6}{3}$$

b) Vinculum
(single terms)

6:3

c) Colon

6 ÷ 3

d) Obelus

6/3

d) Solidus

$$1\ 4\ \overline{)\ 6\ 5\ 8}$$ with 47 above

e) No name

Figure 4-30. Alternative symbols used to represent division.

Performing a division in binary

When it comes to performing a division in binary, we can use a similar process to the one we used for decimal; for example, let's assume that we wish to divide 0010_2 into 0111_2 in binary (2_{10} into 7_{10} in decimal), as shown in Figure 4-31.

In some respects, binary divisions are simpler than their decimal cousins, because the divisor can either be directly subtracted from the portion of the dividend currently under consideration or it can't; that is, we don't have to worry about experimenting with different multiples of the divisor.

In the case of this particular example, we know that we can't divide 10_2 into 0_2, so we set the first digit in our quotient (result) to 0. Similarly, we know that we can't divide 10_2 into 01_2, so we set the second digit in the quotient to 0. However, we can divide 10_2 into 011_2, so we set the third digit in our quotient to 1.

Subtracting 10_2 from 011_2 leaves 01_2, at which point we drop the next 1 from the dividend to form 011_2. Once again, we know that we can divide 10_2 into 011_2, so we set the fourth digit in our quotient to 1. This

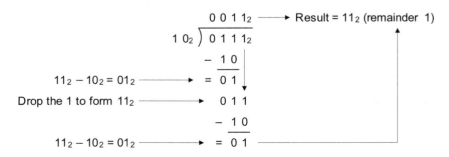

Figure 4-31. Performing a long division in binary.

leaves a value of 01_2, which is too small to be divided by 10_2. Thus, the final result from dividing 0111_2 by 10_2 is 11_2 with a remainder of 1 (that is, 7_{10} divided by $2_{10} = 3_{10}$ with a remainder of 1).

As we shall see, the algorithm we use to make our calculator perform binary divisions only works if both the divisor and dividend have positive values. This means that we will have to convert any negative values into their positive counterparts, perform the division using positive values, and then correct the sign of the quotient if necessary.

Rounding

Finally, let us turn our attention to the concept of rounding. In order to set the scene, let's assume that we have just performed eight decimal division operations using a pencil and paper, and that the results from these divisions are as follows:

(a) 27.0   (b) 27.4   (c) 27.5   (d) 27.6

(e) 28.0   (f)   28.4   (g) 28.5   (h) 28.6

If we have made the decision to work only with integer values, then we will ultimately have to discard any digits to the right of the decimal point. We could simply truncate our answers by "throwing" these digits away, but if we use these results as part of future operations, we can end up with "creeping errors." As an alternative, we can use the digits to the right of the decimal point as part of a rounding algorithm so as to increase the accuracy of the final result.

In fact, there are a plethora of different rounding schemes (of which the truncation approach discussed above is one) that may be used depending on the application. The more common techniques are listed below.

- **Round-down.** Also known as "chopping," this refers to rounding toward 0 or truncating. In this case, the digits to be discarded (those to the right of the decimal point for the purposes of these discussions) are simply ignored:

    (a) 27.0 → 27   (b) 27.4 → 27   (c) 27.5 → 27   (d) 27.6 → 27
    (e) 28.0 → 28   (f) 28.4 → 28   (g) 28.5 → 28   (h) 28.6 → 28

- **Round-up.** This refers to rounding away from 0. In this case, if all of the digits to the right of the decimal point are zero, the integer portion of the result is unchanged; otherwise it will be incremented by 1:

(a) $27.0 \rightarrow 27$ (b) $27.4 \rightarrow 28$ (c) $27.5 \rightarrow 28$ (d) $27.6 \rightarrow 28$

(e) $28.0 \rightarrow 28$ (f) $28.4 \rightarrow 29$ (g) $28.5 \rightarrow 29$ (h) $28.6 \rightarrow 29$

- **Round-half-up.** If the digits to be discarded are *greater than* or *equal to* 0.5, then the integer portion of the result will be incremented by 1; otherwise it will be left unchanged:

  (a) $27.0 \rightarrow 27$ (b) $27.4 \rightarrow 27$ (c) $27.5 \rightarrow 28$ (d) $27.6 \rightarrow 28$

  (e) $28.0 \rightarrow 28$ (f) $28.4 \rightarrow 28$ (g) $28.5 \rightarrow 29$ (h) $28.6 \rightarrow 29$

- **Round-half-down.** If the digits to be discarded are *greater than* 0.5, then the integer portion of the result will be incremented by 1; otherwise (if the digits to be discarded are *equal to* or *less than* 0.5) the integer portion will be left unchanged:

  (a) $27.0 \rightarrow 27$ (b) $27.4 \rightarrow 27$ (c) $27.5 \rightarrow 27$ (d) $27.6 \rightarrow 28$

  (e) $28.0 \rightarrow 28$ (f) $28.4 \rightarrow 28$ (g) $28.5 \rightarrow 28$ (h) $28.6 \rightarrow 29$

- **Round-half-even.** Sometimes referred to as "bankers' rounding," this is possibly the best (most balanced) of the simple rounding techniques, because errors tend to cancel out over the course of multiple operations (see also stochastic rounding below). In this case, if the digits to be discarded are *greater than* 0.5, then the integer portion of the result will be incremented by 1. Or if the digits to be discarded are *less than* 0.5, the integer portion will be left unchanged. If the digits to be discarded are equal to 0.5, the integer portion of the result will be rounded so as to make an even number (that is, if the integer portion of the result is already even then it will be left unaltered, but if the integer portion of the result is odd, it will be incremented by 1 (rounded-up) to form an even number):

  (a) $27.0 \rightarrow 27$ (b) $27.4 \rightarrow 27$ (c) $27.5 \rightarrow 28$ (d) $27.6 \rightarrow 28$

  (e) $28.0 \rightarrow 28$ (f) $28.4 \rightarrow 28$ (g) $28.5 \rightarrow 28$ (h) $28.6 \rightarrow 29$

- **Round-half-odd.** This is similar to the previous mode, the difference being that if the digits to be discarded are equal to 0.5, the integer portion of the result will be rounded so as to make an *odd* number:

(a) 27.0 → 27   (b) 27.4 → 27   (c) 27.5 → 27   (d) 27.6 → 28

(e) 28.0 → 28   (f) 28.4 → 28   (g) 28.5 → 29   (h) 28.6 → 29

This mode is never used in practice because it never rounds to zero, which is often desirable; that is, a result of 0.5 would round to 1 instead of 0.

- **Round-ceiling.** This refers to rounding toward positive infinity ($+\infty$). In this case, the integer portion of the result will remain unchanged if the discarded digits are all zero or if the number is negative; otherwise, it will be incremented by 1 (that is, rounded up):

  (a) +27.0 → +27 (b) +27.4 → +28 (c) +27.5 → +28 (d) +27.6 → +28

  (e) −27.0 → −27 (f) −27.4 → −27 (g) −27.5 → −27 (h) −27.6 → −27

- **Round-floor.** This refers to rounding toward negative infinity ($-\infty$). In this case, the integer portion of the result will remain unchanged if the discarded digits are all zero or if the number is positive; otherwise, it will be incremented by 1:

  (a) +27.0 → +27 (b) +27.4 → +27 (c) +27.5 → +27 (d) +27.6 → +27

  (e) −27.0 → −27 (f) −27.4 → −28 (g) −27.5 → −28 (h) −27.6 → −28

- **Stochastic Rounding.** The term stochastic, which comes from the Greek *stokhazesthai,* meaning "to guess at," refers to a process involving chance or probability by means of a random (or pseudorandom) variable. The idea here is that, if the digits to be discarded are equal to 0.5, the computer will effectively "toss a metaphorical coin into the air" and randomly round the integer portion of the result up or down. Although this approach typically gives the best results over the course of a large number of calculations, it is only employed in highly specialized applications and is rarely found in mainstream usage. This is because the results can vary each time a series of calculations are performed, which, apart from anything else, makes this technique very difficult to test.

> **Note**   This would be a good time to work your way through Lab 4g, in which we create a general-purpose subroutine to perform 16-bit divisions (we will be using this subroutine as part of the integer-based calculator program we are going to build in Chapter 5).

Interactive Laboratories

If you haven't already done so, create and test the testbench program and subroutines described in the following interactive laboratories that you'll find toward the back of this book: We will be using these subroutines to build an integer-based calculator as described in Chapter 5.

Lab 4a: Creating a Testbench Program

Lab 4b: Creating a 16-bit ADD Subroutine

Lab 4c: Creating a 16-bit SUBTRACT Subroutine

Lab 4d: Creating a 16-bit NEGATE Subroutine

Lab 4e: Checking for Overflow in the ADD and SUBTRACT Subroutines

Lab 4f: Creating a 16-bit MULTIPLY Subroutine

Lab 4g: Creating a 16-bit DIVIDE Subroutine

Review

This chapter introduced the concepts of signed versus unsigned binary numbers; adding and subtracting binary numbers; multiplying and dividing binary numbers; and the use of the carry, borrow, and overflow flags.

1) Summarize the concepts of natural (counting), whole, and integer numbers.

2) Document the bit-by-bit addition of the two binary numbers 00101010 and 00110111 in a style similar to that used in Figure 4-3.

3) What range of decimal numbers could be accommodated using a 7-bit sign-magnitude binary format? How about a 9-bit sign-magnitude binary format?

4) What would be the binary equivalents of the decimal values $+44_{10}$ and -44_{10} using an 8-bit sign-magnitude binary format?

5) What range of decimal numbers could be accommodated using a 7-bit signed binary format? How about a 9-bit signed binary format?

6) What would be the binary equivalents of the decimal values $+68_{10}$ and -68_{10} using an 8-bit sign-magnitude binary format?

7) Generate the ones complement equivalents of the following signed binary values: 00110011_2, 11001100_2, 01010011_2, and 10101100_2.

8) Convert the decimal values -150_{10} and -140_{10} into 8-bit signed binary numbers, add them together, and describe the resulting contents of the carry and overflow flags.

9) Convert the decimal values -130_{10} and $+23_{10}$ into 8-bit signed binary numbers and multiply them together using a shift-and-add technique.

10) Convert the hexadecimal values 84_{16} and 24_{16} into their 8-bit signed binary counterparts, and then divide the binary equivalent of 84_{16} by the binary equivalent of 24_{16}. What are the values of the quotient and the remainder? What would be the final value of the quotient if we were using the following rounding techniques: round-down, round-up, round-half-up, round-half-down, round-half-even, round-half-odd, round-ceiling, and round-floor?

CHAPTER
5

CREATING AN INTEGER CALCULATOR

> *A man has one hundred dollars and you leave him with two dollars, that's subtraction.*
> MAE WEST (1893–1980) in *My Little Chickadee* (1940)

In this chapter we will consider:

- Usage models
- The infix, prefix, and postfix notations
- The order of operations
- The precedence of operators
- Direct algebraic logic
- The Polish and reverse Polish notations

A Few Decisions

The purpose of this chapter and its associated laboratories is to create a simple four-function (+, −, *, and /) calculator that can be used as the basis for further experimentation. But, before we plunge into the fray, there are a few decisions that have to be made.

How to represent numbers inside the computer

The first thing we have to establish is how we wish to store and manipulate numbers inside the computer. Based on our discussions in the previous chapters and laboratories, we have made the decision to work with 2-byte signed binary integer values that we will use to represent numbers in the range $-32,767_{10}$ to $+32,767_{10}$.

It's important to note, however, that there are a number of different schemes we could adopt. These alternative approaches are discussed in more detail in Chapter 6 and also on the DIY Calculator website at www.DIYCalculator.com.

In what form(s) to input and output values

Another consideration involves how we wish to enter and display numbers. For example, the testbench program we created to test the math routines in the labs associated with Chapter 4 required us to enter and display values as four hexadecimal digits. By default, however, we typically prefer to work in decimal. Thus, for the purposes of this calculator, we will make the decision to create input and output routines that allow us to enter and display values in decimal. To be more specific, our input routines will allow us to enter values in decimal and they will then convert these values into their signed binary equivalents for storage and manipulation. Similarly, our output routines will take the signed binary values stored inside the calculator and convert them into their decimal equivalents for display.

As discussed in Chapter 6, these preliminary routines could be extended in the future to allow values to be entered and displayed in binary, decimal, and hexadecimal.

Which math functions to implement

The next step in the process involves deciding which math functions we wish to perform. For the purposes of this first-pass calculator, we will offer only the four fundamental binary operators (+, −, *, and /). The reason these are referred to as "binary" operators is that they require two operands.

Our first-pass calculator will also support the unary operator +/−, which will be used to invert the sign of the value currently on display (the term "unary" is employed because this operator requires only a single operand). For example, if the display originally shows a value of +42, then clicking the +/− key will negate this value inside the computer and also display this negated form of −42. Similarly, clicking the +/− key once more will return the value to its positive form and display +42.

As is discussed in Chapter 6, once we have our first-pass calculator up and running, we may decide to extend our selection of math functions to include more interesting operations.

Which usage model to adopt

In this context, the term "usage model" refers to the way in which we intend to interact with our calculator. For example, suppose we wish to perform the operation

$$3 + 6 - 2 = 7$$

How do we intend to communicate our requirements to the calculator? Your knee-jerk reaction may be to create input routines to allow us to enter the following keystrokes:

$$[3] \ [+] \ [6] \ [−] \ [2] \ [=]$$

Known as *infix notation* (because the operators are embedded between the operands), this is certainly a possibility. However, it does raise all sorts of complications for the unwary. Thus, for the purposes of our first-pass calculator, we shall adopt another approach known as *reverse polish notation* (*RPN*).

The remainder of this chapter is devoted to explaining some of the considerations associated with the various possible usage models. As is discussed in Chapter 6, once we have our first-pass calculator up and running, we may decide to modify our input routines to support a different usage model.

Infix, Prefix, and Postfix Notations

In a way, we're jumping a bit ahead of ourselves by introducing these topics here, but "biting the bullet" now will make our subsequent discussions easier to understand. So let's leap into action by stating that there are three fundamental notations we can use to represent mathematical equations:

Infix notation: 1 + 2
Prefix notation: + 1 2
Postfix notation: 1 2 +

In the case of an infix notation, with which we are most familiar, operators like +, −, *, and / are placed between the operands. In this case, we can use the operators to indicate the end of the numbers. Thus, the actual keystrokes we would use with an infix-based input routine would be as follows:

Keystrokes for infix notation: [1] [+] [2] [=] . . .

Note the use of the equals ("=") key to indicate the fact that we've reached the end of the equation and wish to see the final result.

By comparison, in the case of a prefix notation, the operators are placed *before* the operands. This leads to a problem in that we need some way to indicate to the calculator's input routines when we've reached the end of a number; otherwise, the two keystrokes [1] and [2] could be mistakenly assumed to represent the number 12. For this reason, a calculator based on a prefix notation requires an extra "Enter" key to indicate the end of a number. Thus, the actual keystrokes we would use with a prefix-based input routine would be as follows:

Keystrokes for prefix notation: [+] [1] [Enter] [2] [Enter]

In this case, we don't actually require an equals ("=") key because the operation will be performed and the result displayed as soon as the second operand has been entered.

Last, but not least, in the case of a postfix notation, the operators are located *after* the operands. Once again, this requires the use of a special "Enter" key to indicate the end of a number. Thus, the actual keystrokes we would use with a postfix-based input routine would be as follows:

Keystrokes for postfix notation: [1] [Enter] [2] [+]

Once again, we don't actually require an equals ("=") key when using this approach because the operation will be performed and the result displayed as soon as the operator key has been pressed.

Finally, before we move on to the "meat" of this chapter, it's important to note that you can, in fact, mix operators and operands when using both the prefix and postfix schemes as follows:

"Infix notation: 1 * 2 + 3
Prefix notation: + * 1 2 3   or   + 3 * 1 2
Postfix notation: 3 1 2 * +   or   1 2 * 3 +

Using the stack

The notation we are actually going to implement in our first-pass calculator is of the postfix variety, so let's consider the examples presented above in a little more detail. Let's start with the first variant:

1st Postfix example: 3 1 2 * +
Actual keystrokes: [3] [Enter] [1] [Enter] [2] [Enter] [*] [+]

As we previously discussed, the "Enter" key is used to indicate the end of a number. As each value is entered, it is pushed onto the top of the stack. In this case, pressing the "Enter" key after the number "2" is optional because the "+" operator can also be used to indicate the end of the number; but assuming that this "Enter" is used will aid us in our discussions. Thus, the state of the stack after the third number has been entered will be as shown in Figure 5-1(a). (Note that these are high-level representations of the stack that do not reflect the actual number of bytes being used to represent each value.)

When the multiplication (*) operator is keyed in, whatever two numbers are on the top of the stack will be popped off the stack, multiplied together, and the result will be placed back on the stack [Figure 5-1(b)]. Similarly, when the addition (+) operator is keyed in, whatever two numbers are on the top of the stack will be popped off the stack, added together, and the result will be placed back on the stack [Figure 5-1(c)].

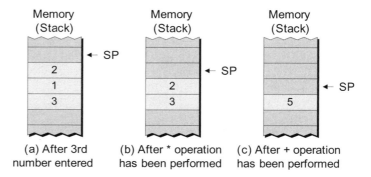

Figure 5-1. The stack during the first postfix example.

Now, let's consider the state of the stack during the course of our second postfix variant, which was as follows:

2nd Postfix example: 1 2 * 3 +
Actual keystrokes: [1] [Enter] [2] [Enter] [*] [3] [Enter] [+]

In this case, pressing the "Enter" key after the number "2" and the number "3" would be optional; once again, however, assuming that these keystrokes are performed will aid us in our discussions. Once the number "2" has been entered, the stack will be as shown in Figure 5-2(a).

When the multiplication (*) operator is keyed in, whatever two numbers are on the top of the stack will be popped off the stack, multiplied together, and the result will be placed back on the stack [Figure 5-2(b)]. When the third number is then entered, the stack will be as shown

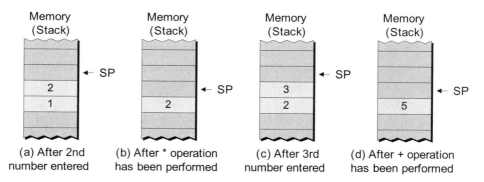

Figure 5-2. The stack during the second postfix example.

in Figure 5-2(c). Finally, when the addition (+) operator is keyed in, whatever two numbers are on the top of the stack will be popped off the stack, added together, and the result will be placed back on the stack [Figure 5-2(d)].

This postfix approach may seem a little counterintuitive at first, so why would anyone wish to go this way? Aha! Therein lies a story...

Order of Operations

The term "order of operations" refers to a convention that dictates which operations should be performed in which order. For example, consider the following expression:

$$1 + 2 * 3 = ?$$

What do you think the result should be from this equation? In fact, most of us would say that the answer should be "seven":

$$1 + 2 * 3 = 7$$

This is because, when we were young innocents at school, we were taught to perform any multiplications and divisions before considering any additions and subtractions. This order of operations is based on a concept known as the *precedence of operators*. The idea here is that some operators are inherently more "powerful" than others. Generally speaking, the precedence of operators is taken to be as follows:

()   Parenthesis (highest precedence)
∧   Exponents
* /   Multiplication and division
+ −   Addition and subtraction (lowest precedence)

In the early days, some authorities stated that all of the multiplications should be performed followed by all of the divisions. Furthermore, some folks decided that all of the additions should be performed followed by all of the subtractions. More recently, it is generally agreed that any multiplications and divisions should be performed from left to right; similarly, any additions and subtractions are now typically performed from left to right.

There are two acronyms associated with this order of operations that are commonly employed as aids to memory:

PEMDAS   Please Excuse My Dear Aunt Sally

Parentheses
Exponents
Multiplication and Division
Addition and Subtraction

BEDMAS   Big Elephants Destroy Mice And Snails

Brackets (another term for parentheses)
Exponents
Division and Multiplication
Addition and Subtraction

The PEMDAS and BEDMAS rules may be applied recursively. For example, a pair of parentheses can contain a complex expression that itself contains one or more pairs of parentheses containing their own expressions, and so forth.

The important thing to note is that all of this is just a convention. Nothing is written in stone. Deep in the mists of time, some people started to adopt a certain way of doing things. As the years went by, more and more folks came to use the same convention because doing so made communication much easier.

Direct Algebraic Logic

The phrase *direct algebraic logic* (*DAL*) may be used to refer to a calculator that is based on an infix notation scheme with limited understanding of the precedence of operators (this was true of early machines and remains the case for the majority of calculators to this day). The idea here is that the +, −, *, and / operators are treated as having equal precedence and are evaluated from left to right (that is, the order in which they are entered). In this case, the result from our original equation would be "nine" because the addition would be performed before the multiplication:

$$1 + 2 * 3 = 9$$

The result is that anyone attempting to use a DAL-based calculator to solve equations of any complexity spends much of their time jotting intermediate results down on a piece of paper. For example, consider how a calculator based on a simple DAL infix notation would evaluate the following equation:

$$1 * 2 + 6 / 2 = 4$$

In this case, the calculator would first perform $1 * 2 = 2$, and then add the 6 to give 8, and then divide this value by 2 to give a final result of 4. If we wished to perform this equation using the standard order of operations, we would be obliged to split it up as follows:

$$1 * 2 = 2$$

$$6 / 2 = 3$$

$$2 + 3 = 5$$

In reality, most complex expressions simply cannot be expressed using a conventional infix notation without the use of parentheses. Thus, some electronic calculators were equipped with "(" and ")" keys, thereby allowing the above expression to be entered as follows:

$$(1 * 2) + (6 / 2) = 5$$

In fact, parenthesis are often used by computer programmers to eliminate any possibility of confusion; for example, consider the following equation:

$$(1 + 2) * 3 = 9$$

This is guaranteed to return a result of "nine" irrespective of the precedence rules being applied.

Interestingly enough, some modern calculators go one step further by employing a concept known as *advanced direct algebraic logic* or *advanced DAL*. In this case, the expression may be keyed in just the way you would write it on a piece of paper, with or without the use of parenthesis; for example:

$$1 * 2 + 6 / 2 \text{ [Enter]}$$

In this case, the input sequence is stored and no operations are performed until the [Enter] key (or possibly the equals "=" key) is pressed. At this point, the calculator analyzes the expression and evaluates it according to the standard PEMDAS/BEDMAS order of precedence, which means that such a machine would return the commonly expected result of "five."

Polish and Reverse Polish Notations

In the early 1920s, the Polish mathematician Jan Lukasiewicz (1878–1956) was working in the field of symbolic logic. At that time, he realized that placing operators in front of operands rather than between them rendered the use of parenthesis (brackets) unnecessary.

As the operators *preceded* the operands, this was referred to as a *prefix* notation. Furthermore, due to the fact that non-Polish speakers found it difficult to wrap their tongues around his surname, Lukasiewicz's technique became known as *Polish notation* (*PN*). (Had his family name been something like "Smith," we would probably refer to this as *Smith notation*.) Although his work predated computers as we know them today, Lukasiewicz did note that his scheme was applicable to standard arithmetic operations.

Some time later, in the mid-1950s, one of Australia's first computer scientists, the philosopher Charles Leonard Hamblin (1822–1985), took Lukasiewicz's ideas one step further. Working at the New South Wales University of Technology (NSWUT), Hamblin started to place the operators *after* the operands. For obvious reasons, this form of postfix notation subsequently became known as *reverse Polish notation* (*RPN*).

When the first electronic calculators were developed, the amount of memory available to them was somewhat limited and their computing power left a lot to be desired. Although RPN may appear confusing at first, it does offer several advantages:

- It can be analyzed quickly and easily by a computer program.
- It provides a very efficient way for a calculator to perform its computations.
- It allows equations to be entered explicitly with no need for parentheses.
- It allows technical users to perform complicated calculations using fewer keystrokes.

As a user interface for calculation, RPN first appeared in the desktop calculators manufactured by Hewlett Packard in the late 1960s, and later in their handheld scientific calculators in the early 1970s. In fact, many of Hewlett Packard's scientific calculators continue to support RPN to this day (although users can now select between using the RPN and DAL input techniques).

Interactive laboratories

OK, this is it, no more dilly-dallying or shilly-shallying because we're poised to create our very own calculator. And so, without further ado, let's race to complete the following interactive laboratories that you'll find toward the back of this book:

Lab 5a: Creating the Calculator Framework

Lab 5b: Adding Some Low-Level Utility Routines

Lab 5c: Creating a Decimal GETNUM ("get number") Subroutine

Lab 5d: Creating a Decimal DISPNUM ("display number") Subroutine

Lab 5e: Implementing a Four-Function Integer Calculator

Review

In this chapter, we first considered some of the decisions we need to make as part of creating a calculator program. In particular, we introduced the concept of the usage model and then discussed the ideas behind the infix, prefix, and postfix notations.

Next, we introduced the concepts of the *order of operations* and the *precedence of operators*. Based on these discussions, we noted the problems associated with the *direct algebraic logic (DAL)* usage model and we then presented the ideas behind the *Polish notation (PN)* and *reverse Polish notation (RPN)* schemas.

1) What do the acronyms PEMDAS and BEDMAS stand for?

2) In the case of the equation "8 * 2 + 15 / 3 – 32 / 8," what would be the result if this equation were to be evaluated using the standard PEMDAS/BEDMAS order of precedence?

3) With regard to the previous question, what would be the result if the same equation were to be evaluated using a calculator based on a simple DAL usage model?

4) Recreate the equation (8 * 2) + (15 / 3) – (32 / 8) in prefix notation.

5) With regard to the previous question, list the minimum number of

keystrokes required to enter this equation into a PN-based calculator (if there were such a beast). (Note: assume that you do not need to press the "Enter" key after an operator.)

6) Recreate the equation (8 * 2) + (15 / 3) – (32 / 8) in postfix notation.

7) With regard to the previous question, list the minimum number of keystrokes required to enter this equation into an RPN-based calculator. (Note, assume that you do not need to press the "Enter" key after a math operator or after a number if that number is followed by a math operator.)

8) With regard to the previous question, create a series of diagrams showing the state of the stack as you progress through the calculation.

9) Recreate the equation {[(1+2) * 6] – 3} using a postfix (RPN) notation.

10) Recreate the equation {1 + [2 * (6 – 3)]} using a postfix (RPN) notation.

CHAPTER
6
MORE FUNCTIONS AND EXPERIMENTS

If a man's wit be wandering, let him study the mathematics.

FRANCIS BACON (1561–1626) in *Essays* (1625)

In this chapter we will ponder:

- Implementing additional functions
- Binary and hexadecimal display and entry
- Reconfiguring the calculator's front panel
- Implementing structured (verbal) numbers
- Creating a binary floating-point calculator
- Creating different flavors of BCD calculators
- Creating code-coverage and profiler applications
- Building a real *DIY Calculator*
- And much, much more!

Introduction

Yes, we know, this is the last chapter in the book. But turn that frown upside down into a smile, because there are still lots of experiments to perform, functions to create, additional programs to play with, and even the possibility of creating a real (physical) version of the DIY Calculator! And, of course, there are a lot of additional resources (with some special content for educators) on the CD-ROM accompanying this book and at the DIY Calculator website itself. Plus, there are many other really interesting books to read. (These additional resources are discussed in more detail in Appendix D.)

Implementing Additional Integer Functions

With regard to the integer calculator we created in Chapter 5, the only math functions we've implemented thus far are +, –, *, /, and +/– (that is: add, subtract, multiply, divide, and negate/change sign). However, the existing front panel contains a variety of other function buttons. Some of these, such as the reciprocal (1/x) and the trigonometric functions sin (pronounced "sign"), cos, and tan, make sense only in the context of a floating-point calculator as discussed later in this chapter. But there are other functions you could use to augment the capabilities of your integer calculator. We explore these below.

Math functions

Examples of some math functions that you might wish to consider implementing in the context of an integer calculator are as follows:

- n!   Calculate the factorial of the value currently being displayed (e.g., if the value currently being displayed is 5, pressing this button should perform the operation $5 \times 4 \times 3 \times 2 \times 1 = 120$).

- x^2 Calculate the square of the value currently being displayed (e.g., if the value currently being displayed is 5, pressing this button should perform the operation $5 \times 5 = 25$).

- x^3 Calculate the cube of the value currently being displayed (e.g., if the value currently being displayed is 5, pressing this button should perform the operation $5 \times 5 \times 5 = 125$).

- Mod Display the modulus (remainder) from an integer division (e.g., the result from the operation 14 Mod 3 will be 2).

Binary and hexadecimal display and entry

Just under the calculator's main display area are three buttons marked **Bin** (binary), **Dec** (decimal), and **Hex** (hexadecimal). You could modify our display routines to use these buttons to convert whatever value is currently being displayed into the appropriate number system. For example, if the value on the main display is 42 in decimal, then clicking the **Hex** button should cause the appropriate LED (the light next to the button) to light up; and it should also cause the display to present the equivalent hexadecimal value, which is 2A. Similarly, clicking the **Bin** button should cause *that* LED to light up and the display to show the equivalent binary value, which is 101010.

As a further modification, you could augment the input routines so as to allow values to be entered in the selected number system.

Reconfiguring the Calculator's Front Panel

One interesting feature associated with the calculator's front panel is that you can modify the annotations on the current buttons and add new annotations to the blank buttons. If you right-click on the **Sin** button, for example, you will be presented with the **Configure Button Attributes** dialog window (Figure 6-1).

The value in the **Code** field is the hexadecimal code that will be loaded into the front panel's latch when you click this button. This value is shown with a gray background to reflect the fact that you cannot change it (at least not using this dialog).

The text in the **Annotation** field reflects the characters that are to be displayed on the button. The pull-down list associated with the **Color** field allows you to select the color of the annotation text. Finally, the text in the **Description** field forms the tool tip that will appear if you leave the mouse cursor "hovering" over this button for a few seconds.

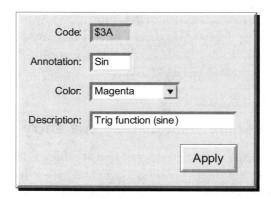

Figure 6-1. Configuring button attributes.

Any changes you make to a calculator button will take effect when you click the **Apply** button. Note, however, that these changes will only persist for the current session. If you wish these changes to become permanent, you must use the **Setup > Save New Buttons** command before you terminate the current session (there is also a corresponding **Setup > Restore Default Buttons** command).

Adding memory functions

One very useful feature would be to augment your calculator with a "memory" in which you can store intermediate results from calculations. For example, you could reconfigure four of the blank buttons to perform the following functions:

- MC   Clear the memory to contain 0.
- MS   Store the value currently on display into the memory (overwrite any existing contents).
- MR   Recall the value in memory and display it on the main display (leave a copy of this value in the memory).
- M+   Add the value currently on display to the current contents of the memory.

Adding logical functions

If you decide to offer binary and hexadecimal number entry and display capabilities as discussed in the previous section, then you might also contemplate adding a selection of logical functions such as AND, OR, XOR, and INV to the front panel buttons.

Adding other functions

In the real world, different calculators provide different functions depending on their target audience. The advantage of the DIY Calculator is that you can configure its interface and program it to perform any function(s) you desire. For example, you might decide to augment your calculator with a "%" button that can be used to easily calculate percentages. Or you could add a variety of statistics and probability functions (again, some of these functions will be of interest only in the context of a floating-point calculator as discussed later in this chapter).

Implementing an Infix Notation Input Schema

The calculator we implemented in Chapter 5 was based on the *reverse Polish notation (RPN)* usage model. As we discussed in that chapter, this is known as a *postfix* scheme because the operators follow the numbers; for example, in order to perform the calculation 6 * 2 + 3 using RPN, we would enter the following:

$$3 \text{ [Enter] } 6 \text{ [Enter] } 2 \text{ [Enter] } * +$$

In this case, we don't require the use of the equals "=" button because the final result from the calculation will be displayed automatically. As an alternative, you could modify our input routines to support an *infix* usage model as discussed below.

A simple infix model

The simplest infix model, which may also be referred to as *Direct Algebraic Logic (DAL)*, is to support operators between numbers as follows:

$$6 * 2 + 3 = 15$$

In this case, we do require the use of the equals "=" button to instruct the calculator to display the final result. Note that when you are implementing this type of scheme, you need to decide what to do if the user clicks on the same operator twice; for example:

$$6 * 2 + + 3 = ?$$

Also, what should occur if the user were to click on two different operators one after the other, as follows:

$$6 * 2 + * 3 = ?$$

As you will soon come to discover, implementing a robust input routine that can handle the user performing unexpected actions is a non-trivial task.

Adding parentheses

One problem with the simple infix model discussed above is that it evaluates operations in the order in which they are entered; that is, from left-to-right. This means that if we were to enter the following,

$$6 + 2 * 3 = 24$$

the calculator would first add the 6 and the 2 to give 8, and then multiply this intermediate value by 3 to give a final result of 24. This is, of course, different from the way in which we would perform such a calculation by hand, in which case we would perform the multiplication before the addition. One way to address this problem would be to augment our input routines to support the use of parentheses. This would allow us to explicitly convey to the calculator what we are trying to achieve as follows:

$$6 + (2 * 3) = 12$$

Initially you may decide to support only one set of parentheses at a time. Later on, you may choose to support nested parentheses; for example:

$$6 + ((2 * 4) / (3 + 1)) = 8$$

A more sophisticated infix model

If you are really daring, you might contemplate offering a more sophisticated infix model, which may also be referred to as *Advanced Direct Algebraic Logic (Advanced DAL)*, that performs operations in exactly the same manner that you would do by hand; for example:

$$6 + 2 * 3 \text{ [Enter]}$$

In this case, the calculator stores the input sequence and doesn't perform any operations until it sees the Enter key, which is used to indicate the end of the equation. At this time, the calculator parses and analyzes the input sequence and performs the various operations using the conventional order of precedence discussed in Chapter 5 (for example, parentheses, then multiplication and division, then addition and subtraction, etc.).

Note that we might choose to use the Enter key and/or the equals "=" key to indicate the end of the equation (we show the Enter key here to emphasize the difference to our previous example).

More Experiments with Integers

Using Bigger Integers (of Fixed Size)

When we created our basic math routines in the labs associated with Chapter 4, we decided to restrict ourselves to work with 2-byte (16-bit) values. But these can represent signed binary numbers only in the range −32,767 to +32,767 (remember that we are ignoring the most negative −32,768 value because we cannot represent its positive counterpart).

One experiment you could perform would be to modify these routines (along with the input and output routines we created in the labs associated with Chapter 5) to work with bigger values. For example, you could elect to work with 3-byte (24-bit) values, which can be used to represent numbers in the range −8,388,607 to +8,388,607; or you might opt for 4-byte (32-bit) values, which can be used to represent numbers in the range −2,147,483,647 to +2,147,483,647 (once again, we are ignoring the most negative values from these 3-byte and 4-byte schemes because we cannot represent their positive counterparts).

Using arbitrarily (dynamically) sized integers

The problem with choosing to work only with numbers of a fixed size (be it 2-byte, 3-byte, 4-byte, or more), is that, at some stage, we will almost inevitably want to perform a calculation whose result exceeds our capability to represent it. One solution is to modify our input, output, and math routines to work with numbers of arbitrary size.

For example, if we entered a value of 42_{10} (that's 00101010_2 in binary) the input routine could determine that this would fit in a single byte, so it could store this (push it onto the stack) as a 2-byte value [Figure 6-2(a)]. By comparison, if we input a value of $1,957_{10}$, the input routine would recognize that this would fit in two bytes, so it would store it as a 3-byte value [Figure 6-2(b)].

Theoretically, we could use this scheme to store numbers ranging from 1 to 255 bytes in size, where the latter can represent numbers so large it makes one's eyes water to think about it.

When we store this type of number, we would start by pushing its least-significant byte onto the stack and work our way up to the most-significant byte. Finally, we would push the extra "size" byte onto the stack. The reason for doing things this way is that, when we come to retrieve one of these numbers, the first byte we pop off the stack will be the "size" byte, which will inform us as to how many additional bytes are used to form the actual number.

Now, suppose we wish to perform an operation such as an addition, in which case we would start by retrieving two numbers from the top of the stack. Let's assume that one of these numbers is four bytes in size while the other is only two bytes. In this case, we would "pad" the smaller number with two bytes of 0s to make the two values the same size before performing the addition. Of course, assuming both values to be positive, it may well be that the result will occupy five bytes.

Similarly, in the case of a multiplication (and depending on the values in question), the result may require more bytes than the two numbers being multiplied (as a worst case, the number of bytes in the result will equal the sum of the bytes in the multiplier and multiplicand). By comparison, in the case of a division, the result may require fewer bytes than the dividend and the divisor.

There are a number of things you have to think about when implementing this type of scheme, as you will doubtless discover for yourself.

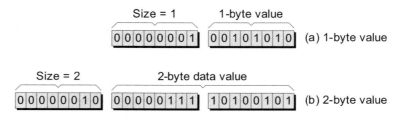

Figure 6-2. Dynamically sized integers.

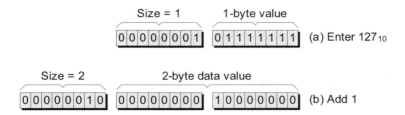

Figure 6-3. The MS bit is always the sign bit.

But just to get you off to the right start, let's suppose that you enter a value of $+127_{10}$, which will fit in a single byte, as shown in Figure 6-3(a).

Now, assume that we add 1 to this value, resulting in $+128_{10}$. In this case, the result will have to be stored in a 2-byte field, as shown in Figure 6-3(b), otherwise we would have to assume that this value represented -128_{10}. Thus, we need to enforce a rule that the most-significant bit of the most-significant data byte is always considered to be the sign bit for this value.

Implementing Structured (Verbal) Numbers

When entering values into a calculator, we often wish to enter a quantity of hundreds or thousands; for example, 700 or 8000. This type of thing is very common when performing financial calculations, so one occasionally runs across a calculator boasting a couple of extra buttons with "00" and "000" annotations (Figure 6-4).

7	8	9
4	5	6
1	2	3
0	00	000

Figure 6-4. Calculator with "00" and "000" buttons.

The idea here is that if you press the "8" button followed by the "000" button, for example, the calculator will immediately display "8000." This can be very useful, and it's possible to extend this scheme in interesting directions. As one example, calculator enthusiast James Redin holds U.S. Patent 5,623,433 on what he refers to as *Structured Numbers* or *Verbal Numerals*. James points out that when we look at a date like 2005, we don't conceptualize it as "two-zero-zero-five," but instead, we think of it the same way we would verbalize it, which is "two thousand and five." Similarly, when we look at a number such as 2,007,000, we think of this as "two million, seven thousand." Entering this number the normal way would require seven keystrokes. James proposes augmenting standard Western/English calculator keypads with three extra keys bearing the annotations "H" (hundred), "T" (thousand), and "M" (million), as shown in Figure 6-5.

In this case, we could enter the 4-key sequence "2" [M] "7" [T], which would result in the following progression appearing on the calculator's display:

Keystrokes	Output Display
[2]	2
[M]	2,000,000
[7]	2,000,007
[T]	2,007,000

The use of structured numbers in this example means that we require approximately half the keystrokes as compared to entering the same value on a standard keypad. Also, the use of structured numbers

Figure 6-5. Calculator with H, T, and M buttons.

saves us from the mental gymnastics associated with figuring out how many zeros to use and where they should be included.

The point is that you could reconfigure some of the DIY Calculator's buttons to represent these additional keys, and then modify our input routines to handle this structured number entry format. (More detailed descriptions of verbal numbers and structured number entry techniques are provided in the *Xnumbers* folder on the CD-ROM accompanying this book; see Appendix D for more details on these additional resources.)

> **Note** The reason we used the "Western/English" qualifier above is that different languages use different word structures to define numbers. For example, in the same way that we have a special word for 1000 ("thousand"), the Japanese have a special word for 10,000, which we would have to verbalize as "ten thousand."

Creating a Binary Floating-Point Calculator

Up until this point, we've only considered working with integer values. In reality, of course, this is very limiting, because it means that we can't represent real numbers like 3.142. If you visit our website at www.DIYCalculator.com, you will find some additional materials in which we discuss the concepts of floating-point representations and define our own simple binary floating-point format. A very interesting project would be to create input, output, and math routines that work with this floating-point format.

Once you have created floating-point routines to perform the fundamental operations (+, −, *, /, and +/−), you could start adding functions like 1/x (reciprocal). You may also decide to program the **Pi** button such that pressing it automatically displays the value of π (to the precision supported by our floating-point format) and pushes a copy of this value onto the top of the stack. Later, you might start experimenting with more complex functions such as **Rx** (square root), and eventually work your way up to the trigonometric functions **sin**, **cos**, and **tan**.

Creating a BCD Calculator

The term *binary coded decimal* (*BCD*) refers to a technique for storing and manipulating numbers inside a computer in which each decimal digit is represented as a 4-bit binary value. Our website at www.DIYCalculator.com features some additional materials in which we introduce the concepts associated with BCD representations. As discussed on the website, there are some advantages to using BCD; thus, it may not surprise you to hear that another interesting project would be to create input, output, and math routines that work in BCD.

Fixed-precision integer BCD

As a starting point, you might choose to create a fixed-precision integer BCD calculator. If you decide that each number will occupy a 4-byte field, where the most significant nybble represents the sign, for example, then you will be able to represent numbers in the range –9,999,999 to +9,999,999 (assuming a sign-magnitude format for the purpose of these discussions).

Arbitrary-precision integer BCD

Alternatively, you might elect to work with arbitrary (dynamically) sized integer BCD values, where the first byte could be used to specify the number of BCD digits in the number.

Structured (verbal) entry and BCD

Irrespective of whether you opt to use fixed or variable precision integer BCD representations, you might decide to support the structured (verbal) number entry techniques mentioned earlier in this chapter.

Fixed-precision fixed-point BCD

As yet another alternative, you could determine to work with a fixed-precision, fixed-point BCD format, in which each number occupied a 7-byte field, for example, with the digits in the least-significant three bytes being considered to be to the right of the decimal point. In this case, you could represent numbers in the range –9,999,999.999999 to +9,999,999.999999 (again, we're assuming a sign-magnitude format for the purpose of these discussions).

Arbitrary-precision fixed-point BCD

Of course, you might prefer to work with arbitrary (dynamically) sized fixed-point BCD values. In this case, the most-significant nybble of the "size" byte could be used to define the number of digits (from 0 to 15) to the left of the decimal point (excluding the sign nybble). Similarly, the least-significant nybble of the "size" byte could be used to define the number of digits (from 0 to 15) to the right of the decimal point. In this case, a number such as +12345.67 would be represented as shown in Figure 6-6.

Floating-point BCD

Last, but certainly not least, you could throw caution to the winds and implement a floating-point BCD calculator. Apart from anything else, if you created a similar format to that used for the binary floating-point implementation presented on the DIY Calculator website (7 digits for the exponent and 16 for the mantissa), then it would be interesting to compare these two implementations in terms of code size (memory utilization) and performance (execution speed).

Of course, the performance metric will be a little tricky. Computers now go so fast that it would be difficult to compare the speed of two functions like the binary and BCD implementations of a floating-point multiply routine using a stopwatch. What we need is some form of profiler application. . .

Creating Code-Coverage and Profiler Applications

Good grief! What a coincidence! We were just talking about this in the previous section! Although we haven't discussed this aspect of things

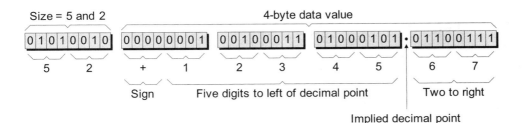

Figure 6-6. A dynamically sized fixed-point BCD representation.

thus far, the DIY Calculator has been designed with "hooks" that allow you to capture interesting data while you are running a program. For example, it can be used to record the number of times each instruction was executed (and the number of times each data location was accessed) over the course of some defined period such as entering and exiting a particular subroutine.

All of this data can be output in the form of an ASCII text file. Also available is a reference file that details the number of virtual clock cycles[1] required to execute each instruction in each addressing mode (in the case of the conditional jump instructions, there are two values depending on whether the test passes or fails).

Our thought was that someone (maybe us, maybe you), could write a program that can analyze this data and present a graphical representation of the code coverage; that is:

- Which instructions were executed?
- Which instructions weren't executed?
- Which data locations were accessed?
- Which data locations were left untouched?

It would also be possible to create a profiler that graphically shows which areas of a program (or subroutine) were used the most and which the least. For example, if you see a "hot spot" that is consuming a large proportion of the processing time, a little judicious tweaking of that portion of the code may yield significant performance improvements for the program as a whole.

The ways in which you can access this data, the format of the files, and some thoughts on the techniques one might use to process the data and present the results are discussed in more detail on the DIY Calculator website at www.DIYCalculator.com.

Building a Real DIY Calculator

The virtual version of the DIY Calculator featured in this book provides a fantastic tool for learning about computers and calculators. However,

---

[1] *The Official DIY Calculator Data Book* (which, as discussed in Appendix D, is included on the CD-ROM accompanying this book) describes the internal architecture of the calculator's CPU, including a breakdown of the clock cycles required to execute each instruction.

there is something strangely satisfying about having access to a real machine, so we've been pondering the idea of creating a real, physical version of the DIY Calculator.

As discussed in Appendix D, the CD-ROM accompanying this book includes a data book that describes the inner architecture of the calculator's *central processing unit* (*CPU*) in excruciating detail. This data book provides more than enough information to create a physical implementation of the calculator, but how should we go about tackling this?

Using discrete devices

One approach would be to use off-the-shelf *integrated circuits* (*ICs*), which are affectionally known as *silicon chips*. In the case of simple ("jelly-bean") devices, for example, one chip may provide four 2-input AND gates, another might provide six inverters, and another may boast four register (memory) elements.

In addition to the CPU, we would also have to add memory devices in which to store programs and input/output (I/O) port devices with which to communicate with the outside world. Then we would have to add the calculator buttons and display device. And we would also have to implement some way for the calculator to "talk" to our personal computer (PC) such that we could download machine-code programs created on the main computer into the physical calculator.

It would certainly be interesting to design and build a system using these basic devices, but it would also be time-consuming, expensive, and prone to error (having said this, the authors have actually considered building a relay-based version of the calculator).

Using an off-the-shelf microprocessor/microcontroller

Another technique would be to take a simple off-the-shelf *microprocessor* (μP) or *microcontroller* (μC) and create a small circuit board featuring this device, some memory, some I/O ports, some buttons (to represent the calculator buttons), and a display device such as a small *liquid crystal display* (*LCD*).

Once again, we would also have to implement some way for the physical calculator and main computer to communicate with each other in order to be able to download machine-code programs into the calculator.

There are now two ways in which we could proceed:

1) We could create a special cross-assembler to run on the main computer. This cross-assembler could be used to take source-code programs written in the DIY Calculator's assembly language and generate two different types of machine-code files: one using the DIY Calculator's instruction set and opcodes; the other using the real μP/μC's instruction set and opcodes. This would allow us to create and debug our programs in the virtual world as usual, and to then download the μP/μC version into the physical calculator.

2) Alternatively, we could use an appropriate assembler to write a special "interpreter" program to run on the μP/μC. We would then create and debug our programs in the virtual world as usual. Once the DIY Calculator's machine-code version of the program was working to our satisfaction, we could download it into the physical calculator. The interpreter program would now be used to interpret our DIY Calculator program on an instruction-by-instruction basis.

Using a field-programmable gate array (FPGA)

We've saved the best for last (hurray!). As its name suggests, a *field-programmable gate array* (*FPGA*) is a silicon chip whose function can be configured (programmed) by the user "in the field." Furthermore, some FPGA types can be reprogrammed with new configurations over and over again.

Depending on the device in question, an FPGA can contain the equivalent of a few hundred thousand logic gates to tens of millions of gates. Also, in addition to this raw logic, the FPGA can contain other elements, such as blocks of memory.

As we know, the virtual DIY Calculator consists of a CPU, some memory, some I/O ports, some buttons, and a display device. The idea here is to program an FPGA to emulate the functions of the CPU, memory, and I/O ports. We would also create a small circuit board containing this device along with the calculator's buttons and displays. The FPGA could also be used to implement a communications interface, such as a serial I/O port or a USB port, that the physical calculator could use to "talk" with the main computer.

When it comes to creating FPGA-based designs, there are a number of different tools and vendors available. For this type of project, however, the design tools from Altium Ltd. (www.altium.com) are incredibly well-suited to the task. First and foremost is Protel 2004, which is a complete circuit board design solution that includes hierarchical schematic (circuit diagram) capture, mixed-signal simulation, signal-integrity analysis, and rules-driven board layout.

This means that we could use Protel 2004 to design the physical calculator's circuit board. However, in addition to all of its board-level capabilities, the reason Protel 2004 is of interest to us here is that it also features a fully integrated FPGA design environment. The system comes equipped with libraries of functions ranging from simple gates, to multi-bit registers, adders, counters, and memory blocks. Also included are communications blocks such as serial I/O and USB ports. Each of these logical functions comes equipped with a corresponding schematic symbol. This means that we could capture the DIY Calculator's CPU, memory, and I/O ports as a schematic of functional blocks. The system can then be used to generate the configuration file used to program the real FPGA device.

Alternatively, if we wished to base our physical implementation on an off-the-shelf microprocessor (such as a PIC, for example) as discussed in the previous section, then Altium also offers a product called Nexar 2004. As with Protel 2004, this little rapscallion allows you to capture an FPGA design as a block-level schematic; but in addition to the regular logical function blocks discussed above, it also allows you to include one or more microprocessor blocks in your design. Each processor comes equipped with appropriate compilers and assemblers, and also a complete source-level debugging environment.

Last, but certainly not least, Altium has an FPGA development board, called the NanoBoard, that works with both Protel 2004 and Nexar 2004. When you connect the NanoBoard to your main computer, you can download your FPGA designs directly into an on-board FPGA and then run and debug these designs in real time. One cool thing about the NanoBoard is that it supports multiple target devices in the form of swappable FPGA daughter boards, which means you aren't tied to a particular FPGA vendor or device.

Further reading

There are several points relating to the creation of a physical version of the DIY Calculator that could be a little tricky unless one thinks them through beforehand. But we don't want to wander off into the weeds here, so if you are interested in learning more, visit our website at www.DIYCalculator.com.

Also, multimedia presentations on Protel 2004, Nexar 2004, the NanoBoard, and associated evaluation kits are provided in the *Altium* folder on the CD-ROM accompanying this book; see Appendix D for more details on these additional resources.

Sharing Your Work With Others

If you do create something interesting as discussed below, then please feel free to send it to us (with appropriate documentation) for inclusion on our website so that others can peruse and ponder your masterpiece. We will, of course, give you full credit for your work (check out the notes at www.DIYCalculator.com for more details).

Subroutines

For example, let's suppose you are looking at a subroutine on our web-site (or a routine presented in this book) and you say to yourself: "I can do the same thing using half of the memory (but it won't go as fast)," or "I can do the same thing twice as quickly (but it will use more memory)," or, if you're really good, "I can do the same thing using half of the memory and make it go twice as fast!"

Well, "talk is cheap," as they say, and "the proof of the pudding is in the eating." If you do create an interesting version of a subroutine (integer, floating-point, or BCD-based), then please feel free to share it with us and we'll make it availabe to other readers via the website.

> **Note** If you reconfigure the buttons on the calculator's front panel as part of your project, then you should also send us a copy of the *newbuttons.ini* file that you'll find in the *C:\DIY Calculator\Config* folder.

Entire calculators

If you are really brave, you may go beyond creating a simple subroutine, and instead implement an entire calculator: for example, your version of a BCD-based floating-point calculator. Once again, please feel free to share anything like this with us for inclusion on our website.

Design files for physical versions of the DIY Calculator

With regard to a physical implementation of the DIY Calculator, it should be noted that the CPU design documented in the data book on the CD-ROM was created so as to be understandable. To put this another way, the original design is not as efficient as one might hope.

Our thoughts are that it would be a good idea to first create the design cycle-for-cycle as documented in the data book. Once this first implementation is working, it would be interesting to explore alternative architectures that can execute the same machine-code files in a more efficient manner.

As usual, if you do anything like this, we would love to hear about it and, if you wish, we will post copies of your design notes and design files on our website for the delectation and delight of other DIY Calculator enthusiasts.

LABS FOR CHAPTER 2

CREATING AND RUNNING PROGRAMS

Lab 2a

Creating a Simple Program

Objective: To introduce the DIY Calculator's assembly language and assembler. Also to create, load, and run a simple program that clears the calculator's main display

Duration: Approximately 25 minutes.

The Memory Map

Before we commence this lab, we need to be aware of the way in which a *central processing unit* (*CPU*) perceives the external world, which is as a series of memory locations. Our CPU's 16-bit address bus can be used to point to $2^{16} = 65,536$ unique locations, which would be numbered from 0 to 65,535 if we were counting in decimal (remember that we typically start counting things from 0 when working with computers).

As was noted in Chapter 1, however, the decimal numbering system is not particularly convenient for our purposes. Thus, we prefer to number these locations from $0000 to $FFFF using the hexadecimal numbering system (where "$" characters are used to indicate hexadecimal values).

A common method for representing the way in which the computer's memory is allocated and organized is by means of a diagram called a *memory map* (Figure L2-1).

In the case of our virtual computer system, addresses $0000 through $3FFF are occupied by *read-only memory* (*ROM*), whereas addresses $4000 through $EFFF are occupied by *random access memory* (*RAM*).

If this were a real calculator, the ROM would contain the predefined routines used to read numbers and function keys from the key-

157

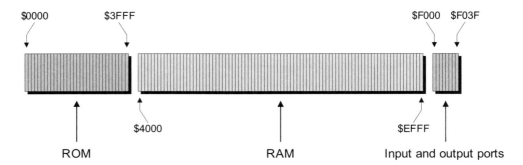

Figure L2-1. The memory map.

pad, perform arithmetic operations on the numbers, and display the results. Thus, when the calculator was first powered up, a *power-on reset* signal would cause the CPU to start executing the program in the ROM. This program would first clear the calculator's display, initialize anything needing to be initialized, and then wait for the user to start pressing some buttons. In such a scenario, the RAM would be used primarily to store temporary data, intermediate values, and pending operations.

With regard to our virtual calculator, however, we don't have any predefined math routines stored in the ROM. During the course of this book we are going to create these routines ourselves, and any programs we generate will be loaded directly into the RAM. Having said this, it will make some things easier to describe and understand if we assume that the ROM contains a simple *monitor program* that performs a few low-level housekeeping tasks.

Toward the top of the memory map are the input and output (I/O) ports that the CPU uses to communicate with the outside world in the form of the calculator's front panel. In fact, the CPU regards these I/O port addresses as standard memory locations and doesn't realize we're using them for our own cunning purposes. Addresses $F000 through $F01F are occupied by a set of 32 input ports (of which the front panel uses only one port at address $F011), whereas addresses $F020 through $F03F are occupied by a set of 32 output ports (of which the front panel employs only two ports at addresses $F031 and $F032). Finally, addresses $F040 through $FFFF are unused in this implementation (Table L2-1).

TABLE L2-1. Summary of memory allocation

Addresses	Function
$0000 to $3FFF	ROM
$4000 to $EFFF	RAM
$F000 to $F01F	Input ports
$F020 to $F03F	Output ports
$F040 to $FFFF	Unused

The Accumulator (ACC)

One other point of which we need to be aware before plunging headfirst into the fray is that, among other things, our CPU contains an 8-bit register called the *accumulator (ACC)*. (The term "register" refers to a group of memory elements, each of which can store a single binary digit.) As its name implies, the accumulator is where the CPU gathers, or accumulates, intermediate results (Figure L2-2).

Observe the way we indicated the size of the accumulator as being [7:0] in this figure. This is referred to as *vector notation.* The accumulator comprises eight bits, and these bits are numbered from 0 to 7, with bit 0 being the *least-significant bit* (*LSB*) on the right-hand side, and bit 7 being the *most-significant bit* (*MSB*) on the left-hand side.

The reason for numbering these bits from 0 to 7 is discussed in

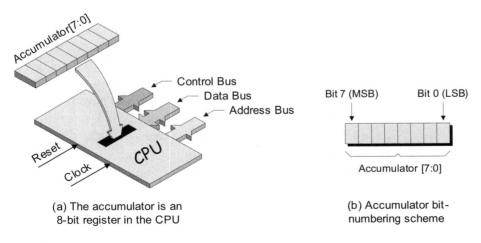

(a) The accumulator is an
8-bit register in the CPU

(b) Accumulator bit-
numbering scheme

Figure L2-2. The accumulator (ACC).

Chapter 4. For our purposes here, we need only note that we could give each of these bits individual names such as acc7, acc6, acc5, acc4, acc3, acc2, acc1, and acc0, where this would be referred to as a *scalar notation*. However, suppose we wish to associate a binary value of %01010011 with bits 7 through 0, respectively. Using a scalar notation, we would have to say something like "acc7 = 0, acc6 = 1, acc5 = 0, acc4=1 . . ." This is laborious and time-consuming to write, and nonintuitive to understand. A better solution is to use a vector notation, in which case we could simply say "acc[7,6,5,4,3,2,1,0] = %01010011," and we can abbreviate this to "acc[7:0] = %01010011," where the colon ":" character is used to indicate a range.

The ways in which we can modify and use the contents of the accumulator will become apparent as we progress. Suffice it to say that:

- The CPU can be instructed to *load* (read) a byte of data from any external memory location (or input port) into the accumulator. This involves taking a *copy* of the data in the memory; the contents of the memory at that location remain undisturbed.

- The CPU can be instructed to perform a variety of arithmetic and logical operations on the data in the accumulator.

- The CPU can be instructed to *store* (write) the contents of the accumulator into a memory location (or output port). This overwrites any existing contents in that memory location, but leaves the contents of the accumulator undisturbed.

The Program We're About to Create

The program we're about to create is very simple and contains just three instructions (Figure L2-3). The first instruction loads a value of $10 into the accumulator (the significance of this value will become apparent shortly). The second instruction writes the contents of the accumulator to the output port that drives the calculator front panel's main display area. And the third instruction directs the CPU to jump to address $0000 (this is the first location in our ROM, which we may assume contains a low-level monitor routine that will cause the CPU to automatically terminate the program and return the system to its *reset mode*).

To the left of Figure L2-3 is a special diagram called a *flowchart*, which is used to illustrate the sequence of operations. The rectangular

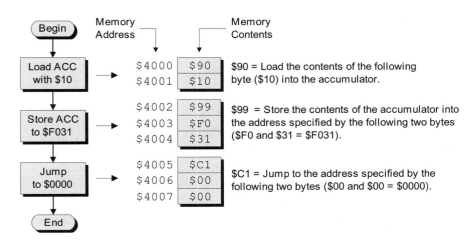

Figure L2-3. Flowchart for a program to clear the main display.

boxes in the flowchart represent the actions to be performed. To the right of each action box is a representation of the equivalent *machine code* (the raw numerical values) that must be placed into our virtual computer's memory. This numerical representation is called "machine code" because this is the form that is understood and executed by the computer (machine).

The first instruction is located at address $4000, which is the first location in the RAM. All of our programs will commence at this location, because this is the address that our monitor program will automatically jump to when we press the **Run** button on the front panel.

The $90, $99, and $C1 values are known as operation codes (*opcodes* for short) and are part of the CPUs instruction set (see also the *More! More!* section at the end of this lab). The $90 opcode at address $4000 instructs the CPU to load the following byte ($10) into the accumulator. This form of instruction is said to use the *immediate addressing mode,* because the data associated with the instruction is located immediately after the opcode (the data is referred to as the *operand*). The reason we wish to load $10 into the accumulator is that this is a special code to clear the calculator's main display (the rationale for this will become apparent in Lab 2c).

The $99 opcode at address $4002 directs the CPU to store a copy of the contents of the accumulator into the location pointed to by the following two bytes ($F0 and $31 = $F031). This form of instruction is said to use the *absolute addressing mode,* because the two-byte operand following the opcode contains a specific (absolute) address. What the CPU doesn't know is that address $F031 doesn't actually point to a location in the memory, but instead points to the output port that drives the front panel's main display (this port is discussed in more detail in Lab 2c).

Finally, the $C1 opcode at address $4005 tells the CPU to perform an unconditional jump to the location pointed to by the following two bytes ($00 and $00 = $0000). This address is the first instruction in the monitor program, which resides in our virtual computer's ROM and which is of no great concern to us here. All we really need to know is that, in the case of our virtual system, a jump to address $0000 will cause the program to automatically terminate and the CPU to enter its *reset mode.*

Entering and Assembling a Program

If we wished, we could write all of our programs directly in machine code and then manually insert them into the system's memory, one byte at a time. However, this approach would be extremely time-consuming, excruciatingly boring, and terribly error-prone to say the least. A vastly more preferable technique is to describe our programs at a higher level of abstraction, and to then use a special computer program to automatically translate these programs into their machine code equivalents. The first step up the evolutionary ladder for programming languages is called *assembly language.* Once we've captured a program in assembly language, we use a special program called an *assembler* to convert it into machine code.

If you haven't already done so, install your DIY Calculator now as discussed in Appendix A. Once you've performed this installation, use the **Start** > **Programs** > **DIY Calculator** > **DIY Calculator** command (or double-click the DIY Calculator icon on your desktop) to invoke the little rascal (Figure L2-4).

In the middle of the main DIY Calculator window is the *calculator front panel* (the "calculator" or "front panel" for short), which can be used to feed data to, display results from, and control the calculator (Figure L2-5). But in order for this front panel to do anything useful we need a program, and creating such a program (even a small one) is, of course, the purpose of this lab.

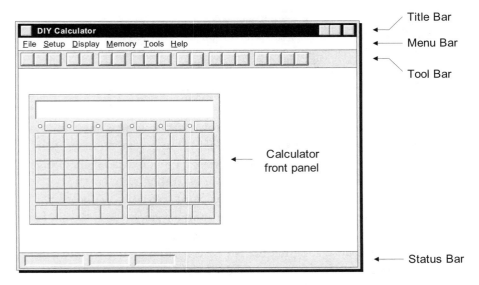

Figure L2-4. The main DIY Calculator window.

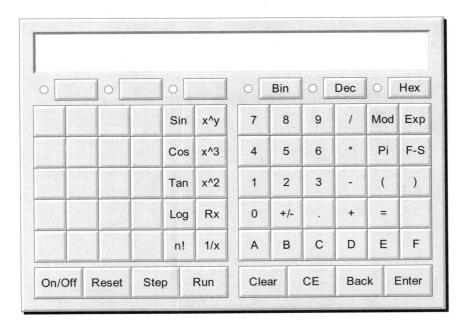

Figure L2-5. The calculator front panel.

> **Note** The assembler is an independent executable application (program) that is launched from within the main DIY Calculator interface. When you click on any part of the DIY Calculator interface after you have launched the assembler, then the main interface will "come forward" and may completely obscure the assembler. Should this occur, you can click the **Assembler / Editor** item in the main Windows® taskbar to bring this application to the foreground.

Use the **Tools** > **Assembler** pull-down menu in the main window (or click the appropriate icon on the tool bar) to invoke the assembler (Figure L2-6). This tool serves two purposes. First, it acts as an editor that allows us to capture programs in our assembly language (this is called the *source program* and, when saved out into a file, it may be referred to as the *source file*). Second, the interface can be used to call the assembler function, which assembles our source code into equivalent machine code. Make sure that the cursor is flashing in the assembler's working area, and then enter the following program in our assembly language:

```
.ORG    $4000       # Set program's origin to address $4000

 LDA    $10         # Load accumulator with clear code

 STA    [$F031]     # Store accumulator to address $F031

 JMP    [$0000]     # Jump to address $0000

.END                # This is the end of the program
```

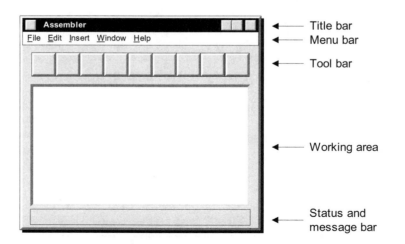

Figure L2-6. The assembler.

> **Note**   Convention dictates that directive staements like .ORG and .END are prefixed by period (full-stop)
> characters. There really is no inherent reason why this should be so, but it does provide a visual cue to the
> reader, thereby making programs slightly easier to understand. Also, these periods can be quickly
> recognized by the assembler, which may help to make it slightly more efficient.

Observe the .ORG and .END statements. These are known as *pseudoin-structions* or *directives* (because they "direct" the assembler). The .ORG is used to specify the start address or *origin* of the program (address $4000 in this case), whereas the .END informs the assembler when it reaches the end of the program.

Also note that anything to the right of a hash "#" character (also known as a *number sign, pound sign,* or *sharp*) is considered to be a comment. Comments may occur anywhere on a source line and are ignored by the assembler. (It's very important to make use of meaningful comments. Just because you know what you intend your program to do today does not mean you will understand it six months—and six thousand lines of code—later!)

Now consider the LDA $10 statement, where LDA is a mnemonic meaning "*load the accumulator.*" This statement instructs the CPU to load the value $10 into the accumulator. Once again, the reason we wish to load $10 into the accumulator is that this is a special code to clear the calculator's main display.

Contrast the immediate addressing mode used by the LDA with the absolute mode employed by the next instruction, the STA [$F031], which instructs the CPU to store the contents of the accumulator to the location at address $F031. (Remember that we're tricking the CPU in this case because, instead of pointing at a memory location, address $F031 actually points to the output port driving the calculator's main display.) As we see, our assembly language uses square brackets [and] to imply that we're talking about the contents of a memory location. The JMP [$0000] instruction also uses the absolute addressing mode to instruct the CPU to perform an unconditional jump to address $0000. Jumping to this address terminates the program and returns control to our monitor program, which is stored in the calculator's ROM.

Once you've entered this program, use the assembler's **File** > **Save As** command to save it to a file called *lab2a.asm.* (By default, the **File** > **Save As** dialog should be pointing to the *C:\DIY Calculator\Work* folder; if not, set the context to this folder before clicking the **Save** button.)

Now comes the exciting part where you assemble your program, which you do by selecting the **File** > **Assemble** pull-down menu in the assembler's menu bar (or by clicking the appropriate icon on the assembler's tool bar). Assuming you haven't made any errors, you'll receive a message along the lines of "*File assembled successfully*" in the assembler's status bar; otherwise you'll have to debug any mistakes and try again (Figure L2-7).

In addition to a number of other actions, the assembler has just created a file called *lab2a.ram*, which contains the machine-code equivalent of your program. Once you've successfully assembled your program, exit the assembler to make room on your screen.

> **Note**   Creating anything other than a very simple program is usually a very iterative process; the assembler locates errors and you fix them. Some assemblers will report a whole slew of errors together, but ours identifies them one at a time (this latter technique is easier for beginners).

> **Note**   The **.ram* output file is only generated if the assembler doesn't detect any errors. This means that if you make a change to a program and try to assemble it, the old **.ram* file won't be overwritten until you've identified and fixed any errors.

Loading and Running a Program

First click the front panel's **On/Off** button to power-up the calculator. Observe that the calculator's main display area fills up with "-" characters, which indicate that it's in an uninitialized state. Similarly, the six little lights next to the **Bin, Dec,** and **Hex** (and related) buttons just under the main display all glow red to indicate that they also are in uninitialized states.

Now use the main DIY Calculator window's **Memory** > **Load RAM** pull-down menu to invoke a dialog offering a list of the programs that are currently available. Select (click on) the *lab2a.ram* file you just created and then click the dialog window's **Load** button to load the contents of this file into the calculator's memory.

Now click the front panel's **Run** button to execute your program (the monitor program causes the CPU to automatically start with whatever instruction it finds at address $4000). As we already know, this program loads the accumulator with the $10 clear code, writes this code to the output port driving the calculator's main display (thereby clearing

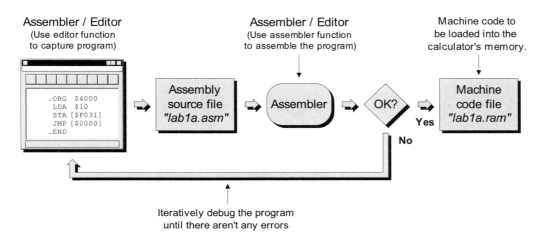

```
Assembler / Editor          Assembler / Editor          Machine code to
(Use editor function       (Use assembler function      be loaded into the
to capture program)        to assemble the program)     calculator's memory.
```

```
.ORG  $4000
LDA   $10
STA   [$F031]
JMP   [$0000]
.END
```

Assembly source file "lab1a.asm" → Assembler → OK? → Yes → Machine code file "lab1a.ram"

No

Iteratively debug the program
until there aren't any errors

Figure L2-7. Using the assembler to create a machine code (.ram) file.

the display), and then jumps to address $0000, which returns the CPU to its *reset mode* (you will see this mode indicated in the main DIY Calculator window's status bar).

Congratulations! You've successfully written, assembled, loaded, and executed your first program. Now that wasn't too bad, was it?

More! More!

These labs have been constructed in such a way that you will be introduced to appropriate opcodes, addressing modes, and assembly language constructs on an "as you need them" basis (similarly for tools like the Assembler, Memory Walker, and CPU Register display). Also, you'll find a more detailed summary of the various addressing modes in Appendix B and a summary of the instruction set, including a table showing the various addressing modes supported by each instruction, in Appendix C.

For your delectation and delight (or if you just want to plunge in deeper), you will discover a wealth of additional information on the instruction set and assembly language on the CD-ROM accompanying this copy of *How Computers Do Math*. See the *Additional Resources* discussions in Appendix D for more details.

Lab 2b
Constant Labels and .EQU Directives

Objective: To use constant labels and .EQU directives to make programs easier to read and understand.

Duration: Approximately 20 minutes.

Making Program Listings More User-Friendly

If you haven't already done so, launch the main DIY Calculator and then use the **Tools > Assembler** pull-down menu (or click the appropriate icon on the main window's tool bar) to invoke the assembler.

Now use the assembler's **File > Open** command to access the *lab2a.asm* file you created in the previous lab, and then use the **File > Save As** command to rename and save this file as *lab2b.asm*. As you will recall, although our original program performed the task for which it was intended, its appearance was not particularly user-friendly:

```
.ORG    $4000       # Set program's origin to address $4000
LDA     $10         # Load accumulator with clear code
STA     [$F031]     # Store accumulator to address $F031
JMP     [$0000]     # Jump to address $0000
.END                # This is the end of the program
```

Using Constant Labels

Even if we happen to know that $10 is the calculator's clear code and $F031 is the address of the output port driving the calculator's main display, this program is still somewhat difficult to read. One way to alleviate this is to replace some of the numeric values with user-friendly labels.

169

These labels can be called anything we want so long as they conform to the rules laid down at the end of this lab. For example, consider two slight modifications to our program as shown below (the changes are highlighted in gray):

```
CLRCODE:  .EQU    $10          # Special code to clear the main display
          .ORG    $4000        # Set program's origin to address $4000
          LDA     CLRCODE      # Load accumulator with clear code
          STA     [$F031]      # Store accumulator to address $F031
          JMP     [$0000]      # Jump to address $0000
          .END                 # This is the end of the program
```

The first modification is to declare a constant label called CLRCODE, and to use an .EQU directive (which stands for "equates to") to assign a value of $10 to this label. Once we've made such an assignment, we can reference this label in the body of the program. When the assembler is assembling the program, it will automatically substitute any constant labels it finds in the body of the program with their numerical equivalents. For example, the assembler will automatically substitute the CLRCODE label associated with the LDA instruction with the data value $10.

In the same way that we can use constant labels to represent data values, we can also use them to depict addresses. Consider two further modifications to our program:

```
CLRCODE:  .EQU    $10          # Special code to clear the main display
MAINDISP: .EQU    $F031        # Address of output port for main display
          .ORG    $4000        # Set program's origin to address $4000
          LDA     CLRCODE      # Load accumulator with clear code
          STA     [MAINDISP]   # Store accumulator to main display
          JMP     [$0000]      # Jump to address $0000
          .END                 # This is the end of the program
```

Here we see a second constant label called MAINDISP, which is assigned a value of $F031 (the address of the output port driving the calculator's main display). This label is subsequently used with the STA instruction in the body of the program.

Note that although the program's source code has been modified, the resulting machine code will be identical to our original program because all the assembler is doing is substituting these labels for their numerical equivalents.

Even though each of these labels is used only a single time in this example, their presence makes the program significantly easier to read and understand, which will become increasingly important as our programs grow in complexity. Using labels also facilitates modifying programs. For example, if we were to use the CLRCODE label multiple times in the body of the program, then we would only have to change the original assignment in the .EQU statement to affect every instance throughout the program.

Testing the New Program

Enter the above changes into the *lab2b.asm* assembly source file, and then assemble the program to generate the corresponding *lab2b.ram* file by selecting the **File > Assemble** pull-down menu in the assembler's menu bar (or by clicking the appropriate icon on the assembler's tool bar). Assuming you don't have any errors and your program assembles successfully, you can now exit the assembler to make room on your screen.

Click the **On/Off** button to power-up the calculator. (If you didn't power the calculator down after the previous lab, click the **On/Off** button two times; the first time to power it down and "scramble" its memory and the second time to power it up again.)

Now use the main window's **Memory > Load RAM** pull-down menu to locate the *lab2b.ram* file you just created and load its contents into the calculator's memory. Next, click the calculator's **Run** button to execute this new version of the program and note that it behaves exactly the same as it did before, which is to load the accumulator with the $10 clear code, write this code to the output port driving the calculator's main display, and then jump to address $0000.

A Few Simple Rules

During future labs you'll come to see that constant labels, along with their address-label cousins, which are introduced in Lab 2c, are exceptionally useful. There are just a few simple rules you should attempt to memorize before proceeding to the next lab:

- Constant labels are declared using .EQU directives, and all such statements must occur *before* the program's .ORG directive.
- Each constant label may be declared only a single time, but they can subsequently be used multiple times within the body of the program.

- Labels can consist of a mixture of alphabetic, numeric, and underscore "_" characters (no spaces or other special characters are allowed), but the first character must be either an alpha or an underscore (not a numeric).

- When a label is declared it is terminated with a colon ":" character, but this character does not form part of the label's name and is not used thereafter.

- The maximum length of a label is eight characters. This includes any underscores, but excludes the colon character used to terminate the label's declaration.

- Labels can include both uppercase and lowercase characters, but the assembler internally converts everything to uppercase, so it will consider labels such as fred, Fred, FrEd, and FRED to be identical.

- Last but not least, assignments to constant labels may employ simple equations and may reference other constant labels. However, if these equations do reference other constant labels, then those labels must be defined *before* they are used.[1] For example:

Valid	**Not Valid**
FRED: .EQU $10	BERT: .EQU (FRED + 2)
BERT: .EQU (FRED + 2)	FRED: .EQU $10
:	:
etc.	etc.
The assignment to BERT *can* reference FRED, because FRED has already been declared.	The assignment to BERT *cannot* reference FRED, because FRED has not yet been declared.

Note that, unlike the constant labels shown here, their address label cousins (as introduced in Lab 2c) can be referenced both *before* and *after* they are used.

---

[1]This is just the way in which our assembler happens to work because it's a simple way of doing things. It would certainly be possible to create an assembler that did support *forward referencing* for constant labels (where "forward referencing" means referencing something that has not yet been declared).

Lab 2C
Driving the Calculator's Main Display

Objective: To learn how to output data to the calculator's main display.

Duration: Approximately 25 minutes.

Viewing the Interface as Discrete Chunks

Conceptually, the calculator's front panel consists of four discrete "chunks" (Figure L2-8). At the bottom-left of the panel are four special buttons that are used to control the virtual computer that we can visualize as lurking "behind" the front panel. The **On/Off** and **Reset** buttons are connected directly to the computer's power supply and reset input, respectively (the **Step** and **Run** buttons will be discussed in more detail later).

The middle of the panel is dominated by the keypad area, whose function and operation are discussed in Lab 2d. Above the keypad is a bank of six *light-emitting diodes* (*LEDs*), whose workings are introduced in Lab 2e, whereas the top of the panel is commanded by the calculator's main display (Figure L2-9).

The Main Display's Character Codes

The calculator's main display is driven by the output port at address $F031 (all of the input and output ports are 8 bits wide to match the system's 8-bit data bus.)

We may consider this display as containing a small amount of internal logic that functions autonomously from the main (virtual) com-

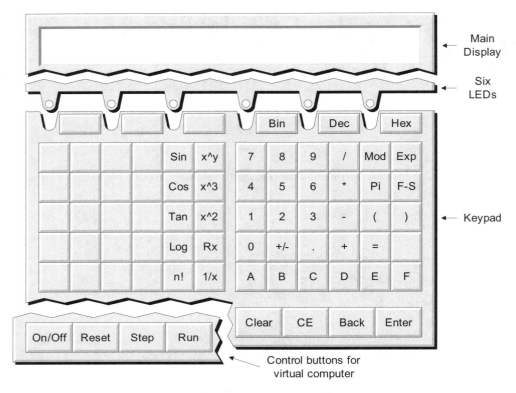

Figure L2-8. The calculator front panel consists of four "chunks."

puter. For example, when the CPU writes an 8-bit value to the port driving the display, any characters that are currently being displayed will be automatically shifted one place to the left and the new character (corresponding to the value written to the port) will appear at the right-hand side of the display. All of this character movement and organization is performed by the display itself, independent of the virtual computer.

Figure L2-9. The calculator's main display.

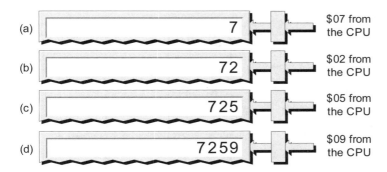

Figure L2-10. Writing a series of codes to the display.

As a simple example (assuming the display to be initially cleared), writing a series of codes such as $07, $02, $05, and $09 to this port will cause the display to show 7, 72, 725, and 7259, respectively (Figure L2-10).

To further increase our programming pleasure, the calculator's main display supports a relatively rich set of alphanumeric, punctuation, and special characters (Figure L2-11).

This character set is very similar to the *American Standard Code for Information Interchange* (*ASCII*), which was introduced in 1963 to facilitate disparate computers "talking" to each other. (Note that ASCII code $20 ("SP") will display a space " " character.) Among other things, supporting this character set means that we can actually present messages like "Hello World" on the display if we so desire (in fact we'll be doing this in Lab 2f).

However, our character set also boasts some variations from standard ASCII, including the fact that we're using codes $00 through $09

> **Note**   ASCII, pronounced "*ask-key,*" started life as a 7-bit code. Thus, the most significant bit in an 8-bit byte was either set to 0 (as in our implementation) or was used to implement a simple form of *error checking* known as *parity checking.*
>
> Error checking is of use when information is being passed from one system (or subsystem) to another. For example, consider the letter "S" for which the 7-bit ASCII code is $53 (or 1010011 in binary). In the case of *even parity,* the most significant bit would be set or cleared so as to maintain an even number of "1"s in an 8-bit byte. Thus the even parity version of the code for the letter "S" would be 01010011. By comparison, 11010011 would be the value used in an odd parity scheme.
>
> The idea is that the transmitting system calculates the parity bit and sends it as part of that byte. For each byte, the receiving system recalculates the parity bit from the first seven bits it sees, and it then compares this bit to the parity bit from the transmitter. Any discrepancy between the two parity bits indicates that some form of corruption occurred during the transmission.

Not Standard ASCII				Standard ASCII											
$00	0	$10	Clr	$20	SP	$30	0	$40	@	$50	P	$60	`	$70	p
$01	1	$11	Bell	$21	!	$31	1	$41	A	$51	Q	$61	a	$71	q
$02	2	$12	Back	$22	"	$32	2	$42	B	$52	R	$62	b	$72	r
$03	3	$13		$23	#	$33	3	$43	C	$53	S	$63	c	$73	s
$04	4	$14		$24	$	$34	4	$44	D	$54	T	$64	d	$74	t
$05	5	$15		$25	%	$35	5	$45	E	$55	U	$65	e	$75	u
$06	6	$16		$26	&	$36	6	$46	F	$56	V	$66	f	$76	v
$07	7	$17		$27	'	$37	7	$47	G	$57	W	$67	g	$77	w
$08	8	$18		$28	(	$38	8	$48	H	$58	X	$68	h	$78	x
$09	9	$19		$29	)	$39	9	$49	I	$59	Y	$69	i	$79	y
$0A	A	$1A		$2A	*	$3A	:	$4A	J	$5A	Z	$6A	j	$7A	z
$0B	B	$1B		$2B	+	$3B	;	$4B	K	$5B	[	$6B	k	$7B	{
$0C	C	$1C		$2C	,	$3C	<	$4C	L	$5C	\	$6C	l	$7C	|
$0D	D	$1D		$2D	-	$3D	=	$4D	M	$5D	]	$6D	m	$7D	}
$0E	E	$1E		$2E	.	$3E	>	$4E	N	$5E	^	$6E	n	$7E	~
$0F	F	$1F		$2F	/	$3F	?	$4F	O	$5F	_	$6F	o	$7F	

Figure L2-11. Character codes supported by the main display.

and $0A through $0F to represent the numeric characters "0" through "9" and the hexadecimal characters "A" through "F," respectively. Also, we're using code $10 ("Clr") as a special instruction that will clear the display, code $11 ("Bell") to cause the interface to "beep" in an annoying fashion, and code $12 ("Back") to cause the display to clear the last (right-most) character and shift any remaining characters one place to the right.[1]

Last, but not least, codes $13 through $1F (along with code $7F) have been left undefined, which means you use them at your peril.

[1]In the case of standard ASCII, codes $00 through $1F were used to represent special control and formatting characters, such as STX ("start of text"), CR ("carriage return"), LF ("line feed"), ACK ("acknowledge"), and EOT ("end of transmission"). As some of these appellations suggest, these characters could be used to control things like printers and old-fashioned Teletype terminals.

The Status Register (SR)

In this lab we're going to create a program that writes to the calculator's main display, but before we proceed there's another register in the CPU that deserves mention: the *status register* (*SR*) (Figure L2-12).

Each bit in the status register is called a *status bit*, but they are also commonly referred to as *status flags* or *condition codes*, because they serve to signal (flag) that certain conditions have occurred. We shall introduce each of these bits and describe how they work as we progress through these labs.

Since we may sometimes wish to load the status register from (or store it to) the memory, it is usual to regard this register as being the same width as the data bus (8 bits in the case of our system). However, our CPU employs only five status flags, which occupy the five least-significant bits of the status register. This means that the three most-significant bits of the register exist only in our imaginations, so their nonexistent contents are, by definition, undefined. Finally, you need to be aware of one more convention as follows:

- A status flag is said to be "set" if it contains a 1, which is used to indicate a "true" condition. For example, if the zero flag is set (contains a 1), this indicates that: "It's TRUE to say that the current value stored in the accumulator (ACC) is zero" (that is, all of the bits in the accumulator contain 0s).

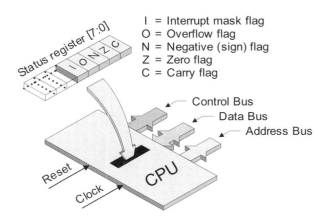

I = Interrupt mask flag
O = Overflow flag
N = Negative (sign) flag
Z = Zero flag
C = Carry flag

Figure L2-12. The status register (SR).

- A status flag is said to be "cleared" if it contains a 0, which is used to indicate a "false" condition. For example, if the zero flag is cleared (contains a 0), this indicates that: "It's FALSE to say that the current value stored in the accumulator is zero" (that is, one or more of the bits in the accumulator contain 1s).

Unfortunately, this can appear somewhat counterintuitive; especially in the case of the zero flag, where a 1 in the flag indicates that the accumulator contains all 0s and a 0 in the flag means that some of the bits in the accumulator are 1. It does take a little effort to wrap your brain around this convention the first time you see it, but you'll find that it really does make sense once you get into the swing of things (especially when we come to consider the other status flags).

We should also note that, although the zero flag is predominantly used to indicate whether or not the accumulator currently contains a zero value as was just discussed, this flag is multifaceted and can be used to indicate other conditions depending on the actual operation being performed. As one example, see the discussions on the CMPA instruction in Lab 2d (and also peruse the *Just When You Started to Feel Happy* section at the end of this lab).

Writing to the Main Display

If you haven't already done so, launch the main DIY Calculator and then invoke the assembler. Now use the assembler's **File > Open** command to access the *lab2b.asm* file you created in the previous lab, and then use the **File > Save As** command to rename and save this file as *lab2c.asm*, which should appear as follows:

```
CLRCODE:  .EQU   $10          # Special code to clear the main display
MAINDISP: .EQU   $F031        # Address of output port for main display
          .ORG   $4000        # Set program's origin to address $4000
          LDA    CLRCODE      # Load accumulator with clear code
          STA    [MAINDISP]   # Store accumulator to main display
          JMP    [$0000]      # Jump to address $0000
          .END                # This is the end of the program
```

As you will recall, this program loads the accumulator with a value of $10, stores this value to the output port driving the main display (thereby clearing the display), and jumps to address $0000, which terminates

the program. But this time our program is going to do something different, because we're going to augment it so as to write some numbers to
the display (Figure L2-13).

The flowchart blocks representing our original program are shown
in white, while any new extensions are shaded gray. Once this new version of the program has cleared the screen, it's going to load the accumulator with a value of $09 (or 9 in decimal). The program now enters a
loop, which consists of copying (storing) the value in the accumulator to
the display, decrementing (subtracting 1 from) the accumulator, and
then testing to see whether to display the accumulator's new contents (if
they're nonzero) or terminate (if they're zero). Note that this test is represented as a diamond shape in the flowchart in order to differentiate it
from the actions, which are shown as rectangles. As we shall see, this will
result in the characters "987654321" appearing on the display.

Modifying the Program

Without any further dilly-dallying or shilly-shallying, modify your program as shown below (the changes are highlighted in gray):

```
CLRCODE:    .EQU    $10       # Special code to clear the main display
MAINDISP:   .EQU    $F031     # Address of output port for main display
            .ORG    $4000     # Set program's origin to address $4000
            LDA     CLRCODE   # Load accumulator with clear code
            STA     [MAINDISP] # Store accumulator to main display
            LDA     $09       # Load the accumulator with $09
LOOP:       STA     [MAINDISP] # Store accumulator to the main display
            DECA              # Decrement the accumulator
            JNZ     [LOOP]    # Jump to LOOP if ACC isn't zero
            JMP     [$0000]   # Jump to address $0000
            .END              # This is the end of the program
```

These modifications include two new instructions. The first is DECA
("decrement accumulator"), which subtracts 1 from the existing contents
of the accumulator. This instruction is said to use the *implied addressing
mode* because it comprises only an opcode without an operand. In this
case, any data required by the instruction, and the destination of any result from the instruction, are *implied* by the instruction itself. (There is
also a corresponding INCA ("*increment accumulator*") instruction that
we can use to add 1 to the existing contents of the accumulator.)

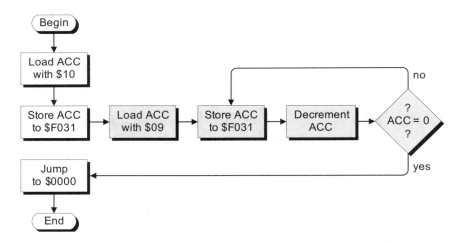

Figure L2-13. Flowchart for a program that writes numbers to the main display.

The second new instruction is JNZ ("jump if not zero"), which is a special form of jump with a condition attached to it. In fact, this is one of a pair of instructions that decide whether or not to jump based on the contents of the Z (zero) status flag. Just for the moment, however, let's forget about the Z flag per se, and instead simply say that:

- The JZ ("jump if zero") instruction will only jump to a specified address if the current value stored in the accumulator (ACC) is zero (that is, all of the bits in the accumulator contain 0s); otherwise the test will fail and the program will simply continue to the next instruction.

- By comparison, the JNZ ("jump if not zero") will only jump to a specified address if the ACC contains a nonzero value (that is, one or more of the bits in the accumulator contain 1s); otherwise the test will fail and, once again, the program will simply continue to the next instruction.

Our modifications also include a label called LOOP. This is a new type of label known as an *address label,* whose value (as "seen" by the assembler) is the memory location of the instruction with which it's associated (the program's second STA instruction in this case). This means that whenever the LOOP label is referenced elsewhere in the program (such as by the JNZ instruction in the above example), the assembler will automatically substitute the address of the memory location containing the STA opcode. (Note that, unlike the constant labels introduced in Lab 2b, *address labels* can be referenced *before* they have been declared. This isn't the case in this particular program, but we'll see lots of examples of this type of *forward referencing* in the not-so-distant future.)

In order to see how the address label actually works, let's consider the actions performed by our new statements. After clearing the main display, the LDA $09 instruction is used to load the accumulator with $09, which is going to provide the loop with an initial value. The subsequent STA [MAINDISP] instruction now stores the current value in the accumulator to the output port driving the main display, thereby causing the character "9" to appear at the right of the display area.

Next, the DECA instruction subtracts 1 from the value in the accumulator, which leaves the accumulator containing $08. As this value is nonzero, this means that the JNZ ("jump if not zero") instruction will cause the program to jump back to the label LOOP, at which point the STA [MAINDISP] will once again store the current value in the accumulator to the output port driving the main display. This will cause the "9" that's already on the display to be shifted one place to the left and an "8" to appear next to it, thereby resulting in a value of "98" on the display.

And so it goes. The program will keep on racing around the loop until the display contains "987654321," at which point executing the DECA instruction will result in the accumulator containing zero. This means that the JNZ instruction will fail and the program will drop through to our original JMP $0000 instruction, which, when executed, will cause the program to terminate.

Testing the Program

To see all of this happen on the real display, first assemble your program to generate the corresponding *lab2c.ram* file. Assuming you don't have any errors and your program assembles successfully, you may now exit the assembler.

Click the **On/Off** button to power-up the calculator. (If you didn't power the calculator down after the previous lab, click the **On/Off** button two times; the first time to power it down and "scramble" its memory, and the second time to power it up again.). Now use the main DIY Calculator window's **Memory > Load RAM** pull-down menu to locate the *lab2c.ram* file you just created and load its contents into the calculator's memory. Next, click the calculator's **Run** button to execute this new version of our program and observe the characters "987654321" appear on the display.

Just When You Started to Feel Happy

We'd just like to say in advance that we're sorry about what we're poised to do to you, which is to (metaphorically) squeeze your brain until your eyes water.

Earlier in this lab, we stated: "Just for the moment, let's forget about the Z flag per se, and instead simply say that...." And then we presented you with the following definitions:

- The JZ ("jump if zero") instruction will only jump to a specified address if the current value stored in the accumulator (ACC) is zero; otherwise the test will fail and the program will simply continue to the next instruction.

- By comparison, the JNZ ("jump if not zero") will only jump to a specified address if the ACC contains a nonzero value; otherwise the test will fail and, once again, the program will simply continue to the next instruction.

Sad to relate, this was a case of "the quickness of the hand deceives the eye." These definitions are certainly true for the program in this lab, but life is a tad more complicated than we've lead you to believe. The problem is that we typically think of the JZ and JNZ instructions in the context of the accumulator. In reality, however, the Z flag can be used to indicate other conditions; for example, consider the CMPA ("compare accumulator") instruction, which is used to compare two numbers (this instruction is introduced in Lab 2d). In this case, the Z flag is set to 1 if the two numbers are equal; otherwise it is cleared to 0.

Thus, more precise definitions of the operation of the JZ and JNZ instructions are as follows:

- The JZ ("jump if zero") instruction will only jump to a specified address if the Z flag contains a 1 (a "true" condition); otherwise the test will fail and the program will simply continue to the next instruction.

- By comparison, the JNZ ("jump if not zero") will only jump to a specified address if the Z flag contains a 0 (a "false" condition); otherwise the test will fail and the program will simply continue to the next instruction.

> **Note**   You might be forgiven for thinking we're wandering off into the weeds here, reminiscent of the now famous statement by U.S. State Department Spokesman Robert McCloskey:
>
> "I know you believe that you understood what you think I said, but I am not sure you realize that what you heard is not what I meant."

So what does all of this really mean? Well, the decision symbol (the diamond) in the flowchart in Figure L2-13 is based on the contents of the accumulator. Another way to think about this would be to base this symbol on the contents of the Z flag (Figure L2-14).

Observe that the "yes" and "no" annotations associated with the decision symbols are swapped between the two figures. This is because a 0 in the Z flag indicates a nonzero value in the accumulator, whereas a 1 in the Z flag indicates a zero value in the accumulator.

Note that the "!" character (sometimes referred to as a "bang" or a "shriek") is often used to indicate an inversion or a negative ("not") state, so the string "ACC != 0" in Figure L2-14 indicates "ACC is not equal to zero." By comparison, a pair of equals signs "==" is used to indicate a logical equivalence (as opposed to an assignment implied by a single equal sign "="), so the string "ACC == 0" in Figure L2-14 is used to indicate that "ACC is logically equal to zero."

Don't panic if all of this seems overly confusing; everyone feels that way when they first grapple with this topic.[2] In fact, the only reason

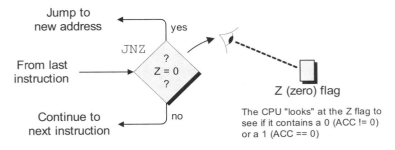

Figure L2-14. The JNZ causes a jump if the Z flag is zero.

we've labored this point into the ground is that the Z flag can be used to indicate a variety of conditions depending on the actual operation being performed. For the moment, however, we can forget all about the inner machinations of the Z flag, and need only remember that, in the case of this program, the JNZ instruction means "Jump to the specified address if the accumulator contains anything other than $00.")

Bonus Question   In the section of this lab entitled *The Main Display's Character Codes,* why do you think we decided to make the main display accept codes $00 through $09 as representing the numeric characters "0" through "9" when these characters are already provided by the standard ASCII codes $30 through $39. Similarly, why did we make the main display accept codes $0A through $0F as representing the hexadecimal characters "A" through "F" when ASCII codes $41 through $46 already provide *these* characters?

---

perform any such conversions.

"F" allows us to output these values directly to the display without having to through $0F as representing the characters "0" through "9" and "A" through ASCII equivalents. Augmenting our calculator display to accept codes $00 we would first have to add ($41 – $0A) to them to convert them into their through %00001111 (representing the hexadecimal digits "A" through "F"), equivalents. Similarly, in order to output values in the range %00001010 "9"), we would first have to add $30 to them to convert them into their ASCII %00001001 (representing the decimal and hexadecimal digits "0" through whenever we wished to output values in the range %00000000 through

Answer   If the calculator display accepted only standard ASCII codes, then

---

[2]Alternatively, if you're feeling frisky and saying to yourself: "Well, that was pretty simple. I really don't know what they were making all the fuss about," then put away these notes, grab a friend, and try articulating the way in which we can view the JN and JNZ instructions working in the context of (a) the ACC and (b) the Z flag.

Lab 2d
Reading from the Calculator's Keypad

Objective: To learn how to read data from the calculator's keypad.

Duration: Approximately 25 minutes.

Keypad Buttons and Codes

In the previous lab, we discovered how to drive the calculator's main display. The next thing we need is some data *to* display, and one source of data is the calculator's keypad (Figure L2-15).

Each of the calculator's buttons has a code associated with it, where these codes are shown as hexadecimal values in parenthesis. For example, $16 and $1A are the codes associated with the "+" and "=" buttons, respectively.

The calculator interface contains an 8-bit latch whose outputs drive our virtual computer's input port at address $F011. (For our purposes here, a *latch* may be considered to be another name for a *register*; that is, a group of memory elements, each of which can store a single bit.) When the calculator is first powered-up using the front panel's **On/Off** button (not shown in Figure L2-15), this latch is automatically loaded with a default value of $FF. This default value remains in the latch until we click one of the calculator's buttons, at which point the code associated with that button is loaded into the latch. Once a button code has been loaded into the latch, it remains there until one of two things occur:

a) We click another button, which causes *its* code to be loaded into the latch.

b) The CPU (directed by one of our programs) reads data from the input port at address $F011. In addition to copying the current value stored in the latch into the accumulator, a special (virtual) electronic circuit automatically causes the latch to be reloaded with its default $FF value.

185

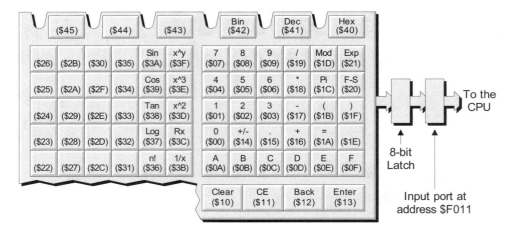

Figure L2-15. The calculator's keypad and associated codes.

The fact that reading from the port causes the latch to be reloaded with $FF is important, because, as we shall see, it provides a way for our programs to detect whether or not a key has been pressed. Also, you will observe that although many of the keys have annotations, some remain blank. The annotated keys are those for which we have already decided the function they are to perform (we'll tell you how to modify these existing annotations or add new ones to the blank keys later). In addition to the buttons marked "0" through "9" and "A" through "F" (which can be used to represent decimal and hexadecimal digits), we have made the following assignments:

Code	Annotation	Intended Function
$10	Clear	Clear the current calculation and the main display
$11	CE	Clear the current value on the main display
$12	Back	Clear the last displayed character
$13	Enter	Indicates the end of a numerical value[1]
$14	+/−	Change the sign of a value from + to − or vice versa[2]

[1]This is used in the case of the *reverse Polish notation* (*RPN*) scheme introduced in Chapter 5.

[2]The actions of the +/− button will eventually be context sensitive in the case of floating point values; that is, we may wish to change the sign of the mantissa and/or the exponent in scientific-notation numbers (this is covered in more detail on the DIY Calculator's website).

$15	.	Decimal point
$16	+	Add
$17	−	Subtract
$18	*	Multiply
$19	/	Divide
$1A	=	Equals
$1B	(	Left parenthesis
$1F	)	Right parenthesis
$1C	Pi	The constant π
$1D	Mod	Displays the modulus (remainder) of a division
$20	F-S	Toggle between scientific notation being on/off
$21	Exp	Used to enter an exponent in scientific notation
$36	n!	Calculate the factorial of the value on display
$37	Log	Calculate the logarithm of the value on display
$38	Tan	Calculate the tangent of the value on display
$39	Cos	Calculate the cosine of the value on display
$3A	Sin	Calculate the sine of the value on display
$3B	1/x	Calculate the reciprocal of the value on display
$3C	Rx	Calculate the square root of the value on display
$3D	x^2	Calculate the square of the value on display
$3E	x^3	Calculate the cube of the value on display
$3F	x^y	Calculate x raised to the yth power
$40	Hex	Enter and display values in hexadecimal
$41	Dec	Enter and display values in decimal (the default)
$42	Bin	Enter and display values in binary

Although the above list sounds rather grand, especially if you read it quickly and aloud, it's worth remembering that none of these buttons actually *do* anything yet! At the moment, the only action that occurs when you click a button is for the code associated with that button to be loaded into the calculator's 8-bit latch. The purpose of this book is to teach you how to create the programs that will cause at least some of these buttons to actually do something useful.

With regard to the unassigned buttons with codes $1E, $22 through $35, and $43 through $45, these buttons have been left free for you to use as you wish. You can reconfigure the front panel to modify the legends on the buttons, change the colors of the legends, and so forth. These aspects of the front panel are introduced in more detail in Chapter 6.

Reading Codes from the Keypad

You should know the drill by now. If you haven't already done so, launch the main DIY Calculator and then invoke the assembler. Now use the assembler's **File > Open** command to access the *lab2b.asm* file you created in an earlier lab, then use the **File > Save As** command to rename and save this file as *lab2d.asm*, which should appear as follows:

```
CLRCODE:   .EQU    $10          # Special code to clear the main display
MAINDISP:  .EQU    $F031        # Address of output port for main display
           .ORG    $4000        # Set program's origin to address $4000
           LDA     CLRCODE      # Load accumulator with clear code
           STA     [MAINDISP]   # Store accumulator to main display
           JMP     [$0000]      # Jump to address $0000
           .END                 # This is the end of the program
```

You should be able to write this particular program in your sleep by this time! All it does is to load the clear code $10 into the accumulator, write this code to the port driving the main display, and then terminate the program by jumping to address $0000. But, as usual, this is just a starting point, because we are going to modify the program such that it waits for a button to be pressed and, if that button corresponds to "0" through "9" (that is, one of the buttons carrying the "0" through "9" annotations), writes this value to the main display (Figure L2-16).

As usual, the flowchart for our original program is shown in white, whereas any extensions are shown in gray. Note that we're going to lose the JMP [$0000] instruction from the end of the original program, because our new program never actually ends. Once the clear code has been written to the main display, the program enters a series of *nested loops*, which means that we have an inner loop "nested" inside one or more higher-level loops.

First, the program loops around, reading values from the input port at address $F011 (the port connected to the calculator's latch) and loading them into the accumulator. The program keeps on doing this

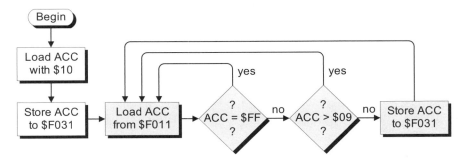

Figure L2-16. Flowchart for a program to read from the keypad and write to the display.

until it detects a value other than $FF, at which point it knows that a button has been pressed. (Remember that the act of reading from this port automatically causes the latch to be reloaded with its original $FF value.)

Next, the program tests to see if the code for this button is greater than $09, in which case it won't perform any actions, but will simply return to wait for another button to be pressed. Only if the code is between $00 through $09, inclusive (corresponding to the '0' through '9' buttons), will the program write this value to the calculator's main display and then return to wait for the next button to be pressed. So now is the time to modify our program as shown below (the changes are highlighted in gray):

```
CLRCODE:  .EQU    $10         # Special code to clear the main display
MAINDISP: .EQU    $F031       # Address of output port for main display
KEYPAD:   .EQU    $F011       # Address of input port for keypad
          .ORG    $4000       # Set program's origin to address $4000
          LDA     CLRCODE     # Load accumulator with clear code
          STA     [MAINDISP]  # Store accumulator to main display
LOOP:     LDA     [KEYPAD]    # Load the accumulator from the keypad
          JN      [LOOP]      # Jump to LOOP if N flag is set
          CMPA    $09         # Compare accumulator to code $09
          JC      [LOOP]      # Jump to LOOP if C flag is set
          STA     [MAINDISP]  # ... else store accumulator to display
          JMP     [LOOP]      # Jump to LOOP
          JMP     [$0000]     # Jump to address $0000
          .END                # This is the end of the program
```

The first change is to introduce a new constant label called KEYPAD, and to assign it the address of the input port driven by the latch connected to the calculator's keypad. After clearing the main display, the LDA [KEYPAD] instruction loads the accumulator from the latch. A new flavor of instruction—JN ("jump if negative")—is now used to jump back to the LOOP label if the N (negative) flag in the status register contains a 1 (Figure L2-17). [Note that there is also a corresponding JNN ("jump if not negative"), which will jump if the N flag contains a 0.]

To be honest this is a bit unfair, because there are a couple of things we haven't told you yet. (Actually, there's a whole lot we haven't told you, but we will—oh yes, we will. Be afraid. Be very afraid.) First of all, it's up to us to decide what the contents of the accumulator represent at any particular point in time. For example, at some stage during our program we might consider the accumulator to contain only positive values, but a little later we could do a complete about-face and decide that it can contain both positive and negative values (all of this is explained in excruciating detail in Chapter 4).

If we assume that the accumulator can contain both positive and negative values, then we can use its *most-significant bit* (*MSB*) to differentiate between them (a 0 or 1 in the MSB indicates a positive or negative value, respectively). The trick here is that the N (negative) status flag typically contains a copy of whatever is in the most-significant bit of the accumulator. As the calculator's default $FF code (which equates to %11111111 in binary) is the only code the front panel's latch will ever contain that has a 1 in its most-significant bit, the JN instruction will only jump back to the LOOP label if no key has been pressed.

If a key *has* been pressed, the JN instruction fails and the program continues to the CMPA ("compare accumulator") instruction, which compares the contents of the accumulator to the value $09. This operation assumes that both of the values being compared are positive values

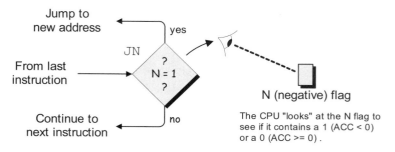

Figure L2-17. The JN causes a jump if the ACC contains a negative value.

(remember that it's up to us to decide what we want the contents of the accumulator to represent at any particular time). The ways in which the CMPA instruction presents the results of its comparison are as follows:

- The C (carry) status flag will be loaded with a 1 if the value in the accumulator is *larger* than the value it's being compared to; otherwise the C flag will be cleared to 0.[3]

- The Z (zero) status flag will be loaded with a 1 if the value in the accumulator is *equal* to the value with which it's being compared; otherwise it will be cleared to 0.

In the case of this particular program, the only keys of interest are those with codes of $00 through $09. If the value in the accumulator is *greater* than $09, the CMPA instruction will cause the C (carry) flag to be set to a 1. In turn, this will cause the JC ("jump if carry") instruction to jump back to LOOP (Figure L2-18). (As usual, there is a corresponding JNC ("jump if not carry") instruction, which will jump if the C flag contains a 0.)

If the JC instruction fails, this means that the value in the accumulator is in the range $00 through $09, in which case the STA [MAINDISP] instruction is used to store this value to the output port driving the calculator's main display. Finally, the JMP ("jump unconditionally") instruction returns the program to LOOP to wait for the next button to be pressed.

Testing the Program

In order to see this work with the real display, first click the assemble icon on the assembler's tool bar to generate the corresponding *lab2d.ram* file and then either minimize or exit the assembler.

Click the **On/Off** button to power-up the calculator. (If the calculator is already powered-up, use the **Memory > Purge RAM** command to "scramble" its memory). Now use the **Memory > Load RAM** command to locate the *lab2d.ram* file you just created and load its contents into the calculator's memory. Next, click the calculator's **Run** button to execute this new version of our program. Observe that, as soon as the display is cleared, the program starts to loop around waiting for you to click a button.

---

[3]The C (carry) flag is perhaps the most multifarious of all the status flags, in that it is used for a multitude of different purposes, as we shall come to discover in future labs.

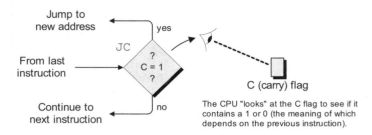

Figure L2-18. The JC causes a jump if the C (carry) status flag contains a 1.

Experiment by clicking the buttons "0" through "9" and note the effect. Also note that even when the display is full, clicking further buttons simply causes the new digits to appear at the right-hand side of the display, whereas the most-significant digits "fall off" the left-hand side of the display. Now click any of the other buttons on the keypad (*excluding* the **On/Off**, **Reset**, **Step**, and **Run** control buttons) and observe that nothing happens. This is because our new program is written in such a way that it ignores them.

Once you are sated with your newfound power, click the **On/Off** button to power-down the calculator, and then consider the question below.

Bonus Question We didn't actually need to use the JN [LOOP] instruction in our program. Why not?

Answer The JN instruction is used to check for the calculator's default $FF code (meaning that no buttons have been pressed), in which case the program jumps back to LOOP. If the JN fails, the CMPA instruction is used to compare the contents of the accumulator to the value $09. If the accumulator's contents are the larger, the carry flag is loaded with a 1, which will cause the JC instruction to jump back to LOOP.

The point is that the CMPA instruction assumes the values it's comparing are positive integers, so it will consider a value of $FF to be larger than a value of $09. Thus, even if the JN instruction were omitted, a value of $FF in the accumulator would still cause the JC instruction to jump back to LOOP, which means that the JN is superfluous in this particular implementation.

Lab 2e | Writing to the Calculator's Six LEDs

Objective:	To learn how to control the six LEDs on the calculator.
Duration:	Approximately 30 minutes.

The Calculator's Six LEDs

As you will recall, there are six buttons on the front panel appearing just below the main display and just above the main keypad area. Three of these buttons have annotations (**Bin, Dec,** and **Hex**), whereas the others are as yet undefined. To the left of each of these buttons is a simple *light-emitting diode* (*LED*), which we can turn on or off as we choose (Figure L2-19).

It's important to note that there is no intrinsic relationship between the **Bin, Dec, Hex,** and similar buttons and these LEDs apart from their being physically close together on the calculator's front panel. The LEDs are driven by an output port at address $F032, and any values written to this output port are totally under the control of our programs. This means that the only relationships between the LEDs and *any* of the keypad's buttons are those *we* decide to implement.

Writing to the Six LEDs

You'll be pleased to hear that this program is going to be short, sharp, and sweet. All we're going to do is to create a loop that flashes the LEDs in sequence (Figure L2-20).

As you will observe, this flowchart assumes we are keeping the two instructions from our earlier programs that are used to clear the main display. In fact there's no particular reason to clear the main display here,

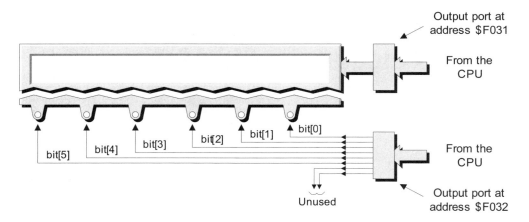

Figure L2-19. The calculator's six LEDs.

but it's something we're used to doing and it makes the display look nice and tidy, so why not?

If you haven't already done so, launch the main DIY Calculator and then invoke the assembler. Use the assembler's **File** > **Open** command to access the *lab2b.asm* file you created in an earlier lab, and then use the **File** > **Save As** command to rename and save this file as *lab2e.asm*, which should appear as follows:

```
CLRCODE:    .EQU    $10         # Special code to clear the main display
MAINDISP:   .EQU    $F031       # Address of output port for main display
            .ORG    $4000       # Set program's origin to address $4000
            LDA     CLRCODE     # Load accumulator with clear code
```

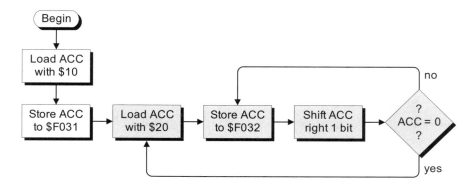

Figure L2-20. Flowchart for a program to flash the six LEDs.

```
         STA     [MAINDISP]  # Store accumulator to main display
         JMP     [$0000]     # Jump to address $0000
         .END                # This is the end of the program
```

In fact we're going to leave the bulk of this program as is. All we really need to do is to declare a new constant called SIXLEDS and assign it to the output port at address $F032, and then replace the existing JMP ("jump unconditionally") instruction at the end of the program with the new instructions required to implement the flowchart in Figure L2-18 (the changes are highlighted in gray):

```
CLRCODE:  .EQU    $10         # Special code to clear the main display
MAINDISP: .EQU    $F031       # Address of output port for main display
SIXLEDS:  .EQU    $F032       # Address of output port for six LEDs
          .ORG    $4000       # Set program's origin to address $4000
          LDA     CLRCODE     # Load accumulator with clear code
          STA     [MAINDISP]  # Store accumulator to main display
LOOPA:    LDA     $20         # Load accumulator with $20
LOOPB:    STA     [SIXLEDS]   # Store accumulator to six LEDs
          SHR                 # Shift accumulator 1 bit to the right
          JNZ     [LOOPB]     # Jump to LOOPB if Z flag not set
          JMP     [LOOPA]     # ..else jump to LOOPA
          JMP     [$0000]     # Jump to address $0000
          .END                # This is the end of the program
```

Everything is familiar until label LOOPA, at which point an LDA ("load accumulator") instruction is used to load the hexadecimal value $20 into the accumulator. This value equates to the pattern %00100000 in binary. Next, a STA ("store accumulator") instruction is used to store this value to the output port driving the LEDs. As it happens, the front panel has been implemented such that an output port bit containing a 1 will turn its corresponding LED on, while a 0 will turn it off. Thus, our %00100000 value will cause only the left-most LED to be activated.

Next, a new instruction, SHR ("shift right"), is used to shift the contents of the accumulator one bit to the right. In fact this is a type of shift known as an *arithmetic shift*,[1] because the *most-significant bit* (*MSB*) in the accumulator is copied back into itself, so shifting the %00100000 value in the accumulator one bit to the right results in its containing a new value of %00010000 (Figure L2-21).

[1]The "shift" instructions, and their "rotate" cousins, are discussed in more detail in Chapters 3 and 4.

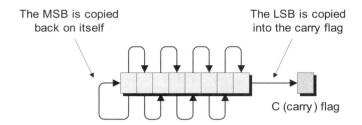

Figure L2-21. The SHR instruction shifts the contents of the accumulator one bit to the right.

Note that when this shift right is performed, the original contents of the *least-significant bit* (*LSB*) in the accumulator conceptually "fall off the end." However, this bit is not completely lost to us, because it is copied into the C (carry) status flag. We aren't making any use of this aspect of the shift instruction in this program, but it will prove useful in the future.

Following the shift, a JNZ ("jump if not zero") instruction is used to jump to the target location if the contents of the accumulator are nonzero. The new value of %00010000 in the accumulator is certainly nonzero, so the program will jump back to the label LOOPB, at which point it will store *this* value to the LEDs. Thus, the sequence of values that will appear on the LEDs is as shown in Figure L2-22.

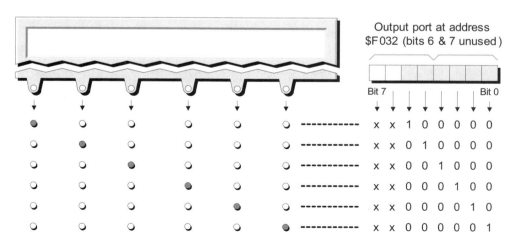

Figure L2-22. The patterns appearing on the six LEDs.

Now consider what happens after the final value has been written to the LEDs. The SHR will cause the %00000001 value in the accumulator to be shifted one bit to the right, leaving the accumulator containing %00000000. In this case, the JNZ instruction will fail, and the program will drop through to the JMP [LOOPA] instruction. This unconditional jump returns the program to LOOPA, at which point the accumulator will be reloaded with $20 and the whole process will start again.

To see this work with the real display, assemble the program to generate the corresponding *lab2e.ram* file and then either minimize or exit the assembler. Now power-up the calculator and note that, in the same way that the main display fills with "-" characters to indicate an uninitialized state, all six LEDs power-up in an active condition to indicate that they too are uninitialized. Load the *lab2e.ram* file we just created into the calculator's memory, and then click the front panel's **Run** button to execute this new version of our program and observe the activity on the six LEDs.

Modifying the System Clock

Depending on the power of your home computer, the patterns on your LEDs are either crawling along like a lethargic snail or flashing before your eyes in a frantic blur (Murphy's Law[2] precludes the third possibility, which would be that your display is changing at just the right speed).

Fortunately, we are prepared for just such an eventuality. Use the DIY Calculator's **Setup > System Clock** command to invoke the appropriate dialog window as shown in Figure L2-23. It should not come as a major revelation to discover that clicking the arrows marked **Slow** and **Fast** followed by the **Apply** button will cause the system clock to slow down or speed up, respectively.[3] Note that this doesn't actually affect the speed of your real computer, only the speed of our virtual system. Also note that the effects of changing the system clock will persist only for the current session; the next time you invoke the calculator, its clock will automatically be returned to its default setting (which is to go as fast as pos-

---

[2]Murphy's Law can be summarized as "Anything that can go wrong will go wrong!" This was first espoused in a more sophisticated form by Captain Edward A. Murphy in 1949 as a useful working assumption in safety-critical engineering.

[3]The **Apply** button applies the new setting and leaves the form on-screen, whereas the **OK** button applies the new setting and then dismisses the form.

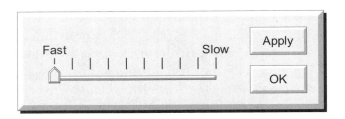

Figure L2-23. The system clock dialog.

sible).

Use this dialog (Figure L2-23) to modify the system clock such that the patterns on the LEDs change approximately once per second; that is, somewhere around the same rate you'd count "Thousand one, thousand two, thousand three. . . ."

Once you've observed that the patterns on the LEDs do indeed follow the sequence described earlier, return the system clock to its fastest setting, click the **OK** button to lock in this setting and dismiss the dialog window, click the **On/Off** button to power-down the calculator, and proceed to the next, and final (Hurray!), laboratory in this section.

Bonus Question   Purely for the sake of discussion, let's assume that our CPU's Z (zero) flag isn't working quite as well as one might hope (perhaps a cosmic ray knocked it out of action). This means that we don't have any faith in the ability of the JNZ ("jump if not zero") instruction in our program to perform its task as planned. Based on our discussions on the SHR ("shift right") instruction illustrated in Figure L2-21, can you think of another conditional jump we could use to replace the JNZ?

■ *Answer* After loading the accumulator with %00100000, our current program loops around, shifting the contents of the accumulator one bit to the right. The JNZ instruction is used to keep on looping until the accumulator finally contains %00000000.

As you will recall, when a SHR ("shift right") is performed, the original contents of the *least-significant bit* (*LSB*) in the accumulator conceptually "fall off the end" and are copied into the C (carry) status flag. Another way of looking at this is that the C (carry) flag will contain a 0 after every shift except the last one, at which point it will contain the 1 that "falls off the end" of the accumulator. This means that, if we feared that the Z flag wasn't working, all we would have to do would be to replace our JNZ [LOOPB] instruction with a JNC [LOOPB] ("jump if not carry") and the program would continue to function as required.

Lab 2f

Using the Memory Walker and Other Diagnostic Displays

Objective: To learn how the Memory Walker, CPU Register, and other diagnostic displays can be used to aid in debugging programs.

Duration: Approximately 40 minutes.

The Need for Diagnostic Tools

Suppose you create a program, assemble it, run it, and it doesn't work as you expected. How are you going to find out what has gone wrong? In fact, there are a variety of different techniques one might use to diagnose the cause of the problem, such as single-stepping through the program one instruction at a time. There are also a number of different display utilities we can use to see what is going on in the CPU's registers, the memory, and the I/O ports. These diagnostic tools and techniques are the focus of this laboratory.

First We Need a Demonstration Program

In order to use the various diagnostic displays supplied with the DIY Calculator, we really need a program to play with. We could use one of the programs from earlier labs, but we've already "been there, done that, read the book, sang the song, saw the play, and bought the T-shirt,"[1] so instead we're going to create a brand new program that will display the message "Hello World" on the calculator's main display.

[1]What? You didn't get the tattoo?

201

One thing of which we need to be aware before we begin is that the CPU contains a special 16-bit register called the "index register," or "X" for short. This register is discussed in more detail in Chapter 3; for our purposes here, we need only note that our new program will use the index register to provide an offset ("index") into a series of characters (Figure L2-24). (Note that we might also describe this series of characters as a *string* of characters or an *array* of characters.)

Purely for the sake of this example, let's assume that we happen to know that the first character in our message is located at memory location $4015. If we load the index register with 0 and add this to the start address of the message; not surprisingly, we end up with the start address again ($4015 + 0 = $4015). Obviously this isn't particularly exciting, but if we now increment the index register and add this new value to the start address of the message, the result points to the second character in our message, and so it goes. As we shall see, this is the basis for our program, the flowchart for which is shown in Figure L2-25.

As usual, this flowchart assumes we are keeping the two instructions from our earlier programs (shown in white) that are used to clear the main display. If you haven't already done so, launch the main DIY Calculator and then invoke the assembler. Use the assembler's **File > Open** command to access the *lab2b.asm* file you created in an earlier lab, use the **File > Save As** command to rename and save this file as *lab2f.asm*, then edit this file as follows (our new additions are highlighted in gray):

```
CLRCODE:   .EQU    $10          # Special code to clear the main display
MAINDISP:  .EQU    $F031        # Address of main display output port
           .ORG    $4000        # Set program's origin to address $4000
           LDA     CLRCODE      # Load accumulator with clear code
           STA     [MAINDISP]   # Store accumulator to main display
           JMP     [$0000]      # Jump to address $0000
           BLDX    0            # Load the index register with 0
LOOP:      LDA     [MESSAGE,X]  # Load accumulator with character
           JZ      [$0000]      # If character = $00 jump to $0000
STORE:     STA     [MAINDISP]   # ... else write character to display
           INCX                 # Increment the index register
           JMP     [LOOP]       # Jump back to LOOP
MESSAGE:   .BYTE $48, $45, $4C, $4C, $4F, $20
           #       H    E    L    L    O  SPACE
           .BYTE $57, $4F, $52, $4C, $44, $00
           #       W    O    R    L    D   NUL
           .END
```

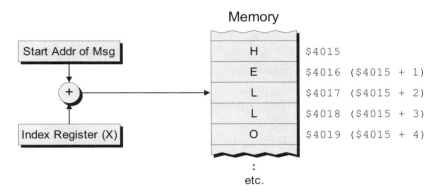

Figure L2-24. Using the index register to provide an offset into a string of characters.

Before you do anything else, use the assembler's **File > Save** command to save this file (never fail to take the opportunity to save your work—trust us on this if nothing else). Now look at the MESSAGE address label toward the bottom of the program. This is followed by a new directive called .BYTE, which instructs the assembler to reserve one or more byte-wide (8-bit) values. This directive is assigned a series of comma-separated values, which the assembler will store in adjacent memory locations.

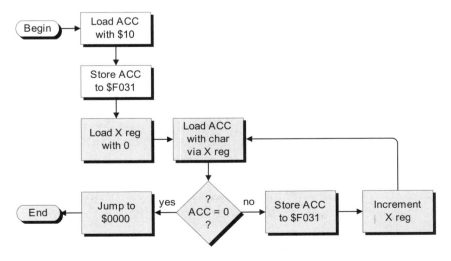

Figure L2-25. Flowchart for a "Hello World" program.

If you cast your mind back to Figure L2-11 in Lab 2c, you will see that these values correspond to the ASCII codes for the letters H, E, L, L, O, which are followed in turn by the code for a space character. Similarly, the second .BYTE directive reserves six byte-wide values and fills them with the ASCII codes for the letters W, O, R, L, D followed by a $00 value (which is sometimes referred to as a NUL or NULL).

Now consider the rest of the program. Everything is reasonably familiar until the BLDX ("big load of the index register") instruction. The reason this is referred to as a "big load" is that our CPU's index register is 16 bits wide (unlike the accumulator which is only 8 bits wide).

The clever part of this program occurs with the LDA [MESSAGE,X] instruction, which uses something called the *indexed addressing mode*. This instructs the assembler to load the accumulator with the contents of a memory location whose address is calculated by adding the address of the MESSAGE label and the contents of the index register (X). As the index register currently contains 0, this results in an address of MESSAGE + 0. This is the first location reserved by the .BYTE directive, which contains the $48 code corresponding to the letter H.

Next, a JZ ("jump if zero") instruction is used to jump to address $0000 if the accumulator contains zero. But, as we've just discussed, the first time around the loop the accumulator contains a value of $48, so this test will fail and the program will continue to the STA [MAINDISP] instruction. This will write the $48 code in the accumulator to the main display, which will now show the letter H.

This is followed by an INCX ("increment index register") instruction,[2] which adds 1 to the current contents of the index register, and then a JMP ("unconditional jump") instruction is used to return to the label LOOP. This is where things start to get interesting, because the index register now contains a value of 1, so this time round the loop the LDA [MESSAGE,X] instruction will load the accumulator with the contents of the memory location at address MESSAGE + 1. This is the second location reserved by the .BYTE directive, which contains the $45 ASCII code for the letter E.

And so the program proceeds around and around the loop, reading the next character from the message and writing it to the calculator's main display. Eventually, the accumulator will be loaded with the $00

---

[2]There is also a corresponding DECX ("decrement index register") instruction, which subtracts 1 from the current contents of the index register.

NUL code at the end of the message. This causes the JZ ("jump if zero") instruction to jump to address $0000, at which point the program will automatically terminate.

Introducing the List File

Assemble the program to generate the corresponding *lab2f.ram* file. Assuming no errors were encountered, the assembler automatically generates another file called *lab2f.lst,* which is referred to as the *list file.* Use the assembler's **Window > View List File** command (or click the corresponding icon on the toolbar), which causes the assembler to display the contents of the list file. If you have a printer connected to your system, now would be a real good time to generate a hard copy of this file to peruse as we proceed.

When you are presented with your first list file, it may initially seem a little complicated, but these little rascals are actually quite easy to read once you know the rules. In fact a list file contains three main sections: the body of the program and two cross-reference tables. First, let's consider the body of the program (Figure L2-26).

```
LINE   ADDR    DATA        LABEL       OPCODE OPERAND
-----  ----  --------    ----------    ------ -------
00001                    CLRCODE :     .EQU   $10
00002                    MAINDISP :    .EQU   $F031
00003                                  .ORG   $4000
00004  4000  90 10                     LDA    CLRCODE
00005  4002  99 F0 31                  STA    [MAINDISP]
00006  4005  A0 00 00                  BLDX   0
00007  4008  92 40 15    LOOP :        LDA    [MESSAGE,X]
00008  400B  D1 00 00                  JZ     [$0000]
00009  400E  99 F0 31    STORE :       STA    [MAINDISP]
00010  4011  82                        INCX
00011  4012  C1 40 08                  JMP    [LOOP]
00012
00013  4015  48 45 4 C   MESSAGE :     .BYTE  $48, $45, $4C, $4C, $4F, $20
       4018  4C 4F 20
00014
00015
00016  401B  57 4F 52                  .BYTE  $57, $4F, $52, $4C, $44, $00
       401E  4C 44 00
```

Figure L2-26. Body of the list file for the "Hello World" program.

During the process of reading our program (also referred to as the *source file*), the assembler assigns a unique number to each line of source code, and these line numbers are displayed in decimal in the left hand column. Our original source code is displayed to the right of the list file. Thus, the list file shows that our program commenced on line 00001 when we declared the constant label CLRCODE and associated it with a value of $10.

Of particular interest is the fact that the machine code relating to each line of source code appears between the line number and the original source code. The first value to the right of the line number is the hexadecimal address of the instruction. Following the address we may see one, two, or three hexadecimal values, of which the first value is always the opcode of the instruction (except in the case of .BYTE directives, along with their .2BYTE and .4BYTE cousins). On line 00011, for example, the JMP instruction is used to jump to label LOOP. This instruction occurs at address $4012, the opcode for the JMP instruction is $C1, and the two data bytes following the opcode are $40 and $08 forming address $4008, which is the address associated with the LOOP label.

Following the body of the program is the *Constant Labels* cross-reference table, which details the constant labels associated with any .EQU directives (Figure L2-27).

```
CONSTANT LABELS CROSS -REFERENCE

  NAME       VALUE     LINE NUMBERS WHERE USED  (* INDICATES DECLARATION )
--------   ---------   ---------------------------------------------------
CLRCODE    00000010    00001 * 00004
MAINDISP   0000F031    00002 * 00005   00009




ADDRESS LABELS CROSS -REFERENCE

  NAME       VALUE     LINE NUMBERS WHERE USED  (* INDICATES DECLARATION )
--------   ---------   ---------------------------------------------------
LOOP       ....4008    00007 * 00011
MESSAGE    ....4015    00007   00013 *
STORE      ....400E    00009 *
```

Figure L2-27. Cross-reference tables for the "Hello World" program.

Reading this cross-reference table from left to right shows that we have a constant label called CLRCODE with a value of $10. This entry also shows that CLRCODE was declared on line 00001 and used on line 00004 ("*" characters are used to indicate the line in which a label is declared). Similarly, we have a constant label called MAINDISP with a value of $F031 that was declared on line 00002 and used on lines 00005 and 00009.

Finally, the *Address Labels* cross-reference table appears at the bottom of the list file. This table shows that the label LOOP is associated with address $4008; it was declared on line 00007 and it is used (referenced) on line 00011. Similarly, the label MESSAGE is associated with address $4015; it was declared on line 00013 and it is used on line 00007. Last, but not least, the label STORE is associated with address $400E; it was declared on line 00009 and it isn't actually used anywhere. So, why did we declare the label STORE in our program if we didn't intend to use it? The answer is that this label made the program a tad more readable, and it also allows us to quickly locate the address of this instruction in the list file. This will prove to be of use shortly; for the moment, however, minimize the assembler to make room on the screen (do *not* exit it, because we'll be needing it shortly).

The Memory Walker Display

Use the main window's **Display** > **Memory Walker** command (or click the appropriate icon on the tool bar) to launch the Memory Walker display, which is used to show the contents of the calculator's RAM.

> **Note**   As soon as you invoke one of the utilities such as the Memory Walker, CPU Register, or I/O Ports displays, it is recommended that you immediately drag the display (using its blue title bar) away from the main calculator front panel and locate it in a clear area of the interface. (You can also resize the Memory Walker in a vertical direction so as to make more rows of data visible.)
>
> The reason we suggest this is that, if you inadvertently click the front panel when one of the other utilities is "on top," the front panel will "come forward" and may completely obscure the other utility.
>
> Should this occur, you can drag the calculator front panel around the screen with its blue title bar until you see a portion of the hidden display. Clicking on any part of that display will bring it to the foreground, at which point you can drag the display to a clear area of the interface.

Assuming that the calculator hasn't been powered-up yet, this display is initially grayed out, indicating that nothing is active, so click the **On/Off** button on the front panel to turn it on. (If the calculator was already running, power it off and on again to return everything to its uninitialized state.)

Applying power causes the Memory Walker display to turn white to show that it's active, but the data column still contains $XX values. These indicate that the RAM locations contain random values because they haven't been loaded with anything meaningful yet.

Now use the **Memory –> Load RAM** command to locate the *lab2f.ram* file we just created and load its contents into the calculator's memory. As you'll see, this data also appears in the Memory Walker display (Figure L2-28).

Compare the values in the Memory Walker display to the contents of the address and data columns in the body of the list file (Figure L2-26) and convince yourself that they are identical. Now, before you do anything else, glance at the main DIY Calculator's status bar (at the bottom of the window) and observe that it's indicating the fact that the calculator is in its *reset mode.*

Now click the **Run** button on the calculator's front panel to execute your program and observe that the main display is cleared, the message "HELLO WORLD" appears, and the program terminates and returns the calculator to its *reset mode.* You should also observe that the Memory

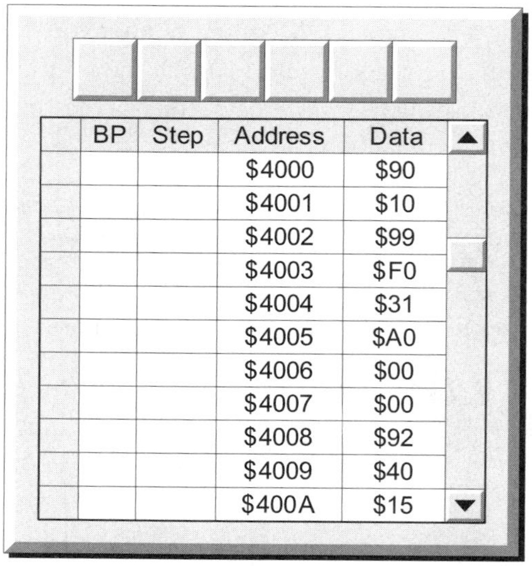

BP	Step	Address	Data	
		$4000	$90	▲
		$4001	$10	
		$4002	$99	
		$4003	$F0	
		$4004	$31	
		$4005	$A0	
		$4006	$00	
		$4007	$00	
		$4008	$92	
		$4009	$40	
		$400A	$15	▼

Figure L2-28. The Memory Walker display.

Walker became grayed out for a short time. This occurs while the calculator is in its *run mode* to indicate that any values being displayed aren't guaranteed to be valid.

So our program works. Hooray! But what would be our options had it failed? One thing we could do is to step through the program one instruction at a time. To see how this works, click the **Step** button on the front panel. This causes the calculator to execute the LDA instruction at address $4000. Observe the chevron characters ("≫") that appear in the Memory Walker's "Step" column pointing to the STA instruction at address $4002. Also observe that the status bar at the bottom of the main window now indicates that the calculator has entered its *step mode*.

When the calculator is in its *reset mode,* both the **Step** and **Run** buttons will start by executing whatever instruction is in address $4000, which is the first location in the calculator's RAM. Once the calculator has entered its *step mode,* however, each subsequent click on the **Step** button will cause it to run the next instruction in the program. Similarly, if the calculator is in its *step mode,* clicking the **Run** button will cause it to continue running from its current location in the program.

Click the **Step** button again, which causes the calculator to execute the STA instruction at address $4002, thereby clearing the calculator's main display, Observe that the chevrons in the Memory Walker now point to the BLDX instruction at address $4005. Click the **Step** button again to execute the BLDX instruction, click again to execute the JZ instruction at address $400B, and click once more to execute the STA instruction at address $400E. As we're currently on the first pass around the loop, this writes the first character in our message—the $48 code corresponding to the letter H—to the display.

Once the calculator is in its *step* mode, it will remain in this mode until you click the calculator's **Run** or **Reset** buttons. Keep on clicking the **Step** button until your curiosity is sated, and then click the **Run** button, which returns the calculator to its *run mode* and executes the rest of the program.

Setting and using breakpoints

Let's assume we've created a program that isn't exactly working as planned and that we're trying to determine what's gone wrong. One option would be to run the program and try to understand what it's doing by puzzling over whatever appears on the calculator's main display.

However, this may not prove to be a particularly useful technique, especially if the program "crashes-and-burns" in a catastrophic manner. Similarly, this approach may falter at the first fence if our program becomes locked up, cycling round in a never-ending loop and sitting there not doing much of anything at all.

As an alternative, we might consider single-stepping through the program using the Memory Walker in order to determine exactly what the program is doing. The problem here is that, depending on the size of our program, we may have to step through hundreds, thousands, or even millions of instructions before reaching the point where the problem lurks. Yet another option would be to start off by running the program, and then dropping back down into the *step mode* in the vague hope that we'll end up somewhere useful. Strange as it may seem, this tactic occasionally works to our favor, particularly if the problem involves the program getting stuck in a loop. Most of the time, however, this approach leaves something to be desired.

One technique that circumvents the above problems is to use one or more *breakpoints*. For the sake of discussion, let's assume that we want to stop our program just before it writes a character to the display. Glancing at our trusty list file, we see that this occurs at the STORE label, which appears at address $400E.

Once we've determined that $400E is the address of interest, we can scroll through the Memory Walker to bring this location into view. Now click the *breakpoint icon* in the Memory Walker's tool bar. This invokes a dialog window that allows us to set or clear breakpoints. Use this dialog to specify a breakpoint at address $400E (just key in the "400E", not the "$") and click the **Apply** button. The characters "BP" appear in the appropriate box in the Memory Walker's breakpoint column to indicate that a breakpoint has been set.

> **Note**   The calculator will only honor breakpoints that occur on the first byte (the opcode byte) of an instruction. If you try to set breakpoints on operand bytes, the calculator will ignore these "pseudo breakpoints" in its frantic desire to reach the next instruction.

Now click the front panel's **Run** button. As usual, the calculator starts executing the program at address $4000 and charges merrily ahead until it runs headfirst into the breakpoint, at which point it comes to a grinding halt and automatically drops down into its *step mode*. As before, chevrons appear in the Memory Walker to indicate where the calculator currently is in the program. If you wanted to, you could now start stepping through the program. Alternatively, if you click the **Run** button again, the simulator will reenter its *run mode* and continue to execute the program until it races around the loop and hits the breakpoint again. This means that every time you click the **Run** button, the program will write another character to the display and then pause.

We can use the Memory Walker to sprinkle breakpoints throughout a program. This means that we can enter the *run mode* and execute the program up to the first breakpoint, single-step for a while as we wish, and then click the **Run** button again to let the calculator proceed to the next breakpoint.

Once we're happy with the portion of code we've been scrutinizing, we can use the breakpoint dialog to clear specific breakpoints, or we can clear them all in one fell swoop by selecting the **Clear All** option followed by the **Apply** button. Last but not least, when we power-down the calculator, it will promptly forget any and all breakpoints from the current session (but do *not* power the calculator down just yet).

The CPU Register Display

Another very useful diagnostic tool is the CPU Register display, which exposes the contents of the CPU's internal registers and status flags. (Don't dismiss the Memory Walker just yet, because we will want to compare the contents of the two displays side by side.) Access this new display using the main window's **Display > CPU Registers** command or by clicking the appropriate icon on the tool bar (Figure L2-29).

Click the front panel's Reset button to ensure that the calculator is in its *reset mode,* and then click the Step button a couple of times and compare what happens in the CPU Register and Memory Walker displays.

As you'll see, the **Accumulator** field displays the current contents of the CPU's 8-bit accumulator while the **Index Register** field reflects the value in the CPU's 16-bit index register. The **Instruction Register** field shows the opcode of the instruction that is being executed, and the **Status Register** fields display the values of the CPU's five status flags. (Note that "X" values are used to indicate register bits that haven't been initialized with anything meaningful yet. Values of $XX and $XXXX are associated with 8-bit and 16-bit registers, respectively.)

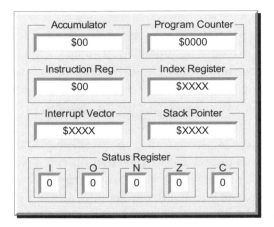

Figure L2-29. The CPU Register display.

Note especially the **Program Counter** field. The program counter is a 16-bit register the CPU uses to point to whatever instruction is to be executed (this register is introduced in more detail in Chapter 3). As you will observe, the value in the program counter tracks the location of the chevrons in the Memory Walker.

Working with the CPU Register display takes a bit of practice, but once you've got the hang of it you will find it to be an incredibly useful tool. The main thing to get used to is the fact that the **Program Counter** field is always left pointing at the first byte of the *next* instruction *to be executed,* whereas all of the other fields reflect the state of their corresponding registers after the *previous* instruction *was executed.*

The I/O Ports Display

As you may recall from Lab 2a, addresses $F000 through $F01F in the calculator's memory map are occupied by a set of 32 input ports (of which the front panel uses only one port at $F011), whereas addresses $F020 through $F03F are occupied by a set of 32 output ports (of which the front panel employs only two ports at addresses $F031 and $F032).

If you wish, you can use the Memory Walker display to see the data values currently associated with these ports (the Memory Walker has two tool bar icons that take you directly to the input and output ports). However, it doesn't take long to realize that this presentation of the data is not quite as useful as one might hope. As an alternative, we can use another diagnostic tool known as the I/O Ports display. Click the front panel's **Reset** button to ensure that the calculator is in its *reset mode,* and then access this new utility via the main window's **Display > I/O Ports** command or by clicking the appropriate icon on the tool bar (Figure L2-30).

Starting from the top, the port at address $F031 is the output port that drives the calculator's main display. The various characters and control codes we can send to the main display were summarized in Figure L2-11 in Lab 2c. The left-hand field associated with this item on the I/O Ports display reflects the last code written to the port ($41 in this example). By comparison, the right-hand field shows the corresponding ASCII character or control code name ("A" in this example):

Code(s)	Character/string displayed in right-hand field
$00 through $0F	"0" through "9" and "A" through "F"
$10	"Clear"
$11	"Bell"
$12	"Back"
$13 through $1F	"---"
$20	"Space"
$21 through $7E	ASCII character
$7F	"---"

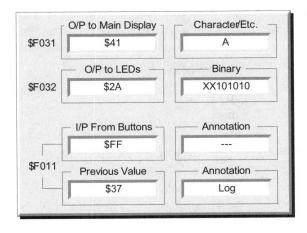

Figure L2-30. The I/O Ports display.

Similarly, in the case of the output port at address $F032 (the port that drives the six LEDs on the front panel), the left-hand field associated with this item on the I/O Ports display reflects the last code written to this port ($2A in this example). By comparison, the right-hand field shows the corresponding 0 and 1 binary values being presented to each bit (XX101010 in this example, where the two "X" values reflect the fact that the most significant two bits in this port aren't used to drive anything).

Things get a tad more interesting when we come to consider the input port at address $F011 (this is the port we use to see which buttons have been pressed on the calculator's front panel). As we discussed in Lab 2d, the front panel has an 8-bit latch that is used to store the code associated with whichever button was last pressed. Thus, the upper-left-hand field associated with this item in the I/O Ports display reflects the value that is *currently* stored in this latch.

The problem is that the act of reading from this port automatically causes the latch to be reloaded with its default value of $FF. Thus, in order to make our lives a little easier, the lower-left-hand field associated with this item reflects the value that was *previously* stored in the latch. Meanwhile, the two right-hand fields associated with this port are used to display the annotation associated with whichever button has been pressed.

In order to see the I/O Ports display in action, click the front panel's **Step** button to execute the LDA instruction at address $4000 (this loads the clear code of $10 into the accumulator). Now click the **Step** button again to execute the STA instruction at address $4002 (this writes the clear code to the calculator's main display). At this time, the left and right fields associated with the output port at $F031 in the I/O Ports display will show the values $10 and "Clr," respectively.

Now try clicking the **Log** button on the front panel and observe what happens on the I/O Ports display.

The Message System Display

In addition to the diagnostic tools described above, there's still one more utility, the Message System display, that's been bashfully hiding in the wings, eagerly awaiting its moment of fame.

Click the front panel's **Reset** button to ensure that the calculator in its *reset mode,* and then access this new utility via the main window's **Display** > **Message System** command or by clicking the appropriate icon on the tool bar.

Now click the **Step** button two or three times to execute the first few instructions in the program. As you will see, the Message System display provides a wealth of information as to all of the steps that were performed during the execution of each instruction (Figure L2-31).

Observe that the messages in this display have different colors and indentations. The left-most text (in black) indicates macroactions, such as acquiring an opcode, decoding an opcode, and so on. By comparison, the first level of indented text (in blue) reflects microactions, such as copying the program counter into the address latch. Finally, the second level of indented text (in red) details any additional points of interest.

New messages will be appended to the bottom of this window every time you click on the **Step** button (this display is automatically deactivated whenever you enter the calculator's *run mode;* otherwise you would be buried up to your ears in messages). Note that you can use the icons in

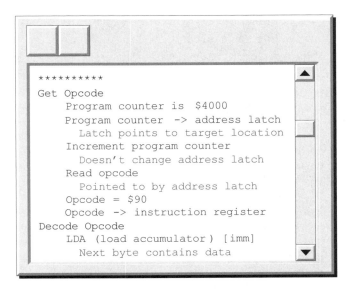

Figure L2-31. The Message System display.

the Message System display's tool bar to print and/or clear the current contents of the display.

The String Generator Utility

Traditionally, creating a routine to write the "Hello World" message to a system's standard output device is the first thing a programmer does to verify a new environment.

A West Coast equivalent to the Hello World program displays the message "Hello Sailor!" (this is supposed to have originated at SAIL, which stands for the Stanford Artificial Intelligence Lab).

Use the assembler's **Window > View Source File** command (or click the corresponding icon on the toolbar) to display the contents of our original source file.

Now use the table in Figure L2-11 from Lab 2c to determine the ASCII codes required to upgrade our program to display the "Hello Sailor!" message (including the use of uppercase and lowercase characters and the exclamation mark). Once you have determined the appropriate codes, modify our program accordingly and then run it to make sure it works.

In future labs we are going to want to display a lot of different character strings; for example, we may wish our calculator to display error messages along the lines of "You can't divide a number by zero!" As you will no doubt have discovered in generating the "Hello Sailor!" version of our program, translating strings into their equivalent ASCII codes by hand is a time-consuming and boring process.

In order to address this issue, the assembler boasts an automatic string generator feature. (Good grief! Do you think we should have mentioned this before? Oh well, it's too late now!) In order to see how this tool works, we must first delete the two message lines from our original program:

~~MESSAGE: .BYTE $??, $??, $??, $??, $??, $20~~
~~   #   H   e   l   l   o   SPACE~~

~~   .BYTE $??, $??, $??, $??, $??, $??, $??, $00~~
~~   #   S   a   i   l   o   r   !   NUL~~

In the real program, the "$??" values shown here will, of course, be the ASCII codes you researched above. Now ensure that the cursor in the as-

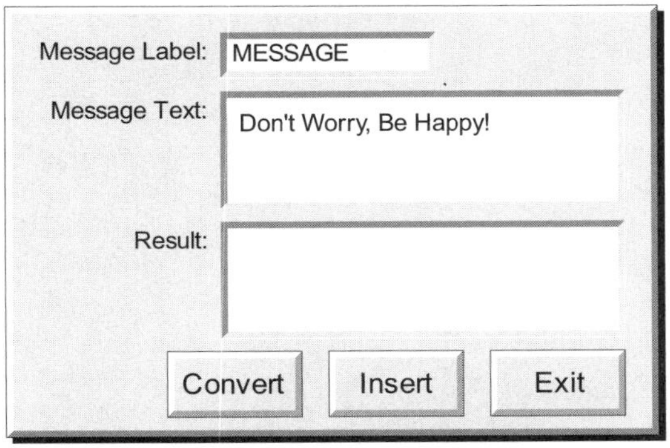

Figure L2-32. The string generator dialog window.

sembler's working area is located at the point where you wish to insert your message, and then use the assembler's **Insert > String** command to access the appropriate dialog window (Figure L2-32).

Enter MESSAGE into the Message Label field (you can, of course, enter any valid label string here; we're just using MESSAGE to match our existing program). Then enter some text such as "Don't Worry, Be Happy!" into the Message Text field. When you're ready to proceed, click the **Convert** button and observe the appropriate label, .BYTE directive, and associated ASCII codes appear in the **Result** field. Assuming you are happy with this result, click the **Insert** button to cause this message to be inserted at the cursor's current location in the source code and then click the **Exit** button to dismiss this dialog.

Bonus Question The program featured in this lab used two .BYTE directives. Were both of these directives necessary, and why didn't the second directive have a label associated with it?

Answer Each .BYTE directive can have as many comma-separated values assigned to it as can fit on one line of assembly source code. This means that the program featured in this lab could have been written using a single directive. The only reason two directives were used was to make the source code a little easier to read. Also, the second directive could be given its own label if required. In the case of this particular program, however, we can view any data associated with the second directive as being an extension of the data associated with the first directive, so it wasn't considered necessary to give the second directive its own label.

LABS FOR CHAPTER 3

SUBROUTINES AND OTHER STUFF

Lab 3a: Using Logical Instructions, Shifts, and Rotates

Lab 3b: Understanding the Program Counter (PC)

Lab 3c: Using the Index Register (X)

Lab 3d: Using the Stack and Stack Pointer (SP)

Lab 3e: Using Subroutines

Lab 3f: Using Recursion

01000010101011010100100100100011011000000 `010001101` 10001100100

Using Logical Instructions, Shifts, and Rotates

Objective: To gain a deeper understanding as to the relationship between binary (and hexadecimal) values and ASCII characters. Also, to enjoy some "hands-on" experience with logical instructions, shifts, and rotates.

Duration: Approximately 50 minutes.

Creating a Skeleton ASM File

We are going to be creating a variety of programs in the following labs, which means it will be well worth our time to create a skeleton (framework) assembly file. This will save us from having to type the same things over and over and over again, so invoke the assembler and enter the following source code:

```
############################################################################
## Start of constant declarations                                        ##
############################################################################
MAINDISP:  .EQU     $F031      # Address of output port for main display
SIXLEDS:   .EQU     $F032      # Address of output port for six LEDs
KEYPAD:    .EQU     $F011      # Address of input port for keypad
CLRCODE:   .EQU     $10        # Special code to clear the main display
BINMODE:   .EQU     %00000100  # LED code to indicate binary mode
DECMODE:   .EQU     %00000010  # LED code to indicate decimal mode
HEXMODE:   .EQU     %00000001  # LED code to indicate hexadecimal mode
############################################################################
## End of constant declarations                                          ##
############################################################################
           .ORG     $4000      # Set program origin
############################################################################
## Start of initialization                                               ##
############################################################################
```

```
INIT:       LDA     CLRCODE     # Load accumulator with clear code
            STA     [MAINDISP]  # Write clear code to main display
            LDA     HEXMODE     # Load accumulator with hex mode code
            STA     [SIXLEDS]   # Write to port driving six LEDs
###########################################################################
## End of initialization                                                 ##
###########################################################################

###########################################################################
## Start of main program body                                            ##
###########################################################################

###########################################################################
## End of main program body                                              ##
###########################################################################

###########################################################################
## Start of subroutines                                                  ##
###########################################################################

###########################################################################
## End of subroutines                                                    ##
###########################################################################

###########################################################################
## Start of global data                                                  ##
###########################################################################

###########################################################################
## End of global data                                                    ##
###########################################################################

            .END                    # That's all folks
```

Now, before you do anything else, use the assembler's **File > Save As** command to save this file as *my-skeleton.asm*. Then try assembling it to make sure that you didn't make any errors. Once you are satisfied that everything is as it should be, use the assembler's **File > New** command to clear the assembler's working area, and then proceed to the next section.

Are Your Codes the Same as My Codes?

As we previously discussed in Lab 2d, each of the buttons on the calculator's keypad area has an 8-bit code associated with it; these codes are shown as hexadecimal values in parentheses. For example, $16 and $1A are the codes associated with the "+" and "=" buttons, respectively (Figure L3-1).

When we click on a button, its associated code is loaded into an 8-bit latch. The code is stored in this latch until the CPU performs a read operation from the input port connected to the latch, at which time the latch is reloaded with its default value of $FF. What we are going to do is to create a program that loops around, waiting for a key to be pressed, and then presents the corresponding code on the main display as a dollar sign "$" followed by two hexadecimal digits. For example, if we click the "=" button, we want "$1A" to be presented on the main display.

"Aha!" you may be thinking to yourself, "This is going to be easy-peasy. All we have to do is read a code from the keypad into the accumulator and then write that code straight back out again to the main display." Maybe you have something like the following in mind:

```
            :
GETKEY:   LDA      [KEYPAD]    # Load ACC with code from keypad
          JN       [GETKEY]    # Jump back if no key pressed
DISPKEY:  STA      [MAINDISP]  # Display code on main display
            :
          etc.
```

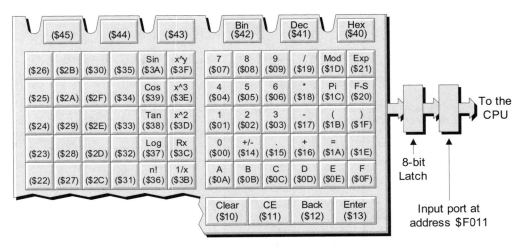

Figure L3-1. The calculator's keypad and associated codes.

Well, sad to relate, things aren't quite this simple. Cast your mind back to the various character and control codes supported by the main display (Figure L3-2).

Let's suppose that we are using the program snippet shown above and that we click on the upper-right-hand button in the main keypad area (the one marked **Exp**) which will result in the code $21 being loaded into the latch. Thus, we want the string "$21" to appear on the main display. However, if we write the hexadecimal code $21 to the main display, from Figure L3-2 we know that the result will be single character: the exclamation mark "!".

Now, you will have either been blessed with an immediate understanding as to the way in which the keypad and display codes are interre-

Not Standard ASCII				Standard ASCII											
$00	0	$10	Clr	$20	SP	$30	0	$40	@	$50	P	$60	`	$70	p
$01	1	$11	Bell	$21	!	$31	1	$41	A	$51	Q	$61	a	$71	q
$02	2	$12	Back	$22	"	$32	2	$42	B	$52	R	$62	b	$72	r
$03	3	$13		$23	#	$33	3	$43	C	$53	S	$63	c	$73	s
$04	4	$14		$24	$	$34	4	$44	D	$54	T	$64	d	$74	t
$05	5	$15		$25	%	$35	5	$45	E	$55	U	$65	e	$75	u
$06	6	$16		$26	&	$36	6	$46	F	$56	V	$66	f	$76	v
$07	7	$17		$27	'	$37	7	$47	G	$57	W	$67	g	$77	w
$08	8	$18		$28	(	$38	8	$48	H	$58	X	$68	h	$78	x
$09	9	$19		$29	)	$39	9	$49	I	$59	Y	$69	i	$79	y
$0A	A	$1A		$2A	*	$3A	:	$4A	J	$5A	Z	$6A	j	$7A	z
$0B	B	$1B		$2B	+	$3B	;	$4B	K	$5B	[	$6B	k	$7B	{
$0C	C	$1C		$2C	,	$3C	<	$4C	L	$5C	\	$6C	l	$7C	|
$0D	D	$1D		$2D	-	$3D	=	$4D	M	$5D	]	$6D	m	$7D	}
$0E	E	$1E		$2E	.	$3E	>	$4E	N	$5E	^	$6E	n	$7E	~
$0F	F	$1F		$2F	/	$3F	?	$4F	O	$5F	_	$6F	o	$7F	

Figure L3-2. Character codes supported by the main display.

lated, or you are currently bouncing back and forth between Figures L3-1 and L3-2 with a growing feeling of panic. If the latter situation is the case, *do not despair*! It takes most folks a little time to wrap their brains around this sort of thing, but once you've grasped the underlying concepts you'll find that everything starts to fall into place.

Let's consider the code $21 associated with the **Exp** button from a slightly different perspective. From our discussions in Chapters 1 and 2, we know that this is an 8-bit value, and that each 4-bit group maps directly onto a hexadecimal digit (Figure L3-3).

This means that we actually wish to write three characters (ASCII codes) to our main display: first a dollar symbol "$"; then the number "2," which corresponds to the most-significant nybble[1] in the accumulator (assuming that we've read the value from the latch into the accumulator); and, finally, the number "1," corresponding to the least-significant nybble in the accumulator.

Writing the dollar symbol will be a no-brainer (we just have to copy the ASCII code $24 to the main display), but what about the other two characters? In the case of the most-significant nybble, we are going to have to shift the contents of the accumulator four bits to the right. This will result in the accumulator containing a value of $02, which will appear as a "2" character when written to the main display [Figure L3-4(a)]. By comparison, in order to access the least-significant nybble in the accumulator, all we have to do is to clear the four most significant bits to 0. This will result in the accumulator containing a value of $01, which will appear as a "1" character when written to the main display [Figure L3-4(b)].

One point we should always remember is that our SHR ("shift right") instruction performs an *arithmetic* shift right, which means that every time we perform a shift the most-significant bit in the accumulator is copied back into itself. Thus, if we were to have a button code whose most significant bit were a logic 1, we would have problems if we used only the four shift operations shown in Figure L3-4a.

For example, assuming we had a button code of 10010001_2 ($91 in hexadecimal), then our four SHR instructions would result in the following intermediate results: 11001000_2, 11100100_2, 11110010_2, and 11111001_2 (the four bits shown in bold are the ones we're interested in).

---

[1] Remember that "nybble" (or "nibble") is the special name computer weenies use for a 4-bit group of binary digits, whereas "byte" refers to an 8-bit group.

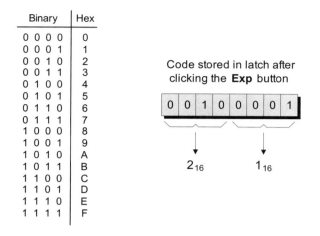

Binary	Hex
0 0 0 0	0
0 0 0 1	1
0 0 1 0	2
0 0 1 1	3
0 1 0 0	4
0 1 0 1	5
0 1 1 0	6
0 1 1 1	7
1 0 0 0	8
1 0 0 1	9
1 0 1 0	A
1 0 1 1	B
1 1 0 0	C
1 1 0 1	D
1 1 1 0	E
1 1 1 1	F

Code stored in latch after clicking the **Exp** button

2_{16} 1_{16}

Figure L3-3. The code associated with the Exp button.

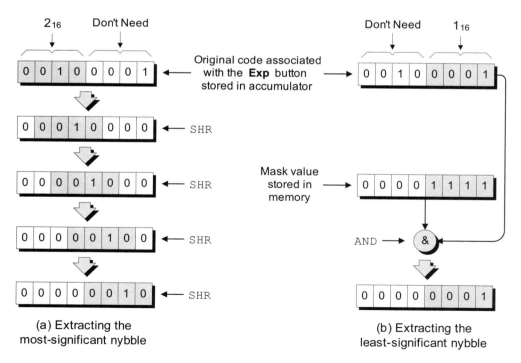

(a) Extracting the most-significant nybble

(b) Extracting the least-significant nybble

Figure L3-4. Extracting the most- and least-significant nybbles.

This means that, following our four shift operations, we would have to use an AND %00001111 instruction to mask out the four most-significant bits and clear them to 0s.

As it happens, none of our button codes has a 1 in the most significant bit, so this is never going to be a problem for us. However, it's good practice to get into the habit of thinking about this sort of thing, so we will perform this AND operation in our programs anyway.

Creating the Hex Display Program

Use the assembler's **File > Open** command to access the *my-skeleton.asm* file you created earlier, and then use the **File > Save As** command to save it as *lab3a-hex-shr.asm*. (If you neglected to build your own skeleton file, you can use the one called *skeleton.asm* that we have already created for you.)[2]

Before we do anything else, let's use a .BYTE directive to reserve an 8-bit (1-byte) temporary location called TEMP8 in the global data area, as shown below:

```
####################################################################
## Start of global data                                          ##
####################################################################
TEMP8:      .BYTE            # 8-bit temp location to store data
####################################################################
## End of global data                                            ##
####################################################################
```

The reason we refer to this area as "global data" is that we intend that any of the locations reserved here may be used by any portion of our program. As we shall see when we start to create subroutines, in some cases we might declare "local data" values as part of a subroutine that are intended to be used only by that particular routine.

Now, let's create the body of our program as follows (in the future, we'll simply say things like "Add the following statements between the *Start Body* and *End Body* comments"):

```
####################################################################
## Start of main program body                                    ##
####################################################################
########## Wait for key to be pressed
```

[2]Oops, did we neglect to mention that before? Sorry!

```
GETKEY:     LDA     [KEYPAD]   # Load ACC with code from keypad
            JN      [GETKEY]   # Jump back if no key pressed
            STA     [TEMP8]    # Store key code in temp location
########## Prepare the main display
CLRDISP:    LDA     CLRCODE    # Load ACC with clear code
            STA     [MAINDISP] # Clear main display
DISPDOLL:   LDA     $24        # Load ACC with ASCII code for '$'
            STA     [MAINDISP] # Write '$' to main display
########## Extract and display the most-significant nybble
DISPMSN:    LDA     [TEMP8]    # Reload ACC with copy of key code
            SHR                # Shift right 1 bit (= 1 bit shift)
            SHR                # Shift right 1 bit (= 2 bit shift)
            SHR                # Shift right 1 bit (= 3 bit shift)
            SHR                # Shift right 1 bit (= 4 bit shift)
            AND     %00001111  # Clear MS 4 bits (not really necessary)
            STA     [MAINDISP] # Copy result to main display
########## Extract and display the least-significant nybble
DISPLSN:    LDA     [TEMP8]    # Reload ACC with copy of key code
            AND     %00001111  # Mask out (clear) MS nybble
            STA     [MAINDISP] # Copy result to main display
########## Do it all again
DONE:       JMP     [GETKEY]   # Jump back and wait for new key
####################################################################
## End of main program body                                      ##
####################################################################
```

The first thing we do (at the GETKEY label) is to loop around, waiting for a key to be pressed. As soon as we do get a valid key, we store it out into our temporary memory location called TEMP8. The reason for performing this store is that we're going to be loading other values into the accumulator, so we need some way to remember which key was pressed.

Next, we prepare the main display (starting at the CLRDISP label) by clearing it and then writing the $24 code for a dollar character ("$") to it. We don't really need to perform the clear operation in the case of the first button we press because we already cleared the main display as part of our initialization routine. When we come to process subsequent buttons, however, we will want to clear the results associated with the previous button from the display.

When we come to extract and display the most-significant nybble (at the DISPMSN label), we start by reloading the key code from our temporary location into the accumulator. Then we shift this value four bits to the right, mask out the most significant four bits (which isn't really necessary, but is good practice as we discussed earlier), and write the resulting code corresponding to a number character to the main display.

Next, when we come to extract and display the least-significant nybble (at the DISPLSN label), we once again reload the key code from our temporary location into the accumulator. Then we mask out the most significant four bits and write the resulting code corresponding to a number character to the main display.

Finally, we jump back to wait for a new key to be pressed.

So let's see if this works. Assemble your program, click the **On/Off** button on the front panel to power-up the calculator, use the **Memory > Load RAM** command to load the *lab3a-hex-shr.ram* file into the calculator's memory and click the **Run** button to set the program running. Observe that the light-emitting diode (LED) associated with the **Hex** legend is lit; this is because we stored the HEXMODE code to the port driving the six LEDs as part of our skeleton file's initialization routine. Now click the **Exp** button on the front panel and the "$21" string will appear on the main display.

Feel free to click some other buttons and ensure that the values appearing on the main display are as expected. When you've finished playing around, click the **On/Off** button to power-down the calculator and scramble its memory again.

Creating a Binary Display Version

Now let's suppose that, as opposed to displaying the key code in hex, we decide that we'd prefer to see it in binary. For example, if we press the **Exp** button on the keypad, we'd like to see the string "%00100001" appear on the main display. Think about how you would go about implementing this for a moment before looking at our solution.

If you don't still have the *lab3a-hex-shr.asm* file open in your assembler, open it now, save it out as *lab3a-bin-v1.asm,* and then make the following changes:

1) The INIT (initialization) routine currently contains the following statement:

```
LDA     HEXMODE     # Load accumulator with hex mode code
```

Replace this statement with the following:

```
LDA     BINMODE    # Load accumulator with bin mode code
```

Of course, this change is cosmetic (some may say superficial), but its all part of the aesthetics of doing things the right way.

2) The statements associated with the DISPDOLL label in the main body of the program currently read as follows:

```
DISPDOLL:  LDA     $24        # Load ACC with ASCII code for '$'
           STA     [MAINDISP] # Write '$' to main display
```

Replace these lines with the following:

```
DISPPERC:  LDA     $25        # Load ACC with ASCII code for '%'
           STA     [MAINDISP] # Write '%' to main display
```

3) Delete all of the statements associated with extracting and displaying the most-significant and least-significant nybbles (at the DISPMSN and DISPLSN labels, respectively), and replace them with the following:

```
########## Display the binary value
## Process bit 7
TEST7:    LDA     [TEMP8]    # Reload ACC with copy of key code
          SHL                # Shift left 1 bit
          STA     [TEMP8]    # Store new value in temp location
          JC      [DISP7_1]  # If carry = 1, jump to display a 1
DISP7_0:  LDA     0          # ... otherwise load acc with 0
          STA     [MAINDISP] # ... and store it to main display
          JMP     [TEST6]    # ... then go and deal with bit 6
DISP7_1:  LDA     1          # Load acc with 1
          STA     [MAINDISP] # ... and store it to main display

## Process bit 6
TEST6:    LDA     [TEMP8]    # Reload ACC from temp location
          SHL                # Shift left 1 bit
          STA     [TEMP8]    # Store new value in temp location
          JC      [DISP6_1]  # If carry = 1, jump to display a 1
DISP6_0:  LDA     0          # ... otherwise load acc with 0
          STA     [MAINDISP] # ... and store it to main display
          JMP     [TEST5]    # ... then go and deal with bit 5
DISP6_1:  LDA     1          # Load acc with 1
          STA     [MAINDISP] # ... and store it to main display

## Process bit 5
TEST5:    LDA     [TEMP8]    # Reload ACC from temp location
          SHL                # Shift left 1 bit
```

```
            STA      [TEMP8]      # Store new value in temp location
            JC       [DISP5_1]    # If carry = 1, jump to display a 1
DISP5_0:    LDA      0            # ... otherwise load acc with 0
            STA      [MAINDISP]   # ... and store it to main display
            JMP      [TEST4]      # ... then go and deal with bit 4
DISP5_1:    LDA      1            # Load acc with 1
            STA      [MAINDISP]   # ... and store it to main display

## Process bit 4
TEST4:      LDA      [TEMP8]      # Reload ACC from temp location
            SHL                   # Shift left 1 bit
            STA      [TEMP8]      # Store new value in temp location
            JC       [DISP4_1]    # If carry = 1, jump to display a 1
DISP4_0:    LDA      0            # ... otherwise load acc with 0
            STA      [MAINDISP]   # ... and store it to main display
            JMP      [TEST3]      # ... then go and deal with bit 3
DISP4_1:    LDA      1            # Load acc with 1
            STA      [MAINDISP]   # ... and store it to main display

## Process bit 3
TEST3:      LDA      [TEMP8]      # Reload ACC from temp location
            SHL                   # Shift left 1 bit
            STA      [TEMP8]      # Store new value in temp location
            JC       [DISP3_1]    # If carry = 1, jump to display a 1
DISP3_0:    LDA      0            # ... otherwise load acc with 0
            STA      [MAINDISP]   # ... and store it to main display
            JMP      [TEST2]      # ... then go and deal with bit 2
DISP3_1:    LDA      1            # Load acc with 1
            STA      [MAINDISP]   # ... and store it to main display

## Process bit 2
TEST2:      LDA      [TEMP8]      # Reload ACC from temp location
            SHL                   # Shift left 1 bit
            STA      [TEMP8]      # Store new value in temp location
            JC       [DISP2_1]    # If carry = 1, jump to display a 1
DISP2_0:    LDA      0            # ... otherwise load acc with 0
            STA      [MAINDISP]   # ... and store it to main display
            JMP      [TEST1]      # ... then go and deal with bit 1
DISP2_1:    LDA      1            # Load acc with 1
            STA      [MAINDISP]   # ... and store it to main display

## Process bit 1
TEST1:      LDA      [TEMP8]      # Reload ACC from temp location
            SHL                   # Shift left 1 bit
            STA      [TEMP8]      # Store new value in temp location
            JC       [DISP1_1]    # If carry = 1, jump to display a 1
```

```
DISP1_0:   LDA    0              # ... otherwise load acc with 0
           STA    [MAINDISP]     # ... and store it to main display
           JMP    [TEST0]        # ... then go and deal with bit 6
DISP1_1:   LDA    1              # Load acc with 0
           STA    [MAINDISP]     # ... and store it to main display

## Process bit 0
TEST0:     LDA    [TEMP8]        # Reload ACC from temp location
           SHL                   # Shift left 1 bit
           STA    [TEMP8]        # Store new value in temp location
           JC     [DISP0_1]      # If carry = 1, jump to display a 1
DISP0_0:   LDA    0              # ... otherwise load acc with 0
           STA    [MAINDISP]     # ... and store it to main display
           JMP    [DONE]         # ... then go and deal with bit 6
DISP0_1:   LDA    1              # Load acc with 1
           STA    [MAINDISP]     # ... and store it to main display
```

This may appear to be a frightening amount of code, but you can obviously speed things up by entering the statements associated with one of the bits, copying and pasting that block of statements seven more times, and then "tweaking" the various labels. (Be careful. This isn't as easy as it might seem, and you are almost bound to make one or more mistakes that will take you ages to track down!)

When you've finished entering and assembling (and *debugging*) the above, load the corresponding *lab3a-bin-v1.ram* file into the calculator and ensure that this new version of the program works as planned.

Additional Exercises

With regard to our original hexadecimal display program, we used the SHR ("shift right") instruction to shift the most-significant nibble four bits to the right. But what if our virtual computer supported only a SHL ("shift left") instruction? Use the assembler to reload the *lab3a-hex-shr.asm* file, save it as *lab3a-hex-shl.asm,* and modify it to use SHL instructions.

Now, suppose that our virtual computer doesn't support either the SHL or SHR instructions, but only the RORC and ROLC rotate-through-carry instructions. Use the assembler to reload the *lab3a-hex-shr.asm* file, save it as *lab3a-hex-rorc.asm,* and modify it to use RORC instructions. Similarly, save a copy called *lab3a-hex-rolc.asm,* and modify it to use ROLC instructions.

Finally, with regard to the binary display program, it's obviously painful to have to explicitly code the same tests and display actions for each and every bit. It would be much nicer to be able to code the tests and display actions a single time, and to then loop around using this block of code for each bit in turn. You may recall that in Lab2f we introduced the concept of the index (X) register. Use the assembler to load the *lab3a-bin-v1.asm* file, save it as *lab3a-bin-v2.asm,* and then see if you can modify this program to use the index register to dramatically reduce the amount of code we have to type in.[3]

Bonus Question   In the original hexadecimal display program we created for this lab (*lab3a-hex-shr.asm*), we employed the fact that the main display accepts codes $00 through $0F and uses them to present the hexadecimal characters "0" through "9" and "A" through "F."

But now let's suppose that we are presented with a different version of the calculator whose main display supports only the standard ASCII codes of $20 through $7F. Assuming that we have an ADD instruction that allows us to add a value to the current contents of the accumulator, how could we modify our original program to achieve the desired result?

Answer   The solution to this conundrum is based on the fact that we know the ASCII codes for "0," "1," "2," . . . are $30, $31, $32, and so on; we also know that the ASCII codes for "A," "B," "C," . . . are $41, $42, $43, and so on.

At the end of the existing DISPMSN ("display most-significant nybble") portion of the program there is a STA [MAINDISP] instruction:

```
########## Extract and display the most-significant nybble
DISPMSN:   LDA     [TEMP8]    # Reload ACC with copy of key code
           SHR                # Shift right 1 bit (= 1 bit shift)
           SHR                # Shift right 1 bit (= 2 bit shift)
           SHR                # Shift right 1 bit (= 3 bit shift)
           SHR                # Shift right 1 bit (= 4 bit shift)
           AND     %00001111  # Clear MS 4 bits (not really necessary)
           STA     [MAINDISP] # Copy result to main display
```

This writes a code of $00 through $0F to the main display in the hope that the display will treat these codes as representing the hexadecimal

---

[3]In order to make your world a slightly happier place, we've included our versions of all of the programs described in these labs in the *C:\DIYCalculator\Data* folder.

characters "0" through "9" and "A" through "F." If we were reduced to using only ASCII character codes, then we would have to replace the existing STA instruction as follows (modifications are shown highlighted in gray):

```
########## Extract and display the most-significant nybble
DISPMSN:   LDA     [TEMP8]     # Reload ACC with copy of key code
           SHR                 # Shift right 1 bit (= 1 bit shift)
           SHR                 # Shift right 1 bit (= 2 bit shift)
           SHR                 # Shift right 1 bit (= 3 bit shift)
           SHR                 # Shift right 1 bit (= 4 bit shift)
           AND     %00001111   # Clear MS 4 bits (not really necessary)
           STA     [MAINDISP]  # Copy result to main display
TEST:      CMPA    $09         # Compare the ACC to $09
           JC      [DISP_A_F]  # If carry is 1, ACC is bigger than 9
DISP_0_9:  ADD     $30         # Add ASCII code for '0' to ACC
           JMP     [WRITEMSN]  # Jump to write character
DISP_A_F:  ADD     $41 - $0A   # Add ASCII for 'A' minus A (see notes)
WRITEMSN:  STA     [MAINDISP]  # Copy result to main display
```

> **Note**   The **ADD $41 - $0A** statement (the second line from the bottom of this code snippet) illustrates the fact that we can employ simple expressions in our assembly code. The assembler will automatically perform this calculation "on-the-fly" and effectively replace our statement with **ADD $37** (because $41 – $0A = $37).
>
> In this case, we could have performed the calculation by hand, and simply said **ADD 37**, but writing it out in this manner will make what we're doing more understandable when someone reads our program in the future.

The way this works is as follows. We start with a CMPA ("compare accumulator") instruction that is used to compare the value in the accumulator with $09. If the value in the accumulator is less than or equal to this value, then we know that we want to display a number character in the range "0" to "9." Alternatively, if the value in the accumulator is greater than $09, then we know that we want to display an alpha character in the range "A" to "F."

.   Thus, following the CMPA we have a JC ("jump if carry") instruction. This will cause a jump to occur if the value in the carry flag is 1, which will be the case if the value in the accumulator is greater than $09. So now we have two different scenarios:

- If the value in the accumulator is less than or equal to $09, we use an ADD instruction to add a value of $30 to the accumulator, and we then write the result to the main display. Remember that $30 is the ASCII code for the "0" character, so if the accumulator contains a value of $00, for example, then $00 + $30 = $30, which corresponds to the ASCII code for "0". Similarly, if the accumulaor originally contains a value of $01, then $01 + $30 = $31, which corresponds to the ASCII code for "1"; and so forth.

- Alternatively, if the value in the accumulator is greater than $09, we use an ADD instruction to add a value of $41 – $0A to the accumulator, and we then write this result to the main display. This one is a bit trickier to explain, but it's based on the fact that $41 is the ASCII code for the "A" character. The problem is that we know the accumulator already contains a value between $0A and $0F, so we have to take this into account. If the accumulator contains a value of $0A, for example, then $0A + $41 – $0A = $41, which corresponds to the ASCII code for "A." Similarly, if the accumulator originally contains a value of $0B, then $0B + $41 – $0A = $42, which corresponds to the ASCII code for "B"; and so forth.

Of course, we would also have to perform a similar treatment at the end of the existing DISPLSN ("display least-significant nybble") portion of the program.

Lab 3b

Understanding the Program Counter (PC)

Objective: To gain a good understanding as to what the program counter is and what it does.

Duration: Approximately 25 minutes.

An Example Program Snippet

As we discussed in Chapter 3, the *program counter* (*PC*) keeps track of where the CPU is in a program, and the contents of the program counter are presented to the outside world via the address bus. In order to get a feel for exactly what it is that the program counter does, let's start by considering the flowchart associated with a small program snippet as shown in Figure L3-5.

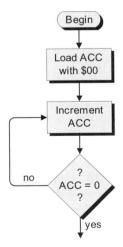

Figure L3-5. Flowchart for example program snippet.

This example is fairly meaningless, of course, because it doesn't actually do anything useful, but it will serve to illustrate some of the key aspects of the way in which the program counter operates.

First, we load the accumulator with $00 and then we increment it, which means that it now contains $01. Next, we test to see whether or not it contains zero, which it doesn't, so we jump back to increment it again. As we keep on cycling around the loop, the accumulator will eventually end up containing $FF (or 255 in decimal). This is the largest number our 8-bit accumulator can hold, so the next time we perform an increment the accumulator will overflow and end up containing $00 again. At this point, our test will show that the accumulator does indeed contain zero, so we'll drop out of the loop and continue on to the rest of the program.

Now let's consider the assembly source code we would have to write to implement our program snippet. For the purposes of these discussions, we'll terminate our program as soon as we drop out of the loop:

```
        .ORG    $4000       # Set program's origin to address $4000
        LDA     0           # Load accumulator with zero
LOOP:   INCA                # Increment the accumulator
        JNZ     [LOOP]      # Jump to LOOP if ACC !=0
        JMP     [$0000]     # Jump to address $0000
        .END                # Terminate the program
```

With regard to the comment associated with the JNZ ("jump if not zero") instruction, you may recall from our discussions in Lab 2c that the "!" character is often used to indicate an inversion or a negative ("not") state, so the string "ACC != 0" indicates "ACC is not equal to zero."

Power-up your DIY Calculator, invoke the assembler, key in this program, and then use the assembler's **File > Save As** command to save this file as *lab3b.asm*.

Assemble the program to generate the corresponding *lab3b.ram* file, and then, assuming no errors were encountered, use the assembler's **Window > View List File** command (or click the corresponding icon on the toolbar) to display the contents of the list file, which should look something like the one shown in Figure L3-6.

Now, just to hammer the point home, let's consider how the ensuing machine code will end up being loaded into the calculator's memory (Figure L3-7).

```
LINE  ADDR   DATA       LABEL     OPCODE  OPERAND
----- ----   --------   --------- ------  -------
00001                             .ORG    $4000
00002 4000   90 00                LDA     0
00003 4002   80         LOOP:     INCA
00004 4003   D6 40 02             JNZ     [LOOP]
00005 4006   C1 00 00             JMP     [$0000]
00006
```

Figure L3-6. List file for example program.

The .ORG directive instructs the assembler to set the origin of the program to address $4000. The body of the program commences with an LDA ("load accumulator"), which loads the accumulator with zero. (Note that LDA 0, LDA $00, and LDA %00000000, with arguments specified in decimal, hexadecimal, and binary, respectively, all have exactly the same effect.) Next, the INCA ("increment accumulator") adds 1 to the contents of the accumulator. Finally, the JNZ ("jump if not zero") jumps back to the label LOOP if the contents of the accumulator are nonzero.

So, after loading the accumulator with zero, the program loops around, incrementing the accumulator until it contains $FF (that's %11111111 in binary or 255 in decimal). The next time round the loop,

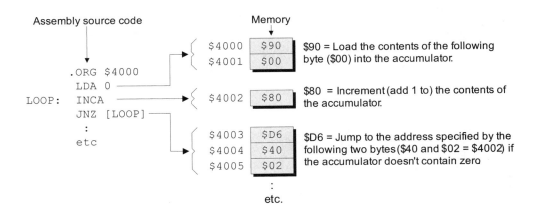

Figure L3-7. The machine code in the calculator's memory.

the INCA will cause the accumulator to overflow, leaving it containing zero. This means that the JNZ will fail (because the accumulator now contains zero), and the program will fall through to whatever instruction it finds at address $4006.

Executing the LDA Instruction

Click the front panel's **On/Off** button to power up the calculator and then use the main window's **Memory > Load RAM** command to load the *lab3b.ram* file into the calculator's memory. Also, use the main window's **Display** pull-down menu to activate the Memory Walker, CPU Register, and Message System displays.

Now, click the **Step** button to execute the first instruction in our program. As we were previously in the calculator's *reset mode,* this automatically causes the program counter to be loaded with a value of $4000, which is used to drive the address bus (Figure L3-8).

The first thing the CPU does is to read whatever opcode is pointed to by the program counter—the LDA ("load accumulator") opcode at address $4000, in this case [Figure L3-8(a)]. The CPU stores the opcode in an 8-bit register called the *instruction register* (IR) until it has finished executing this instruction. Following this read operation, the program counter is automatically incremented to point to address $4001 [Figure L3-8(b)].

As soon as the CPU examines the $90 opcode from address $4000, it recognizes that this is an LDA using the *immediate addressing mode,* so it knows that the next location contains a byte of data. Thus, the CPU reads the data at address $4001 and loads it into the accumulator [Figure L3-8(c)]. Furthermore, at the same time as it's loading

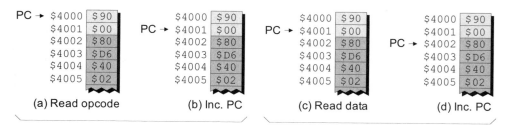

Figure L3-8. Executing the LDA instruction.

this data into the accumulator, the CPU automatically increments the program counter to point to the next opcode at address $4002 [Figure L3-8(d)].

Observe the CPU register display, which shows that the instruction register contains the $90 opcode, the accumulator contains the $00 data value, and the program counter contains $4002. Also, examine the contents of the Message System window to see all of the macro and micro actions that went into executing this instruction.

Executing the INCA Instruction

Click the **Step** button again. The CPU now reads whatever opcode is currently pointed to by the program counter—the INCA ("increment accumulator") opcode at address $4002 [Figure L3-9(a)]—and automatically increments the program counter to point to address $4003 [Figure L3-9(b)].

In this case, as soon as the CPU examines the $80 opcode from address $4002, it recognizes that this is an INCA using the *implied addressing mode,* so it knows that there isn't any data associated with this instruction. Thus, the CPU executes the INCA and immediately turns its attention to whatever the program counter is pointing at now.

Observe the CPU register display, which now shows the instruction register containing the $80 opcode we just executed, the accumulator now contains $01, and the program counter contains $4003. Once again, examine the contents of the Message System window to see all of the macro and micro actions that went into executing this instruction.

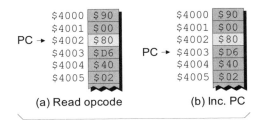

Figure L3-9. Executing the INCA instruction.

Executing the JNZ Instruction

Click the **Step** button once more. As usual, the CPU reads whatever op-code is currently pointed to by the program counter—the JNZ ("jump if not zero") at address $4003 in this case [Figure L3-10(a)]—and automatically increments the program counter to point to address $4004 [Figure L3-10(b)].

When the CPU examines the $D6 opcode from address $4003, it recognizes that this is a JNZ using the *absolute addressing mode,* so it knows that the next two locations contain the most-significant and least-significant bytes of a target jump address. But the CPU also knows that it's faced with a decision: to jump or not to jump.

As this is a JNZ, the test will pass and the CPU will jump if the Z (zero) flag contains a 0 value, thereby indicating that the accumulator contains a nonzero value. Since this is our first time around the loop, the accumulator currently contains $01, so we know that the JNZ instruction is indeed going to jump. In this case, the CPU will make use of a special 16-bit register that we can think of as a *temporary program counter* (*Temp PC*).

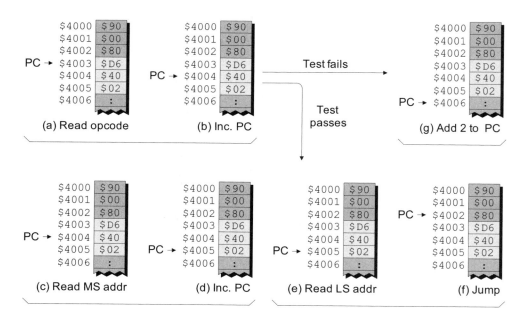

Figure L3-10. Executing the JNZ instruction.

The way to visualize this is that the CPU will read the most-significant byte of the target address ($40) from address $4004 [Figure L3-10(c)], copy it into the most-significant byte of the *temporary* program counter, and increment the *main* program counter [Figure L3-10(d)]. This means that the *temporary* program counter now contains $40XX, where the "XX" values indicate two as-yet undefined hexadecimal digits. Next, the CPU will read the least-significant byte of the target address ($02) from address $4005 [Figure L3-10(e)] and copy it into the least-significant byte of the *temporary* program counter. Thus, the *temporary* program counter now contains $4002. Last, but not least, the CPU will copy the contents of the *temporary* program counter into the *main* program counter, which effectively causes the CPU to jump back to address $4002 [Figure L3-10(f)].

Now, let's assume that we race around the loop again and again until we reach the point where we've just incremented the accumulator such that it has overflowed and contains $00. Once again, the CPU reads the JNZ opcode at address $4003 [Figure L3-10(a)] and increments the main program counter to point to address $4004 [L3-10(b)].

This time the JNZ test will fail, so the CPU will simply add 2 to the contents of the main program counter, leaving it pointing to whatever opcode is to be found at address $4006 [Figure L3-10(g)].

It's important to remember that, as discussed in Chapter 3, there really isn't any way for us to control the program counter per se. When the CPU is first powered-up, the program counter is initialized with some known good value; after this point, the program counter follows the dictates of the program itself. However, it's important for us to have a good understanding as to how the program counter performs its job, because this will help us to appreciate other aspects of the system's machinations when we get to them.

Lab 3c
Using the Index Register (X)

Objective: To become familiar with some of the things we can do with the index register.

Duration: Approximately 40 minutes.

Displaying Button Values in Order

Before we plunge into the fray, we're going to need an example program to provide a starting point. Use the assembler to open the *my-skeleton.asm* file you created in Lab 3a and save it as *lab3c-hex-forward.asm*. Now add the following statements between the "Start" and "End" of the "Main Program Body" comments:

```
########## Wait for key to be pressed
GETKEY:    LDA      [KEYPAD]    # Load ACC from the keypad
           JN       [GETKEY]    # Jump back if no key pressed
           CMPA     $0F         # Compare ACC to $0F
           JC       [GETKEY]    # Jump back if ACC is bigger
           STA      [MAINDISP]  # ... else store ACC to display
           JMP      [GETKEY]    # Go and wait for another key
```

This should be more than familiar to us by now. We commence with an LDA ("load accumulator"), which loads a value from the keypad. If no key has been pressed, the JN ("jump if negative") causes the CPU to jump back to the GETKEY label. If a key has been pressed, we compare its code (the value in the accumulator) to $0F. If the value in the accumulator is larger than $0F, we know that it doesn't correspond to any of the "0" to "9" or "A" to "F" hexadecimal digits. In this case the C (carry) flag will be set to 1, so the JC ("jump if carry") will cause the CPU to jump back to the GETKEY label; otherwise, the test will fail and the STA ("store accumulator") will be used to write the code to the main display. Finally,

245

the JMP ("unconditional jump") will return us to the GETKEY label in order to wait for the next key to be pressed.

Assemble this program, power-up the calculator, load the *lab3c-hex-forward.ram* program into the calculator's memory, and run it. Now click a few of the hexadecimal number buttons one after the other; for example: "2," "4," "6," "8," "A," "C," and "E." Not surprisingly, we'll end up with the string "2468ACE" on the calculator's display.

Displaying Button Values in Reverse Order

Now, suppose that we actually wanted these numbers to be displayed in the reverse order from that in which they were pressed. That is, pressing "2," "4," "6," "8," "A," "C," and "E" would actually result in "ECA8642" appearing on the calculator's display. How might we achieve this?

Well, one approach would be to bring the index register into play. For example, use the assembler's **File > Save As** command to save our original *lab3c-hex-forward.asm* file as *lab3c-hex-reverse-x.asm*. Now, add the following statements between the "Start" and "End" of the "Global Data" comments:

```
TEMP8:     .BYTE                 # Temp  8-bit (1-byte) location
TEMP16:    .2BYTE                # Temp 16-bit (2-byte) location
STORE:     .BYTE      *10        # Reserve 10 x 1-byte locations
```

The first of these statements uses a .BYTE directive to reserve a 1-byte temporary data storage location in the memory; as usual, the address associated with the TEMP8 label will be that of this byte [Figure L3-11(a)]. The second statement employs a .2BYTE directive to reserve a 2-byte location; the address associated with the TEMP16 label will be that of the first byte in this field [Figure L3-11(b)]. The third statement is of especial interest, because this is the first time we've seen the use of the "*" qualifier (in this example, the "*10" instructs the assembler to reserve ten 1-byte locations); the address associated with the STORE label will be that of the first byte in this field [Figure L3-11(c)].

At this time, we don't know precisely where these locations will end up in the memory (that will depend on the rest of the program). However, if we assume that the TEMP8 label is associated with the byte at some

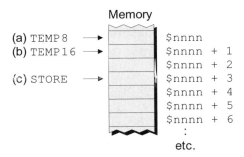

Figure L3-11. Labels are associated with the first location in their respective fields.

address we might call $nnnn in the memory, then we know that the TEMP16 label will be associated with the byte at address $nnnn + 1 (the first byte in the 2-byte field), whereas the STORE label will be associated with the byte at address $nnnn + 3 (the first byte in the 10-byte field).

Now modify the body of our program as follows (as usual, changes are highlighted in gray):

```
##########  Initialize the index register
          BLDX      $0000        # Load index register with zero

##########  Wait for key to be pressed
GETKEY:   LDA       [KEYPAD]     # Load ACC from the keypad
          JN        [GETKEY]     # Jump back if no key pressed
          CMPA      $0F          # Compare ACC to $0F
          JC        [DISPSTUF]   # Jump if ACC is bigger
          STA       [STORE,X]    # ... else store ACC to memory
          INCX                   # ... and increment the index reg
          JMP       [GETKEY]     # Go and wait for another key

##########  Display the numbers
DISPSTUF: LDA       [STORE-1,X]  # Load ACC with a key code
          STA       [MAINDISP]   # Store it to the main display
          DECX                   # Decrement the index register
          JNZ       [DISPSTUF]   # If index reg not 0 get next code
          JMP       [$0000]      # Terminate the program
```

The first change is that we initialize the index register with zero by means of a BLDX ("big load index register") instruction. As we noted earlier, this is referred to as a "big load" because the index register is 16 bits wide (as compared to the accumulator, which is only 8 bits wide).

Our GETKEY routine starts off the same way as before, waiting for a button to be pressed and then comparing that button's code with $0F. However, in the case of this new version of the program, if the value in the accumulator is bigger than $0F (and thus *not* a hexadecimal digit), the JC ("jump if carry") instruction will cause the program to jump to a new routine called DISPSTUF.

Alternatively, if the value in the accumulator is less than or equal to $0F (which means it *is* a hexadecimal digit), then the JC will fail and the program will continue on to the STA [STORE, X] instruction. This stores the value in the accumulator to the memory location whose address is calculated by adding the contents of the index register to the address associated with the STORE label.

The first time around the loop, the index register contains 0 [Figure L3-12(a)], so this key code will be loaded into the first location in our 10-byte field [Figure L3-12(b)]. As soon as this code has been stored, the INCX ("increment index register") is used to add 1 to the current contents of the index register, which therefore ends up containing 1 [Figure L3-12(c)].

The JMP [GETKEY] instruction then returns us to wait for another key to be pressed. Assuming that the next key to be pressed is also a hexadecimal digit, this code will be stored in the memory location whose address is calculated by adding the contents of the index register to the address associated with the STORE label. But, this second time around the loop, the index register contains $0001, so this new key code will be

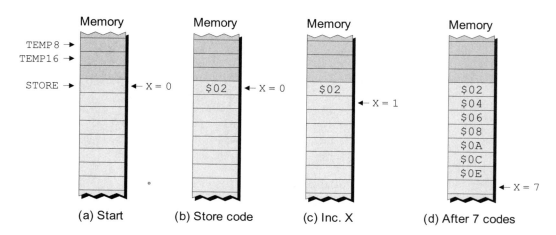

Figure L3-12. One way to view the index register.

loaded into the second location in our 10-byte field. Thus, assuming that we press the keys "2," "4," "6," "8," "A," "C," and "E," the result will be as shown in Figure L3-12(d).

Now, let's assume that the next button we click on the keypad is the "Enter" key with a code of $13. In this case, the test associated with the JC [DISPSTUF] instruction will pass, and the CPU will jump to the start of the DISPSTUF portion of our program:

```
########## Display the numbers
DISPSTUF:  LDA      [STORE-1,X] # Load ACC with a key code
           STA      [MAINDISP]  # Store it to the main display
           DECX                 # Decrement the index register
           JNZ      [DISPSTUF]  # If index reg not 0 get next code
           JMP      [$0000]     # Terminate the program
```

From Figure L3-12(d), we know that adding the address associated with the STORE label to the current contents of the index register ends up pointing to the first "empty" location in our 10-byte field. Thus, the "−1" portion of the LDA [STORE-1,X] instruction is used to load the accumulator with the contents of the last code (the "$0E" value).

Once we've stored this code to the main display, we use a DECX ("decrement index register") instruction to subtract 1 from the current contents of the index register. So long as the index register doesn't contain 0, the JNZ instruction will jump back to the DISPSTUF label in order to display the next code. Finally, when the last code has been written to the display and the index register contains 0, the JNZ instruction will fail and the program will jump to address $0000, which will cause it to terminate.

Assemble this new version of our program. Click the **Reset** button on the front panel to ensure the calculator is halted. Use the **Memory > Purge RAM** command to scramble the calculator's memory. Use the **Memory > Load RAM** command to load our new *lab3c-hex-reverse-x.ram* file. Click the **Run** button, then press the following keys: "2," "4," "6," "8," "A," "C," "E," and "Enter." Ensure that the string appearing on the main display is as we expect it to be.

Accessing the Index Register's Contents

Purely for the sake of discussion, let's assume that we aren't bothered about displaying the codes associated with the keys at all. What we really wish to know is the value in the index register at the time the first non-hexadecimal key is pressed.

First, use the assembler's **File > Save As** command to save our *lab3c-hex-reverse-x.asm* file as *lab3c-hex-x-contents.asm*. The bulk of the program can remain as is; all we have to do is to change the six lines associated with the DISPSTUF portion of the program to read as follows (don't panic, it's not as frightening as it looks):

```
##########  Display the value in the index register
DISPSTUF:   BSTX      [TEMP16]      # Store X reg into 2-byte temp location
            LDA       $24           # Load ACC with ASCII code for '$'
            STA       [MAINDISP]    # Write '$' to main display

##########  Extract/display MS nybble of MS byte
DSMSNMSB:   LDA       [TEMP16]      # Load ACC with MS byte from X reg
            SHR                     # Shift right 1 bit (= 1 bit shift)
            SHR                     # Shift right 1 bit (= 2 bit shift)
            SHR                     # Shift right 1 bit (= 3 bit shift)
            SHR                     # Shift right 1 bit (= 4 bit shift)
            AND       %00001111     # Clear MS 4 bits
            STA       [MAINDISP]    # Copy result to main display

##########  Extract/display LS nybble of MS byte
DSLSNMSB:   LDA       [TEMP16]      # Reload ACC with MS byte from X reg
            AND       %00001111     # Mask out (clear) MS nybble
            STA       [MAINDISP]    # Copy result to main display

##########  Extract/display MS nybble of LS byte
DSMSNLSB:   LDA       [TEMP16+1]    # Load ACC with LS byte from X reg
            SHR                     # Shift right 1 bit (= 1 bit shift)
            SHR                     # Shift right 1 bit (= 2 bit shift)
            SHR                     # Shift right 1 bit (= 3 bit shift)
            SHR                     # Shift right 1 bit (= 4 bit shift)
            AND       %00001111     # Clear MS 4 bits
            STA       [MAINDISP]    # Copy result to main display

##########  Extract/display LS nybble of LS byte
DSLSNLSB:   LDA       [TEMP16+1]    # Reload ACC with LS byte from X reg
            AND       %00001111     # Mask out (clear) MS nybble
            STA       [MAINDISP]    # Copy result to main display
            JMP       [$0000]       # Terminate the program
```

In order to understand what is happening here, let's assume that when we run this new version of the program we are going to click the buttons "2," "4," "6," "8," "A," "C," "E," and "Enter." In this case, from

Figure L3-12d we know that, at the time we click the "Enter" button, the index register will contain 7. Actually, based on the fact that we know that the index register is a 16-bit (2-byte) register, we should really think of its contents as being $0007.

Now, consider what occurs at the start of our new DISPSTUF routine when we execute the BSTX ("big store index register") instruction to store the contents of the index register into the 2-byte location we associated with the label TEMP16 (Figure L3-13).

As before, we don't know precisely where these locations will end up in the memory (that will depend on the rest of the program), but, purely for the purposes of these discussions, we are assuming that the TEMP8 label is associated with the byte at some address we might call $nnnn in the memory, and then referencing all of the other addresses from this point.

Observe that a copy of the most-significant byte of the index register will be loaded into the memory location associated with the TEMP16 label. Also notice that a copy of the least-significant byte of the index register will be loaded into the adjacent memory location.

The rest of the program is very similar to the *lab3a-hex-shr.asm* program we created in Lab 3a. However, in this case we wish to display a 2-byte value in hexadecimal. We commence by writing the ASCII code for a dollar ("$") character to the display. When we come to the DSMSN-MSB ("display the most-significant nybble of the most-significant byte") portion of the program, we use its LDA [TEMP16] instruction to load a

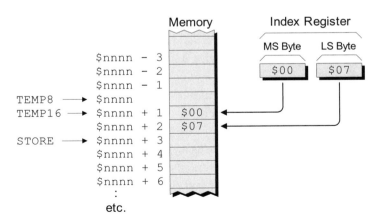

Figure L3-13. The result from the BSTX [TEMP16] instruction.

copy of the most significant byte we saved from the index register into the accumulator. Then we shift it four bits to the right, clear the four most significant bits, and write the resulting code to the display.

In the case of the DSLSNMSB ("display the least-significant nybble of the most-significant byte") portion of the program, we use *its* LDA [TEMP16] instruction to reload a copy of the most significant byte we saved from the index register into the accumulator, and then clear the four most significant bits and write the resulting code to the display.

Now we have to repeat the process to display the most- and least-significant nybbles from the *least-significant byte* we saved from the index register, which we access using the LDA [TEMP16+1] instructions.

Assemble this new version of our program. Click the **Reset** button on the front panel to ensure the calculator is halted. Use the **Memory > Purge RAM** command to scramble its memory. Use the **Memory > Load RAM** command to load our new *lab3c-hex-x-contents.ram* file. Click the **Run** button. Press the keys "2," "4," "6," "8," "A," "C," "E," and "Enter," and observe the string "$0007" appear on the main display.

Using the Index Register as a Counter

As opposed to using the index register to modify an address when using the *indexed addressing mode* (or the *preindexed indirect* and *indirect postindexed* modes that we have not yet introduced), this register often proves useful as a simple counter.

As an example, let's cast our minds back to the *lab3a-bin-v1.asm* program we discussed in the *Creating a Binary Display Version* section of Lab 3a. As you will recall, displaying a button code as eight binary digits required a relatively large amount of code. As we shall see, however, we can use the index register to dramatically reduce the size of the code and make our lives much, much easier.

Let's assume that when any button is pressed on the keypad, we wish to display that button's code as an 8-bit binary value. If the **Cos** button with an associated hexadecimal code of $39 were pressed, for example, we would like to see the equivalent binary value "%00111001" appear on the main display.

In order to achieve this goal, use the assembler to open your *my-skeleton.asm* file, save it as *lab3c-bin-v1.asm,* and then add the following statement between the "Start" and "End" of the "Global Data" comments:

```
TEMP8:    .BYTE                # Temp  8-bit (1-byte) location
```

Now, add the following statements between the "Start" and "End" of the "Main Program Body" comments:

```
########## Get and display key codes in binary
## Wait for a key to be pressed
GETKEY:    LDA     [KEYPAD]     # Load ACC from the keypad
           JN      [GETKEY]     # Jump back if no key pressed
           STA     [TEMP8]      # Store ACC in temp location

## Clear display then display the '%' character
CLRDISP:   LDA     CLRCODE      # Load ACC with clear code
           STA     [MAINDISP]   # Clear main display
DISPPERC:  LDA     $25          # Load ACC with ASCII code for '%'
           STA     [MAINDISP]   # Write '%' to main display

## Initialize index register
LOADX:     BLDX    8            # Load X reg with 8

## Loop around writing
LOOP:      LDA     [TEMP8]      # Reload ACC from temp location
           SHL                  # Shift left 1 bit
           STA     [TEMP8]      # Store new value in temp location
           JC      [DISP_1]     # If carry = 1, jump to display a 1
DISP_0:    LDA     0            # ... otherwise load acc with 0
           STA     [MAINDISP]   # ... and store it to main display
           JMP     [DEALWX]     # ... then go and deal with the X reg
DISP_1:    LDA     1            # Load acc with 1
           STA     [MAINDISP]   # ... and store it to main display
DEALWX:    DECX                 # Decrement the index register
           JNZ     [LOOP]       # If X not zero then do next bit
           JMP     [GETKEY]     # Go back and wait for new key
```

This program should be fairly self-explanatory by now. The interesting part starts once we have a new key code (which we've stored in the TEMP8 memory location) and we load the index register with a value of 8 at the LOADX label.

We then loop around, shifting the value stored in the TEMP8 location 1 bit to the left, displaying a corresponding "0" or "1" on the main display, and decrementing the index register. We do this eight times, once for each bit in the code, until the index register contains zero, at which point we jump back to wait for the next button to be pressed.

As usual, assemble this new program. Click the **Reset** button on the front panel to ensure the calculator is halted. Use the **Memory** > **Purge RAM** command to scramble its memory. Use the **Memory** > **Load RAM** command to load our new *lab3c-bin-v1.ram* file. Click the **Run** button, then press the **Cos** button and observe the string "%00111001" appear on the main display. Now, try clicking some of the other buttons to see their binary codes.

Additional Exercise

Create a new copy of our last program called *lab3c-bin-v2.asm* and modify it to display a dash between the most-significant and least-significant nybbles. For example, pressing the **Cos** button should result in the string "%0011-1001" appearing on the main display.

Lab 3d

Using the Stack and Stack Pointer (SP)

Objective: To become familiar with some of the things we can do with the stack and stack pointer in the context of pushing and popping the accumulator.

Duration: Approximately 25 minutes.

A Simple Stack-Based Program

In the previous lab, we used the index register to create a program that accepted numbers from the keyboard and displayed them in the reverse order to that in which their buttons were pressed. That is, pressing "2," "4," "6," "8," "A," "C," and "E" actually resulted in "ECA8642" appearing on the calculator's display.

In this lab, we are going to do something very similar, but this time we are going to make use of the stack. Use the assembler to open your *my-skeleton.asm* file and save it as *lab3d-hex-reverse-sp.asm*. Now add the following statements between the "Start" and "End" of the "Main Program Body" comments:

```
########## Initialize the stack pointer
          BLDSP    $EFFF         # Load stack pointer with $EFFF

########## Wait for key to be pressed
GETKEY:   NOP                    # "No operation" (see notes)
LOOP:     LDA      [KEYPAD]      # Load ACC from the keypad
          JN       [LOOP]        # Jump back if no key pressed
          CMPA     $0F           # Compare ACC to $0F
          JC       [DISPSTUF]    # Jump if ACC is bigger
          PUSHA                  # ... else push ACC onto the stack
          JMP      [GETKEY]      # Go and wait for another key
```

```
########## Display a number
DISPSTUF:   POPA                    # Pop ACC off the stack
            STA     [MAINDISP]      # Store it to the main display
            JMP     [GETKEY]        # Go and wait for another key
```

This works a tad differently from our previous implementation. We commence by loading the stack pointer with $EFFF, which isn't particularly surprising. What may raise a few eyebrows, however, is the first instruction in the GETKEY portion of the program: the NOP ("no operation"). As its name might suggest, this causes the CPU to do nothing at all, really.

So why might one use a NOP instruction? Well, executing a NOP takes the CPU a finite amount of time, which can make it useful for creating delay loops in a program. In this case, however, we are going to use this instruction to facilitate our use of breakpoints when we come to run the program (see also the *Testing the Program* discussions below).

Following the NOP, we loop around, waiting for a key to be pressed. Once a key has been pressed, we compare it to $0F to see if it is a hexadecimal character in the range "0" to "9" and "A" to "F" or if it is some other key. If it is a hexadecimal character, the JC ("jump if carry") instruction will fail, and the PUSHA ("push accumulator") instruction will push a copy of the accumulator onto the stack.

In the case of a nonhexadecimal key, the JC instruction will pass and the program will jump to the DISPSTUF portion of the program. This uses a POPA ("pop accumulator") instruction to retrieve whatever value is on the top of the stack and copy it into the accumulator. Next, we use a STA ("store accumulator") instruction to write this value to the main display, and then we jump back to wait for another key to be pressed.

Testing the Program

Ensure the calculator is powered-up. Assemble our new program (do *not* dismiss the assembler). Click the **Reset** button on the front panel to ensure that the calculator is halted. Use the **Memory > Purge RAM** command to scramble its memory, then use the **Memory > Load RAM** command to load our new *lab3d-hex-reverse-sp.ram* file.

Now, before we actually run the program, return to the assembler for a moment and use its **Window > View List File** command. Scroll down until you reach the GETKEY portion of our program, which should look something like the following:

```
00035 ########## Wait for key to be pressed
00036 400D 00          GETKEY:    NOP
00037 400E 91 F0 11    LOOP:      LDA     [KEYPAD]
00038 4011 D9 40 0E               JN      [LOOP]
00039 4014 60 0F                  CMPA    $0F
00040 4016 E1 40 1D               JC      [DISPSTUF]
00041 4019 B2                     PUSHA
00042 401A C1 40 0D               JMP     [GETKEY]
```

In the version shown here, the NOP instruction, which occurs on line 00036 in the source code, is associated with memory location $400D. Make a note of the equivalent memory location in your program if it is different for any reason (maybe you've been adding things to your version of our skeleton file).

Next, use the **Display > CPU Registers** and **Display > Memory Walker** commands to activate these two debugging tools. Now click the **Set/Clear Breakpoints** icon in the memory walker, enter the $400D address value[1] (or your equivalent), ensure the **Set Breakpoint** radio button is selected, and click the **Apply** button to set this breakpoint, which appears in the Memory Walker display.

Now click the **Run** button. This will cause the calculator to run until it hits our breakpoint, at which point we will automatically drop back into step mode. Observe the **Stack Pointer** field in the CPU register display, which now shows that the stack pointer has been loaded with $EFFF [Figure L3-14(a)].

Click the **Run** button again. Now the CPU is looping around, waiting for a key to be pressed, so let's put it out of its misery and click the number "2" on the front panel. The result is that this code is pushed onto the top of the stack [Figure L3-14(b)] and the stack pointer is decremented to point to the next free location [L3-14(c)].

Once again, as soon as the calculator hits our breakpoint, it will drop back into the *step mode*. Once again, observe the **Stack Pointer** field in the CPU register display, which now shows that the stack pointer currently contains $EFFE [Figure L3-14(c)]. If you wish, you can also scroll down the Memory Walker display to address

[1]Just the "400D" part; you don't need to key in the "$" character.

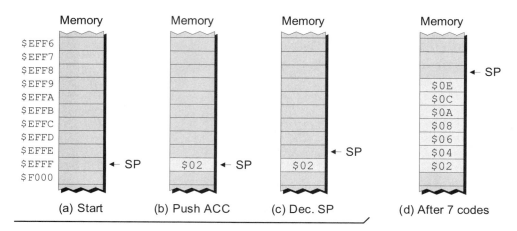

Figure L3-14. Pushing values onto the stack.

$EFFF and see that our $02 value has indeed been stored in this location.

Repeat the above procedure—clicking the **Run** button followed by one of the calculator buttons—for the keys "4," "6," "8," "A," "C," and "E." Each time you hit our breakpoint and drop back into *step mode,* observe the CPU Register display and the stack-related locations in the Memory Walker. After the last key press, the stack will appear as shown in Figure L3-14(d).

OK, now let's go in the other direction. Our starting point is as shown in Figure L3-15(a). Click the **Run** button and then click a nonhexadecimal key, the "=" key, for example, which will result in our to jumping to the DISPSTUF portion of our program. The first instruction here is the POPA ("pop accumulator"), which causes the stack pointer to be incremented to point to the $0E value on the top of the stack [Figure L3-15(b)]; this value is then popped off the stack into the accumulator [Figure L3-15(c)].

Of course, the STA ("store accumulator") instruction following the POPA will cause the $0E value in the accumulator to be written to the calculator's main display, upon which the hexadecimal digit "E" will appear. The program will then jump back to the GETKEY label, which will cause the calculator to hit our breakpoint and drop back down into *step mode.*

Repeat the above procedure—clicking the **Run** button followed by a nonhexadecimal button on the calculator two more times. This will re-

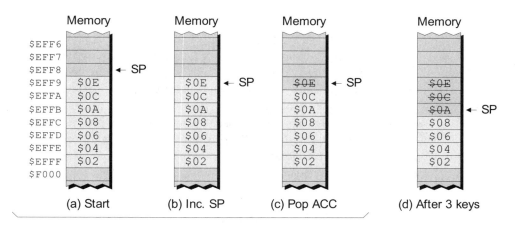

Figure L3-15. Popping values off the stack.

sult in the calculator's main display showing "ECA." Each time you hit our breakpoint and drop back into *step mode,* observe the CPU Register display and the stack-related locations in the Memory Walker. After the last key press, the stack will appear as shown in Figure L3-15(d).

Now try experimenting a little. Perhaps click another hexadecimal key and see what happens. Also, what happens if you pop all of the values off the stack and then you click another nonhexadecimal key?

Bonus Question   What was the advantage of having the NOP ("no operation") as the first instruction in the GETKEY portion of the program (and how could we have managed without it)?

Answer In order to answer this question, let's first remind ourselves how the program looks *with* the NOP instruction:

```
########## Wait for key to be pressed
GETKEY:    NOP                    # "No operation"—see notes
LOOP:      LDA   [KEYPAD]         # Load ACC from the keypad
           JN    [LOOP]           # Jump back if no key pressed
           CMPA  $0F              # Compare ACC to $0F
           JC    [DISPSTUF]       # Jump if ACC is bigger
           PUSHA                  # ... else push ACC onto the stack
           JMP   [GETKEY]         # Go and wait for another key
```

Note especially the use of the LOOP label. This is where the program will loop around, waiting for a key to be pressed on the calculator's front panel. Now let's consider how the program would look *without* the NOP instruction:

```
########## Wait for key to be pressed
GETKEY:    LDA   [KEYPAD]         # Load ACC from the keypad
           JN    [GETKEY]         # Jump back if no key pressed
           CMPA  $0F              # Compare ACC to $0F
           JC    [DISPSTUF]       # Jump if ACC is bigger
           PUSHA                  # ... else push ACC onto the stack
           JMP   [GETKEY]         # Go and wait for another key
```

In the case of this particular program, the GETKEY label is the obvious place to drop a breakpoint. In this new (non-NOP) version of the program, however, we'd hit the breakpoint every time the CPU read a value from the keypad. This means that we'd drop back into *step mode* irrespective of whether or not any-one had pressed a key, which would be something of a pain to us.

However, if we had wished to implement the program in this way for some reason, then the next best solution would have been to set *two* break-points: one on the JMP [GETKEY] instruction at the end of the GETKEY por-tion of the program, and the other on the JMP [GETKEY] instruction at the end of the DISPSTUF portion of the program.

Lab 3e

Using Subroutines

Objective: To create and use a subroutine to display the hex codes associated with a number of bytes in memory.

Duration: Approximately 35 minutes.

A Blast from the Past

If you cast your mind back to Lab 3a, you may recall our creating a program called *lab3a-hex-shr.asm,* which we used to take a byte of data and display the hexadecimal characters associated with its most-significant and least-significant nybbles.

For example, if our byte contained a value of $3A in hexadecimal (or %00111001), we would first isolate the most-significant four bits (0011) and write a value of $03 to the main display (which would present the character "3"). We would then isolate the least-significant four bits (1001) and write a value of $0A to the main display (which would present the character "A").

Assuming that the byte we wished to process was stored in a temporary memory location called TEMP; then the code to do this would look something like the following:

```
########## Extract and display the most-significant nybble
DISPMSN:   LDA     [TEMP]      # Reload ACC with copy of key code
           SHR                 # Shift right 1 bit (= 1 bit shift)
           SHR                 # Shift right 1 bit (= 2 bit shift)
           SHR                 # Shift right 1 bit (= 3 bit shift)
           SHR                 # Shift right 1 bit (= 4 bit shift)
           AND     %00001111   # Clear MS 4 bits
           STA     [MAINDISP]  # Copy result to main display
```

```
########## Extract and display the least-significant nybble
DISPLSN:   LDA     [TEMP]      # Reload ACC with copy of key code
           AND     %00001111   # Mask out (clear) MS nybble
           STA     [MAINDISP]  # Copy result to main display
```

The above was perfectly acceptable when we wished to process only a single byte, but things became a little trickier in Lab 3c, when we created our program *lab3c-hex-x-contents.asm*. In that case, we wished to process two bytes, so we had to replicate all of the statements shown above in order to handle the second byte. But, of course, that was before we knew about subroutines.

Subroutines to the Rescue

Purely for the sake of these discussions, let's suppose that we have four bytes we wish to process; for example, consider the following statement that will reserve four bytes and initialize them with the values shown:

```
TEMP:       .BYTE $23, $5A, $06, $4C
```

Obviously, we don't want to replicate all of our original "extract and display" statements four times. Instead, it will be much more advantageous for us to create a subroutine and to call *this* four times.

Let's start by using the assembler to open your *my-skeleton.asm* file, save it as *lab3e-sub.asm,* and then insert the TEMP line shown above between the "Start Global Data" and "End Global Data" comments.

Next, we need to use a BLDSP ("big load stack pointer") instruction to initialize the stack pointer. In order to do this, add the line highlighted in gray to the end of the existing initialization instructions as follows:

```
####################################################################
## Start of initialization                                        ##
####################################################################
INIT:      LDA     CLRCODE     # Load accumulator with clear code
           STA     [MAINDISP]  # Write clear code to main display
           LDA     HEXMODE     # Load accumulator with hex mode code
           STA     [SIXLEDS]   # Write to port driving six LEDs
           BLDSP   $EFFF       # Load stack pointer with $EFFF
####################################################################
## End of initialization                                          ##
####################################################################
```

Now, enter the following routine between the "Start Subroutine" and "End Subroutine" comments:

```
##########  Subroutine to extract/display MS and LS nybbles of byte
DISPBYTE:   PUSHA                # Push a copy of ACC onto stack

##########  Extract and display the most-significant nybble
DISPMSN:    SHR                  # Shift right 1 bit (= 1 bit shift)
            SHR                  # Shift right 1 bit (= 2 bit shift)
            SHR                  # Shift right 1 bit (= 3 bit shift)
            SHR                  # Shift right 1 bit (= 4 bit shift)
            AND     %00001111    # Clear MS 4 bits
            STA     [MAINDISP]   # Copy result to main display

##########  Extract and display the least-significant nybble
DISPLSN:    POPA                 # Pop copy of original byte off stack
            AND     %00001111    # Mask out (clear) MS nybble
            STA     [MAINDISP]   # Copy result to main display
            RTS                  # Return from subroutine
```

A key point about this subroutine is that, when we enter it, it assumes that the byte we wish to display is already in the accumulator. Thus, the first thing we do is to push a copy of the accumulator onto the stack for later use. We next extract and display the most-significant nybble; this process loses any bits associated with the least-significant nybble, so we pop our copy of the original value back off the stack and then extract and display the least-significant nybble.

Finally, add the following lines of code between the "Start Body" and "End Body" comments:

```
LDA     [TEMP]       # Load 1st byte into ACC
JSR     [DISPBYTE]   # Call display byte subroutine
LDA     [TEMP+1]     # Load 2nd byte into ACC
JSR     [DISPBYTE]   # Call display byte subroutine
LDA     [TEMP+2]     # Load 3rd byte into ACC
JSR     [DISPBYTE]   # Call display byte subroutine
LDA     [TEMP+3]     # Load 4th byte into ACC
JSR     [DISPBYTE]   # Call display byte subroutine
JMP     [$0000]      # Terminate the program
```

In the case of this particular example, for each of the bytes to be displayed, we first load that byte into the accumulator and then call our DISPBYTE subroutine.

Testing the Basic Subroutine

Ensure that the calculator is powered-up. Assemble our new program. Click the **Reset** button on the front panel to ensure that the calculator is halted. Use the **Memory** > **Purge RAM** command to scramble its memory, then use the **Memory** > **Load RAM** command to load our new *lab3e-sub.ram* file.

Now, click the **Run** button to execute this program and observe the character string that appears on the main display: "235A064C."

Creating a Nested Version

One point worth noting about the subroutine we just created is that it extracted and displayed both the most-significant and least significant nybbles. As an alternative, we might decide to create a nested version. First, use the assembler to save a copy of our *lab3e-sub.asm* program with the new name of *lab3e-nested.asm*. Now replace the original version of our DISPBYTE subroutine with the following:

```
########## Subroutine to extract/display MS and LS nybbles of byte
DISPBYTE:   PUSHA                # Push a copy of ACC onto stack
            JSR      [DISPMSN]   # Call sub for MS nybble
            POPA                 # Pop copy of original byte off stack
            JSR      [DISPLSN]   # Call sub for LS nybble
            RTS                  # Return to calling location

########## Subroutine to extract/display the MS nybble
DISPMSN:    SHR                  # Shift right 1 bit (= 1 bit shift)
            SHR                  # Shift right 1 bit (= 2 bit shift)
            SHR                  # Shift right 1 bit (= 3 bit shift)
            SHR                  # Shift right 1 bit (= 4 bit shift)
            AND      %00001111   # Clear MS 4 bits
            STA      [MAINDISP]  # Copy result to main display
            RTS                  # Return to calling location

########## Subroutine to extract/display the LS nybble
DISPLSN:    AND      %00001111   # Mask out (clear) MS nybble
            STA      [MAINDISP]  # Copy result to main display
            RTS                  # Return to calling location
```

As we see, DISPBYTE, DISPMSN, and DISPLSN are all subroutines in their own right. In this case, the body of our program still calls the DISPBYTE subroutine, which in turn calls DISPMSN and DISPLSN as nested subroutines.

Testing the Nested Version

Ensure that the calculator is powered-up. Assemble our new program. Click the **Reset** button on the front panel to ensure that the calculator is halted. Use the **Memory > Purge RAM** command to scramble its memory, then use the **Memory > Load RAM** command to load our new *lab3e-nested.ram* file.

Now, click the **Run** button to execute this program and, once again, observe the string of characters that appears on the main display: "235A064C."

Why Create the Nested Version?

One question you may be asking yourself is: "Why would we bother to create the nested version of our original subroutine?" After all, the two renderings generate the same results, and the nested incarnation required a few more lines of code (which, in this particular example, means it occupies more memory and takes longer to run).

Well, this is only a very simple subroutine but, generally speaking, it's better to break things down into smaller "chunks," because these smaller blocks of code are easier to debug and understand. Furthermore, we can now use the various "chunks" in their own right. For example, if we wished to see both nybbles from the first and second bytes, but only the most-significant nybble from the third byte and only the least significant nybble from the fourth byte, then we could achieve this with our nested implementation by modifying the body of the program as follows:

```
LDA     [TEMP]      # Load 1st byte into ACC
JSR     [DISPBYTE]  # Display MS and LS nybbles
LDA     [TEMP+1]    # Load 2nd byte into ACC
JSR     [DISPBYTE]  # Display MS and LS nybbles
LDA     [TEMP+2]    # Load 3rd byte into ACC
JSR     [DISPMSN]   # Display MS nybble only
LDA     [TEMP+3]    # Load 4th byte into ACC
JSR     [DISPLSN]   # Display LS nybble only
JMP     [$0000]     # Terminate the program
```

Lab 3f

Recursive Subroutines

Objective: To create a recursive subroutine that displays a string of characters in reverse order.

Duration: Approximately 40 minutes.

Creating the Subroutine

As we noted in Chapter 3, recursion can be mind-bogglingly complicated, and it's usually a lot easier to use a more standard alternative such as creating a simple loop that uses a variable as a counter. Purely for the sake of providing an example, however, we are going to create a recursive subroutine and then try to wrap our brains around its multifarious machinations.

Use the assembler to open the *my-skeleton.asm* file you created earlier, save it as *lab3f-reverse.asm,* and then insert the following statements between the "Start Global Data" and "End Global Data" comments:

```
PHRASE:    .BYTE $53, $57, $41, $50, $20
           #    S    W    A    P   SPACE

           .BYTE $50, $41, $57, $53, $00
           #    P    A    W    S   NUL
```

Together, these form the string "SWAP PAWS" terminated by a NUL ($00) character. You can use the assembler's **Insert > String** command to create these statements if you wish; just remember to place the cursor where you want them to appear in your source code *before* you invoke the command.

The body of our program is very simple because it comprises only four instructions, which you should insert between the "Start Body" and "End Body" comments:

267

```
                 BLDX     0              # Load index register with 0
                 BLDSP    $EFFF          # Load stack pointer with $EFFF
OUTSIDE:         JSR      [REVERSE]      # Call the subroutine
FINISH:          JMP      [$0000]        # Terminate the program
```

As we see, the only actions this program performs are to load the index register with zero, initialize the stack pointer to point to memory location $EFFF, call a subroutine called REVERSE, and then jump to address $0000, which will cause the program to automatically terminate. Note that the OUTSIDE and FINISH labels are included only to provide points of reference for our future discussions.

Now, enter the following subroutine between the "Start Subroutine" and "End Subroutine" comments:

```
########## Recursive subroutine to display string in reverse order
REVERSE:         LDA      [PHRASE,X] # Load ACC with a character
                 JNZ      [GOIN]     # If it's not NUL jump to GOIN
RETURN1:         RTS                 # Otherwise return from subroutine

########## Store the character on the stack and go further in
GOIN:            PUSHA               # Push the character onto the stack
                 INCX                # Increment the index register
INSIDE:          JSR      [REVERSE]  # The subroutine calls itself recursively

########## Retrieve and display a character and come out
COMEOUT:         POPA                # Pop a character off the stack
                 STA      [MAINDISP] # Copy to main display
RETURN2:         RTS                 # Return from subroutine
```

Observe that this routine is a good example of one that uses multiple RTS ("return from subroutine") statements.

Boggling the Mind

Now, prepare to have your mind well and truly boggled. Let's walk through our program and ponder what happens on the stack. Assume that the program is at label OUTSIDE and that it's just about to execute the JSR ("jump to subroutine") instruction, which means that the stack is currently empty.

When the program executes this JSR, the least-significant (LS) and most-significant (MS) bytes of the return address (the address of the memory location associated with the FINISH label) will be pushed onto the stack [Figure L3-16(a)], and the program counter will be left pointing at the first instruction in the subroutine at the REVERSE label.

The first thing our subroutine does is to load the accumulator using the indexed addressing mode; that is, to load the value (an ASCII character in this case) from the memory location whose address is calculated by adding the address associated with the PHRASE label and the current contents of the index register (which are $0000). This will be the first $53 data value (the ASCII letter "S").

Next, the JNZ ("jump if not zero") instruction is used to test for a NUL ($00) character. As the accumulator currently contains $53, this test will pass, and the program will jump to the GOIN ("go further in") label. [Observe that even if we had had an empty string containing only a NUL ($00) code, our program would still work in that it would immediately return to the main body of the program. It's very important to consider these "corner case" situations when creating a program.]

When the program arrives at the GOIN label, it uses a PUSHA ("push accumulator") to push a copy of the $53 value onto the stack [Figure L3-16(b)]. The program then uses an INCX ("increment index register") instruction followed by the JSR ("jump to subroutine") at the INSIDE label. This is the point where the subroutine becomes recursive, because it has just called itself.

When the program executes this JSR at the INSIDE label, the least-significant (LS) and most-significant (MS) bytes of the return address

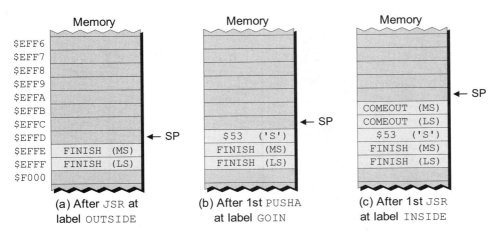

Figure L3-16. The stack during the first few instructions.

(the address of the memory location associated with the COMEOUT label) will be pushed onto the stack [Figure L3-16(c)], and the program counter will, once again, be left pointing at the first instruction in the subroutine.

As before, the subroutine loads the accumulator with the character from the memory location whose address is calculated by adding the address associated with the PHRASE label and the current contents of the index register (which are now $0001). This will be the first $57 data value (the ASCII letter "W").

The JNZ ("jump if not zero") instruction now tests for a NUL ($00) character. As the accumulator currently contains $57, this test will pass, and the program will again jump to the GOIN ("go further in") label. Once again, the program uses the PUSHA ("push accumulator") to push a copy of the $57 value onto the stack [Figure L3-17(a)].

And, once again, the program uses an INCX ("increment index register") instruction followed by the JSR ("jump to subroutine") at the INSIDE label, which causes the least-significant (LS) and most-significant (MS) bytes of the return address (the address of the memory location associated with the COMEOUT label) to be pushed onto the stack [Figure L3-17(b)].

Thus, until we reach a NUL, the subroutine will keep on getting character codes, pushing them onto the stack and calling itself. Now let's assume that our subroutine has pushed the final character in our phrase

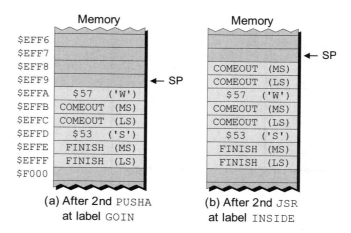

Figure L3-17. The stack during the first few instructions (continued).

onto the stack and has just called itself for the last time. This means that the index register currently contains $0009 and the top of the stack appears as shown in Figure L3-18(a).

As usual, the subroutine loads the accumulator with the character from the memory location whose address is calculated by adding the address associated with the PHRASE label and the current contents of the index register (which are now $0009). This will be the NUL data value (the "$00" code).

Thus, this time round, the JNZ ("jump if not zero") instruction fails and the program drops through to the next instruction, which is the RTS ("return from subroutine") at the RETURN1 label. This causes the top two bytes (forming the address associated with the COMEOUT label) to be popped off the stack and loaded into the program counter [Figure L3-18(b)].

When the program arrives at the COMEOUT label, it uses a POPA ("pop accumulator") to retrieve the data value—the last character we read—from the top of the stack (Figure L3-18c). This is followed by a STA ("store accumulator") that copies this value to the calculator's main display. The program then executes the RTS ("return from subroutine") at the RETURN2 label. As usual, this causes the top two bytes (forming the address associated with the COMEOUT label) to be popped off the stack and loaded into the program counter.

From this point on, the program simply keeps on popping characters off the stack at the COMEOUT label, copying them to the main display,

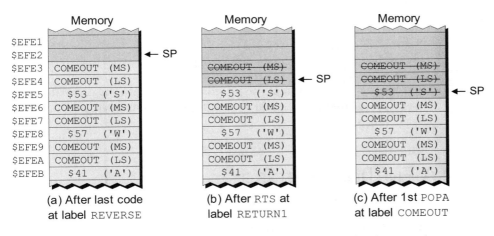

Figure L3-18. The stack after the last valid character has been read.

and then performing the RTS at the RETURN2 label. This RTS keeps on returning the program to the COMEOUT label until we've processed the final character, at which point the final return address stored on the stack is that associated with the FINISH label in the main body of the program [Figure L3-16(a)].

Testing the Subroutine

Ensure that the calculator is powered-up. Assemble our new program. Click the **Reset** button on the front panel to ensure that the calculator is halted. Use the **Memory > Purge RAM** command to scramble its memory, then use the **Memory > Load RAM** command to load our new *lab3f-reverse.ram* file.

Remember that our original character string was "SWAP PAWS" and that our recursive subroutine is going to display these characters in reverse. Click the **Run** button to execute the program and observe the reversed character string appear on the main display.

But wait, the string on the display is "SWAP PAWS" again. How can this be? Did the program fail to work as planned? In fact, the program did perform its task as we described. The problem is that the "SWAP PAWS" string is a palindrome (from the Greek *palindromos,* meaning "running back again"), which means that it reads the same forward or backward (sorry, we couldn't help ourselves).

Let's try again. Replace the space ($20) code associated with the PHRASE label with a NUL ($00) code as follows (the change is highlighted in gray):

```
PHRASE:    .BYTE $53, $44, $41, $50, $00
           #      S    W    A    P   NUL
```

Assemble this new version of the program, load it into the calculator, run it, and observe that the original "SWAP" string is now displayed in reverse as "PAWS."

A Further Exercise

Do you feel confident that you know how recursion works now? Yes? Well, let's put that to the test, shall we? In Chapter 3 we mentioned that the canonical example of a recursive function is a factorial. Consider the following factorials from 1 to 5:

Factorial	Expanded	Result Decimal	Result Hexadecimal
1!	= 1	= 1	= $01
2!	= 2 × 1	= 2	= $02
3!	= 3 × 2 × 1	= 6	= $06
4!	= 4 × 3 × 2 × 1	= 24	= $18
5!	= 5 × 4 × 3 × 2 × 1	= 120	= $78

That's as far as we care to go at the moment, because the next factorial (6! = 720) would be too large to store in a single byte (as we know from Chapter 1, an 8-bit byte can be used to represent positive integers in the range 0 to 255).

For your delectation and delight, we've precreated a special subroutine called FMULT ("factorial multiplier"). Don't worry about how this routine performs its magic, because we'll be covering that in Chapter 4. All we need to know at the moment is that when this subroutine is called, it takes the first two bytes it finds on the top of the stack, multiplies them together, and leaves a single result byte on the top of the stack.[1]

For example, assume that the two bytes on the top of the stack contain $03 and $02 as shown in Figure L3-19(a). If we were to call FMULT at this point, then after this subroutine had completed its task and returned control to the main program, the new value on the top of the stack would be $06, as shown in Figure L3-19(b).

Your mission, should you decide to accept it, is to use the assembler to open up your *my-skeleton.asm* file and save it as *lab3f-factorial.asm*. Next, place the cursor between the *Start Subroutine* and *End Subroutine* comments and then use the assembler's **Insert** > **Insert File** command to insert our FMULT subroutine (you'll find this routing in the *C:\DIYCalculator\Work* folder).

---

[1]As we'll see in Chapter 4, when two binary numbers are multiplied together, the width of the result is equal to the sum of the bits forming the multiplicand and multiplier. Thus, multiplying two 1-byte values together will return a 2-byte result. For the purposes of this application, however, in which we know the result will never be greater than 255, our FMULT routine discards the most-significant byte of the result and returns only the least-significant byte.

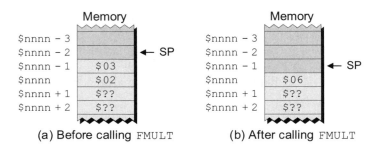

Figure L3-19. Example stack contents before and after calling the FMULT subroutine.

Now, we want you to create a program that loops around, waiting for a button to be pressed in the range "1" to "5" (any other button should be ignored). When a button in the range "1" to "5" is pressed, your program should call a subroutine called FACTOR, which you are going to create. This subroutine will call itself recursively in order to calculate the factorial associated with the button you pressed. Note that this subroutine can call our FMULT routine whenever it wishes to multiply two numbers together.

Once you have calculated the factorial, you can use a version of the DISPBYTE subroutine we created in Lab 3e to display the resulting byte in hexadecimal on the calculator's main display. Last but not least, return to the beginning of the program and wait for another button to be pressed.

Once you've created this program, assemble it, load it into the calculator, run it, and check that it generates the correct results (as shown in the table at the beginning of this section).

LABS FOR CHAPTER 4

INTEGER ARITHMETIC

Lab 4a

Creating a Testbench Program

Objective: To create a testbench program that can be used to input a pair of hexadecimal values, call a 16-bit integer math subroutine, and display the result.

Duration: Approximately 30 minutes.

Creating the Testbench Program

The majority of labs associated with Chapter 4 involve subroutines that perform some task, such as addition, subtraction, multiplication, or division, on a pair of 16-bit numbers. So the first thing we are going to do is to create a testbench program that can read in two 16-bit hexadecimal values and place them on the stack, call one of our math subroutines (which we will be developing in subsequent labs), and display the result.

Copy your original skeleton routine

First of all, invoke the assembler, load the original *my-skeleton.asm* file that you created in Lab 3a, and save it as *skeleton-lab4.asm*. Unless something totally weird has happened to the space–time continuum as we know it, this file should appear something like the following:

```
##################################################################
## Start of constant declarations                              ##
##################################################################
MAINDISP:  .EQU    $F031       # Address of out port for main display
SIXLEDS:   .EQU    $F032       # Address of out port for six LEDs
KEYPAD:    .EQU    $F011       # Address of input port for keypad
CLRCODE:   .EQU    $10         # Special code to clear main display
BINMODE:   .EQU    %00000100   # LED code to indicate binary mode
DECMODE:   .EQU    %00000010   # LED code to indicate decimal mode
HEXMODE:   .EQU    %00000001   # LED code to indicate hexadecimal mode
```

```
########################################################################
## End of constant declarations                                      ##
########################################################################

            .ORG       $4000        # Set program origin

########################################################################
## Start of initialization                                           ##
########################################################################
INIT:       LDA        CLRCODE      # Load accumulator with clear code
            STA        [MAINDISP]   #    and write it to the main display
            LDA        HEXMODE      # Load accumulator with hex mode code
            STA        [SIXLEDS]    #    and write it to port driving LEDs
########################################################################
## End of initialization                                             ##
########################################################################

########################################################################
## Start of main program body                                        ##
########################################################################

########################################################################
## End of main program body                                          ##
########################################################################

########################################################################
## Start of subroutines                                              ##
########################################################################

########################################################################
## End of subroutines                                                ##
########################################################################

########################################################################
## Start of global data                                              ##
########################################################################

########################################################################
## End of global data                                                ##
########################################################################

            .END                    # That's all folks
```

Modifying the initialization routine

First, let's modify the initialization portion of this file to load the stack
pointer (SP) with an initial value of $EFFF (the new statement is shown
highlighted in gray):

```
INIT:       LDA     CLRCODE     # Load accumulator with clear code
            STA     [MAINDISP]  # Write clear code to main display
            LDA     HEXMODE     # Load accumulator with hex mode code
            STA     [SIXLEDS]   # Write to port driving six LEDs
            BLDSP   $EFFF       # Load stack pointer with initial value
```

Now, use the assembler's **File** > **Save** command to save this change. (The moral here is that you should never fail to take an opportunity to save your work. You'll come to understand our paranoia in the fullness of time.)

Reserving global temporary data locations and messages

Next, we'll reserve some locations in which to store temporary address and data values as required, along with any message strings. Let's start by adding the following statements between the "Start" and "End of Global Data" comments.

```
INSTRING: .BYTE *10           # Reserve 10 bytes to store a string
TMPBYTE:  .BYTE               # Reserve a 1-byte temp location
TMP2BYTE: .2BYTE              # Reserve a 2-byte temp location
TMP4BYTE: .4BYTE              # Reserve a 4-byte temp location

########## Start of message strings
```

Now, place your cursor at the beginning of the blank line following the "Start of message strings" comment; then use the assembler's **Insert** > **String** command to access its associated dialog window and use this tool to insert the following message strings (this utility was introduced in Lab 2f):

Label	Message
MSG_000:	ERROR:
MSG_001:	Carry = 1
MSG_002:	Borrow = 0
MSG_003:	Overflow
MSG_004:	Underflow
MSG_005:	Too big
MSG_006:	Too small
MSG_007:	Divide by 0
MSG_008:	Out of range
MSG_009:	-32,768
MSG_010:	Not implemented

With regard to message MSG_000, make sure you include a space following the colon (that is, the string should be "ERROR:"). As you will soon discover, we are going to create a DISPERR ("display error") subroutine that can be called from inside our math subroutines to report error conditions when they are detected.

Adding a GETNUM subroutine

Now, we are going to add three subroutines between the "Start" and "End Subroutine" comments. The first of these subroutines, called GETNUM ("get number"), allows you to enter a number by clicking four hexadecimal digits on the calculator's keypad (you will indicate when you've finished by clicking the "Enter" key). This routine then uses these digits to form a 16-bit binary number that it places on the stack.

The GETNUM subroutine is shown below in its entirety. This is followed by a detailed description as to the way in which this routine performs its magic.

```
########## Start of GETNUM subroutine
########## Get four hex digits, use them to form a 16-bit binary
########## value, and store that value on the stack

########## Pop the return address off the stack and save it
GETNUM:     POPA                        # Pop the MS byte of return address
            STA     [_GN_RADD]          #   off the stack and store it
            POPA                        # Pop the LS byte of return address
            STA     [_GN_RADD+1] #      off the stack and store it

########## Load a series of key codes into 'INSTRING' until we see
########## the 'Enter' key (we assume other keys are 0-9 and/or A-F
########## and we don't perform any error checking)
            BLDX    $0000               # Load the index register with 0
_GN_LOOP:   LDA     [KEYPAD]            # Load accumulator from keypad
            JN      [_GN_LOOP]          # Jump back if no key pressed
            CMPA    $13                 # Compare to code for 'Enter" key
            JZ      [_GN_STR]           # If the same, jump to process
string
            STA     [MAINDISP]          # Copy this key code to main display
            STA     [INSTRING,X]        # Also store this code in string
            INCX                        # Increment the index register
            JMP     [_GN_LOOP]          # Jump back and wait for another key

########## Build a 16-bit binary number and push it in the stack
_GN_STR:    LDA     [INSTRING+2] # Load accumulator with 3rd character
```

```
            SHL                   #    in `INSTRING' and shift it left
            SHL                   #    four bits ...
            SHL                   #      :
            SHL                   #      :
            OR       [INSTRING+3] # OR result with 4th character and
            PUSHA                 #    push this LS byte onto the stack

            LDA      [INSTRING]   # Load accumulator with 1st character
            SHL                   #    in `INSTRING' and shift it left
            SHL                   #    four bits ...
            SHL                   #      :
            SHL                   #      :
            OR       [INSTRING+1] # OR result with 2nd character and
            PUSHA                 #    push this MS byte onto the stack

########## Write a space to the main display
_GN_SPC:    LDA      $20          # Load acc with ASCII code for space
            STA      [MAINDISP]   # Write it to the main display

########## Return gracefully from this subroutine
_GN_RET:    LDA      [_GN_RADD+1] # Get LS byte of return address from
            PUSHA                 #    temp location and push onto stack
            LDA      [_GN_RADD]   # Get MS byte of return address from
            PUSHA                 #    temp location and push onto stack
            RTS                   # Return from this subroutine


_GN_RADD: .2BYTE                  # 2-byte temp location used to store
                                  #    the return address for this routine
########## End of GETNUM subroutine

# - - - - - - - - - - - - - - - - - - - - - - - - - - - - - - - #
```

Before we begin, let's take a moment to observe that all of the labels in this routine, with the exception of the name of the subroutine itself, start with the same four-characters (_GN_), where the "GN" is short for "GETNUM." This helps to prevent the problems that can easily arise if we attempt to use identical labels in multiple subroutines.

Also, observe the final real statement in this routine at the _GN_RADD label, where "RADD" is short for "Return Address." This statement is used to reserve two bytes of data in which we are going to temporarily store the subroutine's return address. (In some cases, we will make use of the global temporary data storage elements we reserved in the previous section. In other cases, it is preferable to use storage elements that are local to the subroutine in question, because this prevents any future nested subroutine calls from overwriting the same global locations.)

There is one last point to make before we plunge into the heart of this subroutine—observe the final comment line we used to terminate the subroutine:

```
# – – – – – – – – – – – – – – – – – – – – – – – – – – – – – – – #
```

As you will come to see, when we have a large number of subroutines (each of which is, ideally, replete with comments, it can be a little tricky to spot where one stops and another begins. Using a special comment like this provides a nice visual indication as to the boundary between adjacent routines.

Storing the return address

Now, let's consider the state of play just before we call the GETNUM subroutine. The stack pointer (SP) will be pointing to whatever location in the memory is currently considered to represent the top of the stack (we'll reference this location as being at address $nnnn for the purpose of these discussions). Meanwhile, the two bytes in memory that are associated with the _GN_RADD label will contain unknown values Figure L4-1(a)].

As you will recall, when a JSR ("jump to subroutine") instruction is used to call a subroutine like GETNUM, the return address (the address of the instruction immediately following the JSR) is automatically pushed onto the current top of the stack. As part of this process, the

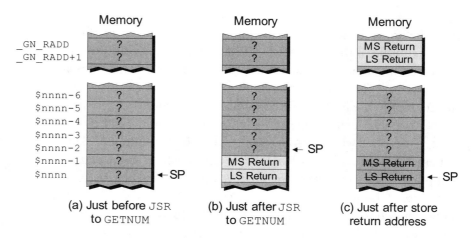

Figure L4-1. The state of play when GETNUM is called.

stack pointer is modified to point to the next free location on the top of the stack [Figure L4-1(b)].

Sad to relate, we now have a bit of a problem, because the mission of the GETNUM routine is to build a 2-byte (16-bit) number and store it such that, when we finally exit GETNUM, this number appears on the top of the stack ready to be used by a subsequent routine. However, if we were to simply push a 2-byte number onto the top of the stack as it appears in Figure L4-1(b), then when we eventually use a RTS ("return from subroutine") to exit GETNUM, this instruction would mistakenly attempt to use our number as its return address with disastrous consequences. This why the first four instructions in GETNUM are used to pop the return address off the top of the stack and store it in our 2-byte temporary location at label _GN_RADD, as shown in Figure L4-1(c) (the act of popping these two bytes off the stack returns the stack pointer to its original position).

Reading and temporarily storing the key codes

Now we want to read in a series of four hexadecimal digits. The way we've decided to implement this part of the subroutine is to loop around, reading the codes associated with buttons being pressed on the key pad and storing them as a string of characters until the user clicks on the "Enter" key. Purely for the purposes of this portion of our discussions, let's assume that the user is going to click the following sequence of buttons:

"3," "A," "6," "C," "Enter"

Now, consider the global 10-byte temporary location called INSTRING that we reserved earlier. Initially, this contains unknown values as shown in Figure L4-2(a).

We commence by loading the index register (X) with 0, and then we start a loop at label _GN_LOOP. Initially, we perform an inner loop, loading the accumulator with values from the input port connected to the calculator's keypad, waiting for a key to be pressed.

When a key is pressed, we use a CMPA ("compare accumulator") instruction to compare the contents of the accumulator to $13, which is the code associated with the "Enter" key. If the value in the accumulator is equal to this code, the Z (zero) status flag will be set (loaded with a logic 1), in which case the JZ ("jump if zero") instruction will cause us to

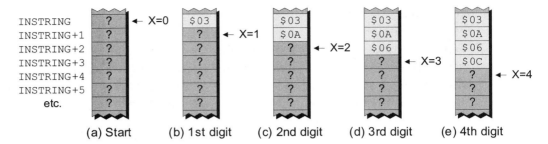

Figure L4-2. Storing key codes as they are clicked.

jump to label _GN_STR. Otherwise, we will copy the value in the accumulator to the output port driving the calculator's main display, and also store into the next free location in our string.

The STA [INSTRING, X] instruction will store the value in the accumulator into the memory location whose address is computed by adding the address associated with the INSTRING label to the contents of the index register. The first time a key is pressed (the "3" key in our example), the index register contains 0, so the $03 code associated with this key is stored in memory location INSTRING+0 [Figure L4-2(b)]. The index register is incremented to contain 1, and the program jumps back to _GN_LOOP to wait for the next key to be pressed.

The second time a key is pressed (the "A" key in our example), the index register contains 1, so the $0A code associated with this key is stored in memory location INSTRING+1 [Figure L4-2(c)]. And so it goes, until the user eventually clicks the "Enter" key, which causes the program to jump to label _GN_STR.

Converting the key codes into a binary number

Now, this will take a little bit of thought in order to understand exactly what we're trying to do (actually doing it is easy-peasy). We currently have four bytes representing key codes stored in INSTRING, and we want to take these key codes and convert them into a 2-byte (16-bit) binary number (we'll refer to this number as NUM for the purpose of these discussions). We can represent this pictorially as shown in Figure L4-3.

The only other trick to remember is that we wish to store the ensuing 2-byte number on the stack such that the most-significant (MS) byte

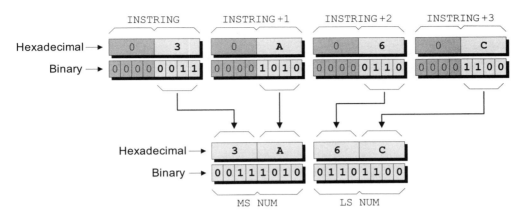

Figure L4-3. Converting the key codes into a binary number.

is "on top" of the least-significant (LS) byte, which means we have to process the LS byte first.

Bearing all of this in mind, consider what happens starting at the _GN_STR label. We start by loading the accumulator with the third key code (the one stored at INSTRING+2), as shown in Figure L4-4(a). Then, we shift this code four bits to the left as shown in Figure L4-4(b), and OR it with the fourth key code (the one stored at INSTRING+3), as shown in Figure L4-4(c).

Now remember that after storing the return address in a safe place at the beginning of this subroutine, we left the stack pointer as shown in

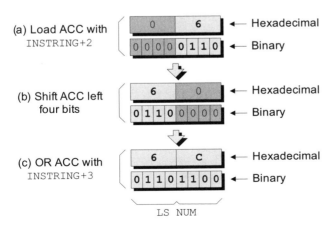

Figure L4-4. Forming the LS byte of the binary number.

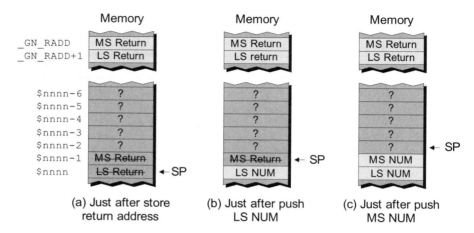

Figure L4-5. Pushing the 2-byte binary number onto the stack.

Figures L4-1(c) and L4-5(a). Thus, after we've formed the LS byte of our 2-byte number as shown in Figure L4-4(c), we push it onto the stack as shown in L4-5(b). (We don't know about you, but even we are starting to get confused with all of these figure numbers!)[1]

Now, we form the MS byte of NUM in the same way: we load the accumulator with the first key code (at INSTRING), shift this code four bits to the left, OR the result with the second key code (INSTRING+1), and push the result onto the stack as shown in Figure L4-5(c).

Retrieving the return address

Now we're really racing toward the end of this subroutine. When we get to the _GN_SPC label, we write a space character to the main display (as you will see, this will make things look "pretty" when we come to use this testbench program). Next, at the _GN_RET label, we retrieve the LS byte of the return address from its temporary location and push it onto the stack [Figure L4-6(b)]. Similarly, we retrieve the MS byte of the return address and push *it* onto the stack [Figure L4-6(c)].

Finally, we execute the RTS ("return from subroutine") instruction, which automatically pops the return address from the top of the stack, as shown in Figure L4-6(d), and returns us to wherever the GETNUM subroutine was called from. The point is that the stack now contains our

---

[1]But look on the bright side; at least *you* don't have to draw the figures!

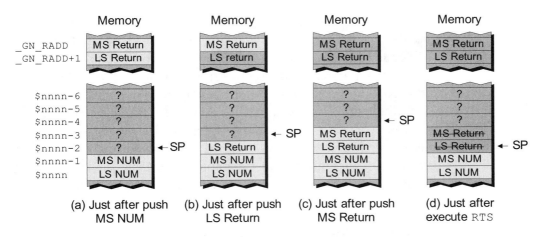

Figure L4-6. Pushing the return address onto the stack.

2-byte number, which is waiting to be used by one of our math subroutines, as we will see in future labs.

Note that when we retrieve the LS and MS bytes of the return address in Figures L4-6(b) and L4-6(c), we leave their values showing (but grayed-out) in the _GN_RADD+1 and _GN_RADD locations, respectively. This is because the act of copying these values into the accumulator doesn't destroy the original versions stored in our temporary memory locations. (Truth to tell, popping the return address out of the two locations on the top of the stack doesn't corrupt these values either, but accessing them would be a bit tricky, so we've shown these locations with their values lined through.)

Once you are confident that you understand how this subroutine works, enter it between the "Start" and "End Subroutine" comments and then use the assembler's **File > Save** command to save your work.

Improving on our subroutine

It wouldn't surprise us if you've noticed that, in many respects, our GETNUM routine leaves something to be desired. It performs no error-checking, it doesn't warn us if the user enters an incorrect number of characters, and it doesn't even chastise the user if he or she presses a nonhexadecimal key. For example, what would happen if the user were to enter the sequence "3," "A," "*," "C," "Enter"? Alternatively, what would happen if the user clicked only three keys before pressing the

"Enter" button? What if they clicked five? Even worse, what if they clicked eleven or more, which would exceed the 10-byte capacity of IN-STRING and start overwriting subsequent memory locations?

The point is that creating "bullet-proof" input routines takes a lot of thought and effort, and we decided not to spend too much time on this aspect of things in the case of this simple testbench program. But once you've got everything working, you might decide to revisit this routine and augment it with some additional tests and warnings of your own or replace it in its entirety with something better. Generally speaking, it is immaterial how this routine performs its machinations as long as it leaves a 2-byte binary number on the top of the stack when it is done.

Adding a DISPNUM subroutine

For the moment, let's assume that we have performed some math function such as adding two 2-byte binary numbers together, and we've left the 2-byte result on the top of the stack. The purpose of the DISPNUM subroutine is to retrieve this 2-byte result from the stack and convert it into a sequence of character codes that are presented to the calculator's main display.

The DISPNUM subroutine is shown below in its entirety. This is followed by a detailed description as to the way in which this routine works.

```
########## Start of DISPNUM subroutine
########## Retrieve a 16-bit binary value from the stack, convert
########## it into 4 hex digits, and display these digits

########## Pop the return address off the stack and save it
DISPNUM:    POPA                    # Pop the MS byte of return address
            STA     [_DN_RADD]      # off the stack and store it
            POPA                    # Pop the LS byte of return address
            STA     [_DN_RADD+1]    # off the stack and store it

########## Write an equals sign ('=') and space to the main display
_DN_EQ:     LDA     $3D             # Load acc with ASCII code for '='
            STA     [MAINDISP]      #   and store it to the main display
            LDA     $20             # Load acc with ASCII code for ' '
            STA     [MAINDISP]      #   and store it to the main display

########## Break-out and display the MS byte (1st and 2nd digits)
_DN_MSB:    POPA                    # Pop the MS byte off the stack
            PUSHA                   #   and push a copy back for later
            SHR                     # Then shift it right by 4 bits
            SHR                     #   :
```

```
             SHR                       #   :
             SHR                       #   :
             AND      $0F              # Clear MS 4 bits to 0
             STA      [MAINDISP]       #   and store 1st digit to main display
             POPA                      # Pop the MS byte off the stack again
             AND      $0F              # Clear MS 4 bits to 0
             STA      [MAINDISP]       #   and store 2nd digit to main display

########## Break-out and display the LS byte (3rd and 4th digits)
_DN_LSB:     POPA                      # Pop the LS byte off the stack
             PUSHA                     #   and push a copy back for later
             SHR                       # Then shift it right by 4 bits
             SHR                       #   :
             SHR                       #   :
             SHR                       #   :
             AND      $0F              # Clear MS 4 bits to 0
             STA      [MAINDISP]       #   and store 3rd digit to main display
             POPA                      # Pop the LS byte off the stack again
             AND      $0F              # Clear MS 4 bits to 0
             STA      [MAINDISP]       #   and store 4th digit to main display

########## Return gracefully from this subroutine
_DN_RET:     LDA      [_DN_RADD+1]     # Get LS byte of return address from
             PUSHA                     #   temp location and push onto stack
             LDA      [_DN_RADD]       # Get MS byte of return address from
             PUSHA                     #   temp location and push onto stack
             RTS                       # Return from this subroutine

_DN_RADD:    .2BYTE                    # 2-byte temp location used to store
                                       #   the return address for this routine
########## End of DISPNUM subroutine

# - - - - - - - - - - - - - - - - - - - - - - - - - - - - - - - - - #
```

In this case, all of the labels inside this routine begin with the character sequence _DN_ (for "display number"), and the 2-byte temporary location reserved at the end of the routine that is used to store the return address is called _DN_RADD.

For the purposes of these discussions, let's assume the 2-byte number on the top of the stack that we wish to display has the hexadecimal value $9C7E, and that we'll refer to this value as RES (for "result"). Now, let's consider the state of play just before we call the DISPNUM subroutine. The stack pointer (SP) will be pointing to whatever location in the memory is currently considered to represent the top of the stack. Meanwhile, the two bytes in memory that are associated with the _DN_RADD label will contain unknown values [Figure L4-7(a)].

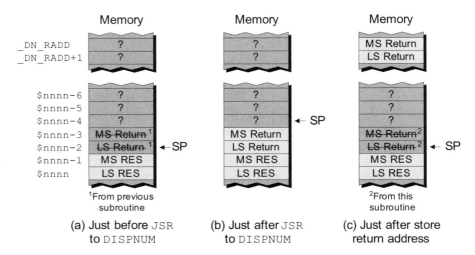

Figure L4-7. The state of play when DISPNUM is called.

Once again, when a JSR ("jump to subroutine") instruction is used to call DISPNUM, the return address is automatically pushed onto the current top of the stack and the stack pointer is modified to point to the next free location on the new top of the stack [Figure L4-7(b)]. As usual, the first four instructions in DISPNUM are used to pop the return address off the top of the stack and store them in our 2-byte temporary location at label _DN_RADD, as shown in Figure L4-7(c).

While we're here, let's quickly note that the next four instructions, starting at label _DN_EQ, are used to write an equals sign ("=") and a space character (" ") to the main display so as to make the output look more aesthetically pleasing.

Now, let's consider just what it is that we need to do in order to display the 2-byte number on the top of the stack (Figure 4-8), and then we'll discuss how to do it.

As we see, we wish to extract four 4-bit nybbles from our 2-byte result, and then convert each nybble into an 8-bit character code to be copied to the main display. We start at _DN_MSB by processing the most-significant byte of the result. First we pop the MS RES byte off the stack into the accumulator as shown in Figure L4-9(a) (note that, in the next line of the program, we push a copy of MS RES back onto the stack because we'll need it again in a moment).

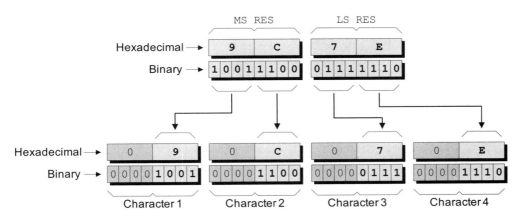

Figure L4-8. Converting the binary number into character codes.

Next, we shift the contents of the accumulator four bits to the right. However, due to the fact that our virtual computer's SHR ("shift right") instruction performs an arithmetic shift, the most-significant bit (a "1" in this case) keeps on getting copied back into itself, resulting in the accumulator containing a value of $F9 [Figure L4-9(b)]. Thus, we next AND the contents of the accumulator with a value of $0F so as to clear the most-significant four bits to 0 [Figure L4-9(c)], and then we write the resulting character code to the main display.

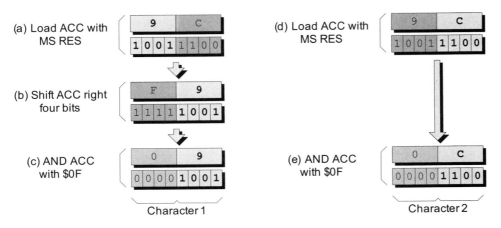

Figure L4-9. Forming the character codes corresponding to MS RES.

Now, we pop the MS RES byte off the stack into the accumulator a second time [Figure L4-9(d)]. In this case, all we have to do is AND the contents of the accumulator with a value of $0F so as to clear the most-significant four bits to 0 [Figure L4-9(e)], and then we write *this* character code to the main display.

We then process the least-significant byte of the result in a similar manner at label _DN_LSB. Finally, we reach the _DN_RET label, which is where we retrieve the return address from its temporary location, push it onto the stack, and use a RTS ("return from subroutine") instruction, which returns the program to whichever point the DISPNUM routine was called from.

Now, enter this routine somewhere between the "Start" and "End Subroutine" comments (preferably, just after the GETNUM routine) and, as usual, use the assembler's **File** > **Save** command to save your work.

Adding a DISPERR subroutine

Last, but not least, we wish to create a DISPERR ("display error") sub-routine that we can use to warn us if our program detects that something has gone wrong. Cast your mind back to the *Reserving global temporary data locations and messages* section earlier in this lab, in which we stored the following messages:

Label	Message
MSG_000:	ERROR:
MSG_001:	Carry = 1
MSG_002:	Borrow = 0
MSG_003:	Overflow
MSG_004:	Underflow
MSG_005:	Too big
MSG_006:	Too small
MSG_007:	Divide by 0
MSG_008:	Out of range
MSG_009:	-32,768
MSG_010:	Not implemented

Now, let's assume that our program detects that it has generated a value that is too large. In such an event, the message associated with label MSG_005 would be reasonably appropriate. In this case, the program could call our DISPERR routine as follows:

```
     BLDX      MSG_005      # Load X reg with start of msg 005
     JSR       [DISPERR]    # Call the error message routine
```

That is, we use a BLDX ("big load index register") instruction to load the index register (X) with the address of the first character in the desired error message, then we call the DISPERR subroutine. Now let's look at the DISPERR routine itself:

```
########## Start of DISPERR subroutine
########## Display an error message and then quit the program

########## First clear the display
DISPERR:   LDA      CLRCODE      # Load ACC with clear code
           STA      [MAINDISP]   #    and copy it to the main display

########## Display the word "Error: " (note the space)
_DE_ERR:   BSTX     [TMP2BYTE]   # Store the value in the X register
           BLDX     MSG_000      # Load X reg with start of msg 000
_DE_LUPA:  LDA      [0,X]        # Load a character from the msg
           JZ       [_DE_MSG]    # If it's a NUL jump to next bit
           STA      [MAINDISP]   #    otherwise copy it to main display
           INCX                  # Increment the index register
           JMP      [_DE_LUPA]   # Jump back for next character

########## Now display the main error message then terminate
_DE_MSG:   BLDX     [TMP2BYTE]   # Reload the X reg's original value
_DE_LUPB:  LDA      [0,X]        # Load a character from the msg
           JZ       [$0000]      # If it's a NUL terminate the program
           STA      [MAINDISP]   #    otherwise copy it to main display
           INCX                  # Increment the index register
           JMP      [_DE_LUPB]   # Jump back for next character
########## End of DISPERR subroutine

# - - - - - - - - - - - - - - - - - - - - - - - - - - - - - - #

########## This is where the math subroutine will go
```

Observe that we are not concerned with storing the return address in this case, because once the DISPERR routine has finished displaying the required message, it will terminate the program.

First, we clear the calculator's main display by loading the accumulator with the display's clear code and copying it to the display. Now, we are at label _DE_ERR, at which point we copy the current value in the index register—the address of the first character in the message we wish to display—into our temporary global location at label TMP2BYTE.

We then load the index register with the address of the first character in the message associated with label MSG_000 (this message is "ERROR: "). Now we enter a loop at label _DE_LUPA that retrieves the character pointed to by the index register and checks to see if it's a NUL character. If it is a NUL, we jump to label _DE_MSG; otherwise, we copy the character to the main display, increment the index register, and jump back to the beginning of the loop. Thus, this loop will cause the main display to show the following string:

ERROR:

Once this portion of the task is completed and the program reaches label _DE_MSG, it reloads the index register with the start address of the original message. Next, we enter a loop at label _DE_LUPB that retrieves the character pointed to by the index register and checks to see if it's a NUL character. If it is a NUL, we jump to address $0000 and terminate the program; otherwise, we copy the character to the main display, increment the index register, and jump back to the beginning of the loop.

Thus, in this particular example, the main display will end up showing the following string of characters:

ERROR: Too big

You know the drill: enter this routine after the DISPNUM routine and use the assembler's **File** > **Save** command to save your work.

Creating the body of the program

Good grief! As you'll see, building this testbench program has actually taken much more effort than creating the math subroutines themselves, but we're almost finished. All we have to do is to enter the main body of the program as follows:

```
########## Get two 16-bit numbers and push them onto the top of the
########## stack; perform some action on them and display the result
MAINLOOP:   JSR    [GETNUM]    # Get 1st 16-bit number (NUMA)
            JSR    [GETNUM]    # Get 2nd 16-bit number (NUMB)
            ### THIS IS WHERE WE WILL CALL THE MATH ROUTINE
            JSR    [DISPNUM]   # Display 16-bit number from stack

########## Wait for any key to be pressed, then clear the main
########## display and do it all again
```

```
WAITKEY:    LDA     [KEYPAD]     # Load accumulator from keypad
            JN      [WAITKEY]    # Jump back if no key pressed
            LDA     CLRCODE      # Load accumulator with clear code
            STA     [MAINDISP]   #   and store it to the main display
            JMP     [MAINLOOP]   # Jump back and do it all again
```

We'll commence by considering the statements associated with the MAINLOOP label. First we call our GETNUM subroutine, which, as we know, allows us to key in four hexadecimal codes that are converted into a two-byte number that will be pushed onto the stack. Assuming that the user clicks the following sequence of buttons,

"3," "A," "6," "C," "Enter"

then the state of the stack immediately after we return from this first call to the GETNUM routine will be as shown in Figure L4-10(a).

Now, we call GETNUM for a second time. Assuming that the user clicks the following sequence of buttons,

"6," "2," "1," "2," "Enter"

then the state of the stack immediately after we return from this second call to the GETNUM routine will be as shown in Figure L4-10(b).

At this point, we haven't actually created any of our math subroutines. However, purely for the sake of discussion, let's assume that we now call a subroutine called _ADD[2] that retrieves our two 2-byte numbers from the top of the stack, adds them together, and pushes the 2-byte result back onto the stack as shown in Figure L4-10(c) ($3A6C + $6212 = $9C7E).

Next, we call our DISPNUM routine, which retrieves a 2-byte number from the top of the stack, converts it into corresponding display codes, and presents these codes to the calculator's main display. In the case of our testbench program, this routine does *not* leave a copy of the value to be displayed on the stack (this will change when we come to create a real calculator program). Thus, the state of the stack following the DISPNUM routine is as shown in Figure L4-10(d).

Finally, when we reach the WAITKEY label, the program waits for any key on the calculator's front panel to be clicked, then it clears the

---

[2]Note that we have to call our subroutine something like _ADD (with a leading underscore character) because we already have an instruction called ADD (without the underscore).

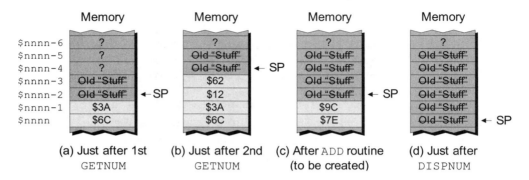

Figure L4-10. The state of the stack at key points in the program.

main display and jumps back to MAINLOOP to do the whole thing all over again.

We're nearly there. Enter these lines of code between the "Start" and "End of main program body" comments and, you guessed it, use the assembler's **File** > **Save** command to save your work.

Testing the Testbench Program

Ensure that the calculator is powered-up. Assemble (and debug) your new program. Click the **Reset** button on the front panel to ensure that the calculator is halted. Use the **Memory** > **Purge RAM** command to scramble its memory, then use the **Memory** > **Load RAM** command to load the new *skeleton-lab4.ram* file.

Now click the **Run** button to execute the program and observe the main display being cleared. The program is waiting for you to input your first number, so click on the sequence of keys we discussed earlier:

"3," "A," "6," "C," "Enter"

Remember that, in addition to building a 2-byte number and placing it on the stack, the GETNUM routine also copies the digits back to the main display, and then it writes out a space character for good measure, so the main display will now show

3A6C

Next, click on the second sequence of keys we discussed earlier as follows:

"6," "2," "1," "2," "Enter"

Once again, the GETNUM routine will copy these back to the main display and follow them with a space character, so for a fraction of a second the main display will show

3A6C 6212

If we had actually created a math subroutine called something like _ADD, this is the point where it would be called from the main program; but we have not yet created that subroutine so, instead, we immediately call our DISPNUM routine, which commences by outputting an equals-space ("= ") pair of characters followed by whatever 2-byte number it finds on top of the stack. As that 2-byte number is currently the last number we entered, the main display will now show

3A6C 6212 = 6212

The program is now waiting for you to click another key. When you do so, it clears the display and then jumps back to the beginning to await your entering of two more numbers.

Experiment by using a few different number sequences. What happens if on the second number you enter only three digits before clicking the "Enter" key? What happens if you enter 5?

Bonus Question What will happen if you click the "0" key four times for your first number followed by one or more nonhexadecimal keys when forming your second number? For example,

> "0," "0," "0," "0," "Enter"   {First number}
>
> "+," "−," "*," "/," "Enter"   {Second number}

Try this for yourself, note the result, and try to explain what you see to your own satisfaction *before* perusing and pondering the answer on the following page.

■ *Answer* The codes associated with the "+," "–," "*," and "/" keys are $16, $17, $18, and $19, respectively. As each button is clicked, the GETNUM routine reads the code associated with that button and immediately copies it back to the main display. However, if you look at Figure L2-11 in Lab 2c, you will see that codes $16, $17, $18, and $19 are not actually supported by the main display. As these codes are undefined, in theory we don't know what will happen (in practice, however, we designed the main display such that it will show a pound "#" character if it's presented with an invalid character).

Now let's consider what will happen when the GETNUM routine uses the second pair of characters to form the LS byte of our binary number. The $18 code (binary 1000 1000) associated with the "*" key will be loaded into the accumulator and shifted four bits to the left, leaving the accumulator containing $80 (binary 1000 0000). This will then be OR-ed with the $19 code (binary 0001 1001) associated with the "/" key, thereby leaving the accumulator containing $99 (binary 1001 1001).

Similarly, when the first pair of characters is used to form the MS byte of our binary number, the $16 code (binary 0001 0110) associated with the "+" key will be loaded into the accumulator and shifted four bits to the left, leaving the accumulator containing $60 (binary 0110 0000). This will then be OR-ed with the $17 code (binary 0001 0111) associated with the "–" key, thereby leaving the accumulator containing $77 (binary 0111 0111).

Thus, clicking in a sequence of "+," "–," "*," and "/" will result in GET-NUM placing a 2-byte value of $7799 on the stack. Of course the DISPNUM routine doesn't care where these values came from, so the main display will now show

$$0000 \ \#\#\#\# = 7799$$

Lab 4b

Creating a 16-bit ADD Subroutine

Objective: To create a general-purpose subroutine that can add two 2-byte (16-bit) unsigned binary numbers together.

Duration: Approximately 45 minutes.

Before We Start

In Lab 4a, we created a testbench program that we are going to use to verify the functionality of our math routines. For the purposes of these discussions, the interesting portion of this testbench program was the first part of its main body, as follows:

```
MAINLOOP:   JSR     [GETNUM]    # Get 1st 16-bit number (NUMA)
            JSR     [GETNUM]    # Get 2nd 16-bit number (NUMB)
            ### THIS IS WHERE WE WILL CALL THE MATH ROUTINE
            JSR     [DISPNUM]   # Display 16-bit number from stack
```

The purpose of this lab is to create a subroutine called _ADD[1] that will add two 2-byte (16-bit) *unsigned* binary numbers together (this routine will be modified to deal with *signed* binary numbers in Lab 4e). We will investigate the creation of the _ADD routine shortly, but for the moment, let's simply assume that we will eventually modify the testbench routine to call this routine as follows (the change is highlighted in gray):

```
MAINLOOP:   JSR     [GETNUM]    # Get 1st 16-bit number (NUMA)
            JSR     [GETNUM]    # Get 2nd 16-bit number (NUMB)
            JSR     [_ADD]      # Add NUMA and NUMB together
            JSR     [DISPNUM]   # Display 16-bit number from stack
```

[1] The reason for calling this subroutine _ADD is that we can't use ADD without the underscore because we already have an instruction with that name.

We commence by calling our GETNUM routine twice so as to obtain two 2-byte numbers on the top of the stack. We shall refer to these numbers as NUMA and NUMB. This means that the state of the stack when we return from the second call to GETNUM will appear as shown in Figure L4-11(a).

The task in store for our _ADD routine is to retrieve these two 2-byte numbers from the stack, add them together, and store the 2-byte result (which we'll refer to as RES) back on the stack as shown in Figure L4-11(b). Finally, we call the DISPNUM routine, which retrieves this result from the stack and presents it to the main display as four hexadecimal characters [Figure L4-11(c)].

Just to remind ourselves, the number to which another number is to be added is called the *augend,* a number that is added to another number is called the *addend,* and the result from adding two or more numbers together is called the *sum.* For example, if we were to enter $3A6C and $6232 as the two numbers to be added, we can represent the addition operation as follows:

Augend +	Addend =	Sum	(Official terminology)
$3A6C +	$6232 =	$9C9E	(Example values)
NUMA +	NUMB =	RES	(Names used in these discussions)

The 16-bit _ADD Subroutine Itself

The _ADD subroutine is shown below in its entirety. This is followed by a detailed description as to the way in which this routine functions.

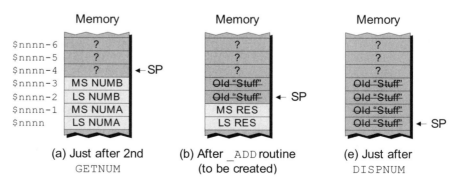

Figure L4-11. The state of the stack at key points in the program.

Observe the big block of comments at the top of the routine. It isn't really necessary for you to replicate all of these comments, but it would be good practice to do so. The reason for providing this level of information is to ease the task of other programmers who may wish to incorporate your routine into their code (and also to aid your own memory when you come to revisit this routine in the future).

```
###########################################################################
# Name:      _ADD                                                         #
#                                                                         #
# Function: Add two 16-bit UNSIGNED binary numbers together and           #
#           return a 16-bit result in the range 0 through 65,535          #
#                                                                         #
# Entry:    Top of stack                                                  #
#           Most-significant byte of return address                       #
#           Least-significant byte of return address                      #
#           Most-significant byte of second number   (addend)             #
#           Least-significant byte of second number  (addend)             #
#           Most-significant byte of first number    (augend)             #
#           Least-significant byte of first number   (augend)             #
#                                                                         #
# Exit:     Top of stack                                                  #
#           Most-significant byte of result          (sum)                #
#           Least-significant byte of result         (sum)                #
#                                                                         #
# Modifies: Accumulator (also index register if error)                    #
#                                                                         #
# Size:     Program = 50 bytes                                            #
#           Data    =  5 bytes                                            #
###########################################################################

########## Get return address from stack and store it
_ADD:       POPA                    # Retrieve MS byte of return
            STA      [_AD_RADD]     #   address from stack and store it
            POPA                    # Retrieve LS byte of return
            STA      [_AD_RADD+1]   #   address from stack and store it

########## Get addend and augend from stack
_AD_GNUM:   POPA                    # Retrieve MS byte of addend from
            STA      [_AD_NUMB]     #   stack and store it
            POPA                    # Retrieve LS byte of addend from
            STA      [_AD_NUMB+1]   #   stack and store it
            POPA                    # Retrieve MS byte of augend from
            STA      [_AD_NUMA]     #   stack and store it
            POPA                    # Retrieve LS byte of augend from
                                    #   stack & leave it in the ACC
```

```
########## Perform the addition
_AD_DOIT:   ADD     [_AD_NUMB+1] # Add LS byte of addend to ACC
            PUSHA                #   and push LS sum onto stack
            LDA     [_AD_NUMA]   # Load ACC with MS byte of augend
                                 # from temp location
            ADDC    [_AD_NUMB]   # Add MS byte of addend to ACC w
            PUSHA                # carry and push MS sum onto stack

########## Make sure there isn't a carry out from the MS addition
_AD_CHK:    JNC     [_AD_RET]    # If carry flag = 0 jump to return
            BLDX    MSG_001      # Load X reg with addr of message
            JSR     [DISPERR]    # Jump to display error subroutine
                                 # (which terminates the program)

########## Return to main program
_AD_RET:    LDA     [_AD_RADD+1] # Load ACC with LS byte of return
                                 #   address from temp location and
            PUSHA                #   push it back onto the stack
            LDA     [_AD_RADD]   # Load ACC with MS byte of return
                                 #   address from temp location and
            PUSHA                #   push it back onto the stack
            RTS                  # That's it, exit the subroutine

########## Reserve temp locations for this subroutine
_AD_RADD:   .2BYTE               # Reserve 2-byte temp location for
                                 #   the return address
_AD_NUMA:   .BYTE                # Reserve 1-byte temp location for
                                 #   the MS byte of the augend
_AD_NUMB:   .2BYTE               # Reserve 2-byte temp location for
                                 #   the addend
########## This is the end of the _ADD subroutine

# - - - - - - - - - - - - - - - - - - - - - - - - - - - - - - - - #
```

In this case, all of the labels inside this routine begin with the character sequence _AD_ (for "add"). Observe the 2-byte temporary location at the end of the routine called _AD_RADD, that is used to store the return address. Also, observe the 1-byte temporary location, called _AD_NUMA, that will be used to store the *most-significant* (*MS*) byte of the augend, and the 2-byte temporary location, called _AD_NUMB, that is used to store both the most-significant and *least-significant* (*LS*) bytes of the addend.

Storing the return address

As we previously discussed, our testbench program is going to arrange for two 2-byte numbers to be stored on the top of the stack [Figure L4-12(a)]. As usual, when we first call the _ADD subroutine, its return address will be

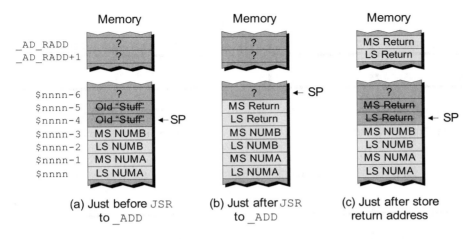

Figure L4-12. The state of play when _ADD is called.

automatically placed on the top of the stack [Figure L4-12(b)]. Thus, the first thing we do at label _ADD is to retrieve the return address and squirrel it away in our temporary _AD_RADD location [Figure L4-12(c)].

Retrieving the numbers to be added from the stack

Now, consider the sequence of instructions starting at the _AD_GNUM label (where "GNUM" is short for "get numbers") as illustrated in Figure L4-13.

We start by popping the MS byte of NUMB off the top of the stack and copying it into the memory location associated with label _AD_NUMB [Figure L4-13(a)]. Next, we pop the least-significant (LS) byte of NUMB off the top of the stack and copy it into the memory location at _AD_NUMB+1 [Figure L4-13(b)].

Now, we pop the MS byte of NUMA off the top of the stack and copy it into the memory location associated with label _AD_NUMA [Figure L4-13(c)]. Finally, we pop the LS byte of NUMA off the top of the stack and leave this value in the accumulator [Figure L4-13(d)].

Actually performing the addition

As we've popped both of our 2-byte numbers off the stack, the stack is now empty, as shown in Figure L4-14(a).

The task of actually performing the addition, which commences at label _AD_DOIT, is surprisingly easy. We already have the LS byte of

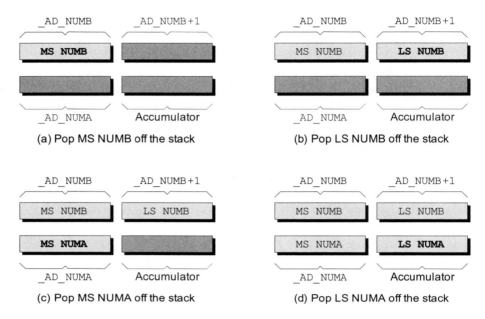

Figure L4-13. Retrieving the numbers to be added from the stack.

NUMA (the augend) in the accumulator, so we use an ADD [NUMB+1] (a standard add instruction) to add the LS byte of NUMB (the addend) to the contents of accumulator, and then we push the result onto the stack, as shown in Figure L4-14(b).

This addition will, of course, load the C (carry) flag with a 0 or 1 depending on the values being added. Next, we use a LDA [NUMA] instruction load the MS byte of NUMA into the accumulator. Then we use an ADDC [NUMB] ("add with carry") instruction to add both the MS byte of NUMB and the current contents of the carry flag to the accumulator, and then we push the result onto the stack as shown in Figure L4-14(c).

Testing for errors

Before we consider the portion of the _ADD subroutine starting at label _AD_CHK, we need to remember that the main testbench program discussed in Lab 4a was equipped with a series of error message strings associated with labels MSG_000, MSG_001, MSG_002, and so on. Furthermore, the testbench program includes a DISPERR ("display error") subroutine.

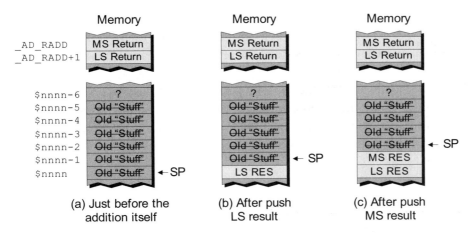

Figure L4-14. The state of the stack as we perform the addition.

For the purposes of this lab, we're assuming that we're working with16-bit unsigned binary numbers. This means that the range of numbers we can support is $0000 through $FFFF (or 0 through 65,535 in decimal). If the final result from our addition is larger than this, the carry flag will contain a logic 1, thereby indicating an error condition.

Thus, we perform a test on the carry flag at label _AD_CHK. If the carry flag contains a 0, which tells us that everything is OK, the JNC ("jump if not carry") instruction will cause the program to jump to label _AD_RET. Otherwise, we load the index register with the value associated with label MSG_001 and then use a JSR ("jump to subroutine") to jump to our DISPERR routine.

> **Note** Ideally, our general-purpose math subroutines are going to be created in such a way as to facilitate translating them into another assembly language for use on a real microprocessor or microcontroller.
>
> Thus, with the exception of this error check, the rest of the **_ADD** subroutine is totally self-contained—it uses its own local temporary storage locations and its only interaction with the outside world is via the stack.
>
> If we did wish to port this routine to another language, we would have several options, as follows: (a) cut this error test out (not recommended), (b) keep the test but embed the message string and associated display code inside this routine, or (c) replicate the external environment in the form of the message strings and DISPNUM subroutine.

Retrieving the return address

Assuming that our error test didn't detect any problems, we arrive at the _AD_RET label with the result from the addition on the top of the stack

[Figure L4-15(a)]. We retrieve the LS byte of the return address from its temporary location and push it onto the stack [Figure L4-15(b)], and then we retrieve the MS byte of the return address and push *it* onto the stack [Figure L4-15(c)].

Finally, we execute the RTS ("return from subroutine") instruction. This automatically pops the return address from the top of the stack, as shown in Figure L4-15(d), and returns us to the testbench program, which will then call the its DISPNUM subroutine to display the result.

Creating and Debugging the _ADD Routine

Invoke the assembler, enter the _ADD subroutine as shown above, and save it in a file called *int-add-2-byte-v1.asm*. In the not-so-distant future, we are going to insert this file into our main testbench program, but first we need to make sure that we've entered it correctly (without any "finger-slips") and that it will assemble without errors.

If you try clicking the assemble button, however, the assembler will immediately inform you that you need a .ORG statement. And if you fix this, you'll run into another problem. In fact, we need to "wrap" the _ADD routine with a couple of statements as follows:

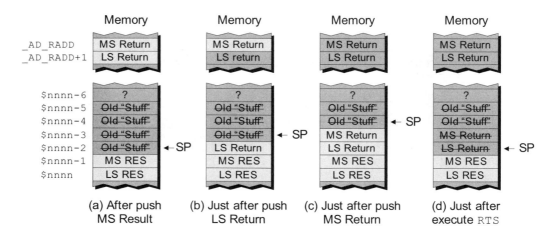

Figure L4-15. Pushing the return address onto the stack.

```
            .ORG        $4000
{The _ADD subroutine goes here}
MSG_001:    .BYTE
DISPERR:    RTS
            .END
```

That is, in order to make the assembler happy, we need a .ORG statement
to kick everything off, a .END statement to terminate things, and two
labels (MSG_001 and DISPERR) to satisfy the external references made
by the _ADD routine.

Now, try clicking the assemble button again. Any errors you detect
from here on are your own, so track them down and debug them until
you receive a "Your file was assembled without any errors" message.

Once you have fully debugged your _ADD routine, strip out the
"wrapping statements" we added above, and use the assembler's **File >
Save** command to save this file one last time.

> **Note**   If you wish, you can launch multiple instantiations of the assembler. Every time you use the **Tools >
> Assembler** command (or click the assembler toolbar icon), a new copy of the assembler will appear. This can
> be very useful for "power users" who wish to edit different files in different assembler windows, but be careful
> not to open the same file in multiple copies of the assembler (otherwise things will become very confusing very
> quickly).

Modifying the Testbench to Call the _ADD Routine

Use the assembler's **File > Open** command to open up the *skeleton-
lab4.asm* file you created in Lab 4a, and then use the **File > Save As** command to save this as *lab4b.asm*.

Modify the body of the testbench program to call the _ADD routine
as follows (the change is highlighted in gray):

```
MAINLOOP:   JSR      [GETNUM]      # Get 1st 16-bit number (NUMA)
            JSR      [GETNUM]      # Get 2nd 16-but number (NUMB)
            JSR      [_ADD]        # Add NUMA and NUMB together
            JSR      [DISPNUM]     # Display 16-bit number from stack
```

Now, scroll down this file until the end of the subroutines to find the
comment line that reads as follows:

```
########## This is where the math subroutine will go
```

Delete this comment and leave your cursor at the beginning of this line, then use the assembler's **Insert > Insert File** command, locate the *int-add-2-byte-v1.asm* source file you just created in the ensuing dialog, and click the **Open** button to insert the contents of that file into this version of the testbench program.

Use the **File > Save** command to save this modified version of the testbench, and then click the assemble button to assemble it. If all goes well, there shouldn't be any problems; otherwise, it's up to you to isolate any bugs and fix them (sorry, we don't have a magic wand for this).

Using the Testbench to Test the Routine

Click the **On/Off** button on the front panel to power-up the calculator, use the **Memory > Load RAM** command to load the *lab4b.ram* file into the calculator's memory, and click the **Run** button to set the program running.

After clearing the display, the program loops around waiting for you to input your first number. As a first test, click on the following sequence of keys:

"3," "A," "6," "C," "Enter"   "6," "2," "1," "2," "Enter"

As we've discussed (ad infinitum), the program will add these numbers together and present the result on the main display, which will now show the following:

3A6C 6212 = 9C7E

The program now waits for you to click any key, at which point it will clear the main display and wait for you to enter another two numbers. Experiment with different values to make sure that this routine works as expected. As another test, try clearing the display and entering the following sequence of keys:

"F," "F," "F," "E," "Enter"   "0," "0," "0," "1," "Enter"

This should result in the following characters appearing on the main display:

FFFE 0001 = FFFF

The result, FFFF, is the largest value we can hold in a 2-byte (16-bit) field. Now, try clearing the display and entering the following sequence of keys:

"F," "F," "F," "E," "Enter"   "0," "0," "0," "2," "Enter"

The result is too big to fit in our 2-byte field, so the C (carry) flag will be set to 1, the error check should call the DISPERR subroutine, and the following message should appear on the main display:

ERROR: Carry = 1

Further Modifications

There are a variety of little "tweaks" you could make to improve the aesthetics of our testbench program; for example:

- When the program is ready for you to input a number, it could write a question mark "?" character to the main display. [When you start entering that number, you could write a $12 ("Back") code to the display; this will cause the display to clear the last (right-most) character—the "?"—and shift any remaining characters one place to the right.]

- You could write "$" characters to the main display so as to make it obvious that the numerical values are in hexadecimal.

- You could write a "+" character to the display between inputting the first and second numbers so as to make clear the math operation that is being performed. Taking this and the previous point together, the display resulting from adding two numbers (say, $1234 and $4321) together would now appear as follows:

$1234 + $4321 = $5555

Lab 4C

Creating a 16-bit SUBTRACT Subroutine

Objective: To create a general-purpose subroutine that can subtract one 2-byte (16-bit) unsigned binary number from another.

Duration: Approximately 35 minutes.

The 16-bit _SUB Subroutine Itself

The _SUB subroutine is very similar to the _ADD routine we created in Lab 4b. This new routine is presented below in its entirety, followed by notes on the differences between this routine and its _ADD counterpart.

```
################################################################
# Name:      _SUB                                             #
#                                                             #
# Function: Subtracts one 16-bit UNSIGNED binary numbers from another #
#           and return a 16-bit result in the range 0 through 65,535 #
#                                                             #
# Entry:     Top of stack                                     #
#            Most-significant byte of return address          #
#            Least-significant byte of return address         #
#            Most-significant byte of second number  (subtrahend) #
#            Least-significant byte of second number (subtrahend) #
#            Most-significant byte of first number   (minuend) #
#            Least-significant byte of first number  (minuend) #
#                                                             #
# Exit:      Top of stack                                     #
#            Most-significant byte of result      (difference) #
#            Least-significant byte of result     (difference) #
#                                                             #
# Modifies: Accumulator (also index register if error)       #
#                                                             #
# Size:      Program = 50 bytes                               #
#            Data    =  5 bytes                               #
################################################################
```

311

```
########## Get return address from stack and store it
_SUB:      POPA                   # Retrieve MS byte of return
           STA      [_SB_RADD]    #   address from stack and store it
           POPA                   # Retrieve LS byte of return
           STA      [_SB_RADD+1]  #   address from stack and store it

########## Get subtrahend and minuend from stack
_SB_GNUM:  POPA                   # Retrieve MS byte of subtrahend
           STA      [_SB_NUMB]    #   from stack and store it
           POPA                   # Retrieve LS byte of subtrahend
           STA      [_SB_NUMB+1]  #   from stack and store it
           POPA                   # Retrieve MS byte of minuend from
           STA      [_SB_NUMA]    #   stack and store it
           POPA                   # Retrieve LS byte of minuend from
                                  #   stack & leave it in the ACC

########## Perform the subtraction
_SB_DOIT:  SUB      [_SB_NUMB+1]  # Sub LS byte of subtrahend from ACC
           PUSHA                  #   and push LS difference onto stack
           LDA      [_SB_NUMA]    # Load ACC with MS byte of minuend
                                  #   from temp location
           SUBC     [_SB_NUMB]    # Sub MS byte of subtrahend from ACC
           PUSHA                  # w (borrow) and push MS diff to stack

########## Make sure there isn't a borrow out from the MS subtraction
_SB_CHK:   JC       [_SB_RET]     # If carry flag = 1 jump to return
           BLDX     MSG_002       # Load X reg with addr of message
           JSR      [DISPERR]     # Jump to display error subroutine
                                  # (which terminates the program)

########## Return to main program
_SB_RET:   LDA      [_SB_RADD+1]  # Load ACC with LS byte of return
                                  #   address from temp location and
           PUSHA                  #   push it back onto the stack
           LDA      [_SB_RADD]    # Load ACC with MS byte of return
                                  #   address from temp location and
           PUSHA                  #   push it back onto the stack
           RTS                    # That's it, exit the subroutine

########## Reserve temp locations for this subroutine
_SB_RADD: .2BYTE                  # Reserve 2-byte temp location for
                                  #   the return address
_SB_NUMA: .BYTE                   # Reserve 1-byte temp location for
                                  #   the MS byte of the minuend
_SB_NUMB: .2BYTE                  # Reserve 2-byte temp location for
                                  #   the subtrahend
########## This is the end of the _SUB subroutine

# ------------------------------------------------- #
```

The differences between the _SUB routine and its _ADD cousin from the previous lab are as follows:

- The labels are prefixed by _SB_.
- When we actually perform the subtraction at label _SB_DOIT, the ADD instruction is replaced with a SUB, whereas the ADDC ("add with carry") is replaced by a SUBC ("subtract with carry/borrow").
- With regard to the error test at label _SB_CHK, a 1 in the C (carry flag) indicates that there were no problems, whereas a 0 indicates a borrow condition (in which case the index register is loaded with the address associated with label MSG_002). In turn, this indicates that we have tried to subtract a bigger number from a smaller one, which results in a negative value. For the purposes of this lab, however, we are assuming the use of unsigned binary numbers, which means that we don't have any way of representing negative values.

> **Note**   When we subtract a larger number from a smaller number, the result is a borrow-out of the higher-order bit, and we say that an *underflow* condition has occurred.
>
> Thus, as opposed to flagging our error using the string associated with the **MSG_002** label, we could have used the string associated with the **MSG_004** label had we so desired (refer to Lab 4a for details on the various message strings).

Creating and Debugging the _SUB Routine

Invoke the assembler, enter the _SUB subroutine as shown above, and save it in a file called *int-sub-2-byte-v1.asm.*

As before, in order to perform preliminary testing on this routine, we need a .ORG statement to kick everything off, a .END statement to terminate things, and two labels (MSG_002 and DISPERR) to satisfy the external references made by the _SUB routine. Thus, we need to "wrap" our routine with a couple of statements as follows:

```
          .ORG      $4000

{The _SUB subroutine goes here}

MSG_002:  .BYTE
DISPERR:  RTS
          .END
```

Now, try clicking the assemble button and track down and debug any errors until you receive a "Your file was assembled without any errors" message. Once you reach this point, strip out the "wrapping statements" we added above and use the assembler's **File > Save** command to save this file one last time.

Modifying the Testbench to Call the _SUB Routine

Use the assembler's **File > Open** command to open up the *skeleton-lab4.asm* file you created in Lab 4a, and then use the **File > Save As** command to save this out as *lab4c.asm*.

Modify the body of the testbench program to call the _SUB routine as follows (the change is highlighted in gray):

```
MAINLOOP:    JSR      [GETNUM]      # Get 1st 16-bit number (NUMA)
             JSR      [GETNUM]      # Get 2nd 16-but number (NUMB)
             JSR      [_SUB]        # Subtract NUMB from NUMA
             JSR      [DISPNUM]     # Display 16-bit number from stack
```

Now, scroll down the file until the end of the subroutines to find the comment line that reads as follows:

```
########## This is where the math subroutine will go
```

Delete this comment and leave your cursor at the beginning of this line, then use the assembler's **Insert > Insert File** command, locate the *int-sub-2-byte-v1.asm* source file you just created in the ensuing dialog, and click the **Open** button to insert the contents of that file into this version of the testbench program.

Use the **File > Save** command to save this modified version of the testbench, and then click the assemble button to assemble it.

Using the Testbench to Test the Routine

Click the **On/Off** button on the front panel to power-up the calculator, use the **Memory > Load RAM** command to load the *lab4c.ram* file into the calculator's memory, and click the **Run** button to set the program running.

As usual, once the display has been cleared, the program will loop around, waiting for you to input your first number. As a first test, click on the following sequence of keys:

"8," "3," "5," "1," "Enter"   "6," "2," "2," "0," "Enter"

Unless something totally unforeseen occurs, the program will subtract the second number from the first and present the result on the main display, which will now show the following:

8351 6220 = 2131

The program now waits for you to click any key, at which point it will clear the main display and wait for you to enter another two numbers. Experiment with different values to make sure that this routine works as expected.

In order to test the error-checking mechanism, clear the display and enter the following sequence of keys:

"0," "1," "2," "3," "Enter"   "0," "1," "2," "3," "Enter"

This should result in the following characters appearing on the main display:

0123 0123 = 0000

The result, 0000, is the smallest value we can hold in a 2-byte (16-bit) unsigned field. Now, clear the display once more and enter the following sequence of keys:

"0," "1," "2," "3," "Enter"   "0," "1," "2," "4," "Enter"

The result (minus 1) is too small to fit in our 2-byte field, so the C (carry) flag will be cleared to 0, the error check should call the DISPERR subroutine, and the following message should appear on the main display:

ERROR: Borrow = 0

Lab 4d
Creating a 16-bit NEGATE Subroutine

Objective: To create a general-purpose subroutine that can negate a 2-byte (16-bit) signed binary number (that is, convert a positive value into its negative counterpart, and vice versa).

Duration: Approximately 30 minutes.

The 16-bit _NEG Subroutine Itself

The easiest way to negate a value is to subtract it from zero, so that's the approach we will take with our _NEG subroutine as shown below:

```
##########################################################################
# Name:     _NEG                                                         #
#                                                                        #
# Function: Negates a 16-bit signed binary value (changes a positive     #
#           value into its negative equivalent and vice versa) and       #
#           returns a 16-bit result in the range -32,767 through         #
#           +32,767. Note that this routine assumes that it will         #
#           not be presented with a value of -32,768, and therefore      #
#           does not perform an error check for this input value)        #
#                                                                        #
# Entry:    Top of stack                                                 #
#           Most-significant byte of return address                      #
#           Least-significant byte of return address                     #
#           Most-significant byte of number to be negated                #
#           Least-significant byte of number to be negated               #
#                                                                        #
# Exit:     Top of stack                                                 #
#           Most-significant byte of result                              #
#           Least-significant byte of result                             #
#                                                                        #
# Modifies: Accumulator                                                  #
#                                                                        #
```

```
# Size:      Program = 37 bytes                                         #
#            Data    =  4 bytes                                         #
########################################################################

########## Get return address from stack and store it
_NEG:        POPA                   # Retrieve MS byte of return
             STA      [_NG_RADD]    #   address from stack and store it
             POPA                   # Retrieve LS byte of return
             STA      [_NG_RADD+1]  #   address from stack and store it

########## Get 2-byte number to be negated from stack
_NG_GNUM:    POPA                   # Retrieve MS byte of number to be
             STA      [_NG_NUM]     #   negated from stack and store it
             POPA                   # Retrieve LS byte of number to be
             STA      [_NG_NUM+1]   #   negated from stack and store it

########## Perform the subtraction
_NG_DOIT:    LDA      $00           # Load ACC with zero
             SUB      [_NG_NUM+1]   # Sub LS byte of number from ACC
             PUSHA                  #   and push LS result onto stack
             LDA      $00           # Load ACC zero
             SUBC     [_NG_NUM]     # Sub MS byte of number from ACC w
             PUSHA                  # (borrow) and push MS result to stack

########## Return to main program
_NG_RET:     LDA      [_NG_RADD+1]  # Load ACC with LS byte of return
                                    #   address from temp location and
             PUSHA                  #   push it back onto the stack
             LDA      [_NG_RADD]    # Load ACC with MS byte of return
                                    #   address from temp location and
             PUSHA                  #   push it back onto the stack
             RTS                    # That's it, exit the subroutine

########## Reserve temp locations for this subroutine
_NG_RADD:    .2BYTE                 # Reserve 2-byte temp location for
                                    #   the return address
_NG_NUM:     .2BYTE                 # Reserve 2-byte temp location for
                                    #   the number to be negated
########## This is the end of the _NEG subroutine

# - - - - - - - - - - - - - - - - - - - - - - - - - - - - - - - - - - #
```

In fact, this routine is very similar to the _SUB routine we created in the previous lab, the main differences being: (a) we only have to retrieve and store one 2-byte number from the stack, (b) we subtract this number from $0000, and (c) we don't perform any error checking.

One key point associated with this routine is that a 2-byte (16-bit) unsigned binary number can be used to represent values in the range

−32,768 to +32,767. However, this means that is isn't possible to negate a value of −32,768, so this routine assumes that it will only ever be asked to negate values in the range −32,767 to +32,767.

Creating and Debugging the _NEG Routine

Invoke the assembler, enter the _NEG subroutine as shown above, and save it in a file called *int-neg-2-byte.asm*.

In order to perform preliminary testing, we need to "wrap" this routine with an .ORG statement to kick everything off and an .END statement to terminate things as follows:

```
    .ORG      $4000

{The _NEG subroutine goes here}

    .END
```

Now, click the assemble button and track down and debug any errors until you receive a "Your file was assembled without any errors" message. Once you reach this point, strip out the "wrapping statements" we added above and use the assembler's **File > Save** command to save this file one last time.

Modifying the Testbench to Call the _NEG Routine

Use the assembler's **File > Open** command to open up the *skeleton-lab4.asm* file you created in Lab 4a, and then use the **File > Save As** command to save this out as *lab4d.asm*.

Modify the body of the testbench program to delete one of our calls to the GETNUM subroutine and to add a call the _NEG routine as follows (these changes are highlighted in gray):

```
MAINLOOP:   JSR      [GETNUM]      # Get a 16-bit number
            JSR      [GETNUM]      # Get 2nd 16-but number (NUMB)
            JSR      [_NEG]        # Negate the value
            JSR      [DISPNUM]     # Display 16-bit number from stack
```

Now, scroll down the file until the end of the subroutines to find the comment line that reads as follows:

```
########## This is where the math subroutine will go
```

Delete this comment and leave your cursor at the beginning of this line, then use the assembler's **Insert > Insert File** command, locate the *int-neg-2-byte.asm* source file you just created in the ensuing dialog, and click the Open button to insert the contents of that file into this version of the testbench program.

Use the **File > Save** command to save this modified version of the testbench, and then click the assemble button to assemble it.

Using the Testbench to Test the Routine

Click the **On/Off** button on the front panel to power-up the calculator, use the **Memory > Load RAM** command to load the *lab4d.ram* file into the calculator's memory, and click the **Run** button to set the program running.

Once the display has been cleared, the program will loop around, waiting for you to input a single 4-digit hexadecimal number. As a first test, click on the following sequence of keys:

"0," "0," "0," "0," "Enter"

The program will subtract this value from zero, and 0 − 0 = 0, which means that the main display should now show the following:

0000 = 0000

The program now waits for you to click any key, at which point it will clear the main display and wait for you to enter another number. Experiment with different values to make sure that this routine works as expected; for example, clear the display and try the following sequence of keys:

"0," "0," "0," "1," "Enter"

This should result in the following characters appearing on the main display:

0001 = FFFF

This is just what we expect, because $FFFF equates to –1 in the signed binary number format. Similarly, if you clear the display and enter the following sequence of keys,

"F," "F," "F," "E," "Enter"

this should result in the following characters appearing on the main display:

FFFE = 0002

Again, this is what we expect, because $FFFE equates to –2 in the signed binary number format. Don't forget that, should you wish to check the results by hand, all you have to do is generate the *twos complement* of the number to be negated. As we discussed in Chapter 4, the easiest way to do this is to copy the bits up to and including the first 1, and then invert the remaining bits. For example, consider how we would use this technique to generate the twos complement of $FFFE as shown in Figure L4-16.

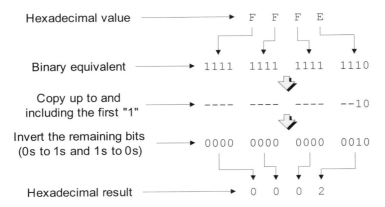

Figure L4-16. Generating twos complement values by hand to check the results.

Lab 4e
Checking for Overflow in the ADD and SUBTRACT Routines

Objective: To modify our general-purpose _ADD and _SUB routines to work with signed binary numbers; that is, to check the O (overflow) flag.

Duration: Approximately 35 minutes.

Testing for –32,768

Based on the discussions of signed binary numbers presented in Chapter 4, we know that a 2-byte (16-bit) signed binary number can be used to represent values in the range –32,768 through +32,767 (Figure L4-17).

The point is that both the _ADD and _SUB routines can easily generate a value of –32,768. For example, if we were to perform the operation $8001 + $FFFF [which equates to –32,767 + (–1) in decimal], the result would be $8000 (that is, –32,768 in decimal).

Now this isn't an issue with the _ADD and _SUB routines per se, but it does present us with something of a problem with regard to the _MULT ("multiply") and _DIV ("divide") routines we are going to create in Labs 4f and 4g, respectively. This is because their core algorithms work only with positive values, which means that the first thing they do is convert any negative quantities into their positive counterparts, perform the multiplication or division operation, and then correct the sign of the result.

The problem, of course, is that if these routines are presented with a value of –32,768, there is no way we can negate it, because the maximum positive value our 2-byte field can support is +32,767. The solution is to make a decision that we will work only with numbers in the range $8001 through $7FFF (that is –32,767 through +32,767 in decimal). In turn, this means that once we've performed an operation such as an ad-

	Binary			Hex	Decimal
0000	0000	0000	0000	0000	+0
0000	0000	0000	0001	0001	+1
0000	0000	0000	0010	0002	+2
	:			:	:
0111	1111	1111	1110	7FFE	+32,766
0111	1111	1111	1111	7FFF	+32,767
1000	0000	0000	0000	8000	-32,768
1000	0000	0000	0001	8001	-32,767
1000	0000	0000	0010	8002	-32,766
	:			:	:
1111	1111	1111	1101	FFFD	-3
1111	1111	1111	1110	FFFE	-2
1111	1111	1111	1111	FFFF	-1

Figure L4-17. Signed binary numbers in a 2-byte field.

dition or subtraction using our _ADD or _SUB routines, respectively, we need to check to ensure that the result is *not* $8000 (−32,768 in decimal).

Devising a subroutine to check for −32,768

If we wished, we could add tests for −32,768 directly into the _ADD and _SUB routines themselves. As an alternative, we can create a special CHK2NEG ("check if too negative") subroutine to perform this task for us as shown below:

```
########## Start of CHK2NEG subroutine
########## Retrieve a 16-bit binary value from the stack and test it.
########## If -32,768, call DISPERR, else return number intact.

########## Pop the return address off the stack and save it
CHK2NEG:    POPA                    # Pop the MS byte of return address
            STA     [_CN_RADD]      #   off the stack and store it
            POPA                    # Pop the LS byte of return address
            STA     [_CN_RADD+1] #   off the stack and store it

########## Pop the number to be tested off the stack
_CN_GNUM:   POPA                    # Pop MS byte of number off the
            STA     [_CN_NUM]       #   stack and store it
            POPA                    # Pop LS byte of number off the
            STA     [_CN_NUM+1] #   stack and store it

########## Push a copy of the number back onto the stack
_CN_PNUM:   PUSHA                   # Push LS byte of number onto stack
            LDA     [_CN_NUM]       # Retrieve MS byte of number
            PUSHA                   #   and push it onto the stack
```

```
########## Check the MS byte to see if it's equal to $80
_CN_CKMS:   CMPA     $80              # Compare contents of ACC to $80
            JNZ      [_CN_RET]        # If not equal we're OK so return

########## Check the LS byte to see if it's equal to $00
_CN_CKLS:   LDA      [_CN_NUM+1]      # Load ACC with LS byte of number
            CMPA     $00              # Compare contents of ACC to $00
            JNZ      [_CN_RET]        # If not equal we're OK so return

########## Call the error message display subroutine
_CN_DERR:   BLDX     MSG_009          # Load X reg with addr of message
            JSR      [DISPERR]        # Jump to display error subroutine
                                      # (which terminates the program)

########## Return gracefully from this subroutine
_CN_RET:    LDA      [_CN_RADD+1]     # Get LS byte of return address from
            PUSHA                     #   temp location and push onto stack
            LDA      [_CN_RADD]       # Get MS byte of return address from
            PUSHA                     #   temp location and push onto stack
            RTS                       # Return from this subroutine

########## Reserve temp locations for this subroutine
_CN_RADD:   .2BYTE                    # 2-byte temp location used to store
                                      #   the return address for this routine

_CN_NUM:    .2BYTE                    # 2-byte temp location to store
                                      #   the number we're checking

########## End of CHK2NEG subroutine

# - - - - - - - - - - - - - - - - - - - - - - - - - - - - - - - #
```

As we see, there is nothing too surprising here. The first thing we do when we enter this subroutine at label CHK2NEG is to pop the return address off the stack and store it in a temporary location.

Next, at label _CN_GNUM ("get number") we pop the 2-byte number to be tested off the stack and store it in a temporary location. This leaves the least-significant (LS) byte in the accumulator, which is just where we want it when we arrive at label _GN_PNUM ("put number"). This is where we push a copy of the 2-byte number back onto the stack, because if the number passes our tests, then we want it to be available to whichever part of the program requested this check when we eventually return from this subroutine.

The act of pushing a copy of the number back onto the stack leaves a copy of the MS byte of the number in the accumulator. By some strange quirk of fate, this is just what we want when we arrive at label _CN_CKMS ("check MS byte"), at which point we compare the value in the accumulator to $80. If the values are *not* equal (in which case the Z (zero) flag will be cleared to its "false" state), we know there isn't a problem, so we use the JNZ ("jump if not zero") instruction to immediately jump to the _CN_RET ("return") label.

Alternatively, if the MS byte of the number does equal $80, then we potentially have a problem depending on whether or not the value in the LS byte equals $00. Thus, at label _CN_CKLS ("check LS byte"), we load a copy of the LS byte into the accumulator and compare it to $00. Once again, if the values are *not* equal (in which case the Z (zero) flag will be cleared to its "false" state), we know there isn't a problem, so we use the JNZ ("jump if not zero") instruction to jump to the _CN_RET ("return") label. Otherwise, we fall through to label _CN_DERR ("display error"), at which point we load the index register (X) with the address of an appropriate error message and jump to the DISPERR ("display error") subroutine, which will display the error message and terminate the program.

And, of course, if there is no error, then when we arrive at label _CN_RET ("return"), we push a copy of the return address back onto the stack and then use an RTS ("return from subroutine") instruction to take us back to the main program.

Creating and debugging the routine

Invoke the assembler, enter the CHK2NEG subroutine as shown above, and save it in a file called *int-check-32768.asm*.

As before, in order to perform preliminary testing on this routine, we need a .ORG statement to kick everything off, a .END statement to terminate things, and two labels (MSG_009 and DISPERR) to satisfy the external references made by the CHK2NEG routine. Thus, we need to "wrap" our routine with a couple of statements as follows:

```
          .ORG      $4000

{The CHK2NEG subroutine goes here}

MSG_009:  .BYTE
DISPERR:  RTS
          .END
```

Now, try clicking the assemble button and track down and debug any errors until you receive a "Your file was assembled without any errors" message. Once you reach this point, strip out the "wrapping statements" we added above and use the assembler's **File** > **Save** command to save this file one last time.

Modifying the _ADD Routine

Use the assembler's **File > Open** command to open up the *int-add-2-byte-v1.asm* file you created in Lab 4b, and then use the **File > Save As** command to save this as *int-add-2-byte-v2.asm*.

Now, scroll down through this routine until you reach the _AD_CHK label, which is where we originally checked the carry flag to see if there was a carry-out from the addition of the two MS bytes. Of course, that was when we were considering the values to be added as *unsigned* binary numbers. Now that we are considering the values to be *signed* binary numbers, we need to modify this test to check the overflow flag, and also to check for a negative result of −32,768, as discussed in the previous section. So, make the changes shown below and save this file for future use:

```
########## Make sure there isn't a carry out from the MS addition
_AD_CHK:    JNC       [_AD_RET]     # If carry flag = 0 jump to return
            BLDX      MSG_001       # Load X reg with addr of message
            JSR       [DISPERR]     # Jump to display error subroutine
                                    # (which terminates the program)

########## Make sure there isn't an overflow from the MS addition
_AD_CHKO:   JNO       [_AD_CHKN]    # If no overflow jump to next test
            BLDX      MSG_003       # Load X reg with addr of message
            JSR       [DISPERR]     # Jump to display error subroutine
                                    # (which terminates the program)

########## Call the CHK2NEG subroutine to test for -32,768
_AD_CHKN:   JSR       [CHK2NEG]     # Call the CHK2NEG subroutine
```

Observe that we've deleted nine bytes from our original code and added twelve bytes back in. Now, consider the following comment in the header at the beginning of the _ADD routine:

```
Size: Program = 50 bytes
```

Based on the changes discussed above, you need to change the value associated with this comment from 50 to 53 bytes.

Testing the Modified _ADD Routine

Use the assembler's **File > Open** command to open up the *lab4b.asm* file you created in Lab 4b, and then use the **File > Save As** command to save this as *lab4e-add.asm*.

Scroll down the file until you find the start of the original _ADD subroutine, delete this routine in its entirety, and leave the cursor where the routine used to be.

Now use the **Insert > Insert File** command to insert the *int-add-2-byte-v2.asm* file that contains the new version of our _ADD subroutine. Then use the **Insert > Insert File** command once more to insert the *int-check-32768.asm* file containing the CHK2NEG subroutine we created at the beginning of this lab.

Click the assemble button. Hopefully, there won't be any errors, but if there are, track them down and debug them until you receive a "Your file was assembled without any errors" message.

Click the **On/Off** button on the front panel to power-up the calculator, use the **Memory > Load RAM** command to load the *lab4e-add.ram* file into the calculator's memory, and click the **Run** button to set the program running.

After clearing the display, the program loops around, waiting for you to input your first number. As a first test, let's try adding two small numbers together, so click on the following sequence of keys:

"0," "0," "0," "6," "Enter"   "0," "0," "0," "2," "Enter"

The main display should now show the following:

0006 0002 = 0008

This shouldn't be too surprising to us, because 6 + 2 does indeed equal 8. Now let's see what happens if we add a small negative number to a larger positive one as follows. Click any key to clear the display, and then click on the following sequence of keys:

"0," "0," "0," "6," "Enter"   "F," "F," "F," "E," "Enter"

As $FFFE equates to –2 when we're dealing with signed binary numbers, the main display should now show the following:

0006 FFFE = 0004

Again, this is what we expect, because 6 + (–2) does equal 4. Let's get a little more daring, and try adding a larger negative number to a smaller positive one. Click any key to clear the display, and then click on the following sequence of keys:

"0," "0," "0," "6," "Enter"   "F," "F," "F," "8," "Enter"

As $FFF8 equates to –8, the main display should now show the following:

0006 FFF8 = FFFE

As we discussed a little earlier, the result ($FFFE) equates to –2, which is what we expect because 6 + (–8) does equal –2. This means that we can change the sign of the result without triggering the overflow flag if we perform a legitimate operation. Now let's see what happens if the sign changes as a result of an illegitimate operation. Click any key to clear the display, and then click on the following sequence of keys:

"7," "F," "F," "F," "Enter"   "0," "0," "0," "2," "Enter"

In this case, the $7FFF equates to 32,767, which is the largest positive value we can hold in our 2-byte signed binary number. Adding 2 to this will give a result of $8001, which equates to –32,767, and which will trigger an overflow condition. When our test in the _ADD subroutine detects that the O (overflow) flag has been set, it will call the DISPERR subroutine, which will display the following error message and terminate the program:

ERROR: Overflow

As a final test, rerun the program by clicking the **Run** button and then click on the following sequence of keys:

"8," "0," "0," "1," "Enter"   "F," "F," "F," "F," "Enter"

As we discussed at the beginning of this chapter, $8001 + $FFFF = $8000 [which equates to –32,767 + (–1) = –32,768 in decimal]. This is, of course, a perfectly legitimate operation, so the overflow flag won't be triggered. However, the _ADD subroutine calls the CHK2NEG subroutine, and when *it* detects this value it will call the DISPERR subroutine, which will display the following error message and terminate the program:

ERROR: –32,768

Modifying the _SUB Routine

Use the assembler's **File** > **Open** command to open up the *int-sub-2-byte-v1.asm* file you created in Lab 4c, and then use the **File** > **Save As** command to save this as *int-sub-2-byte-v2.asm*.

Now, scroll down through this routine until you reach the _SB_CHK label, which is where we originally checked the carry flag to see if there was a borrow-out from the subtraction of the two MS bytes. Once again, that was when we were considering the values to be added as *unsigned* binary numbers. Now that we are considering the values to be *signed* binary numbers, we need to modify this test to check the overflow flag, and also to check for a negative result of –32,768. So, make the changes shown below and save this file for future use:

```
########## Make sure there isn't a borrow out from the MS subtraction
_SB_CHK:    JC      [_SB_RET]    # If carry flag = 1 jump to return
            BLDX    [MSG_002]    # Load X reg with addr of message
            JSR     [DISPERR]    # Jump to display error subroutine
                                 # (which terminates the program)

########## Make sure there isn't an overflow from the MS subtraction
_SB_CHKO:   JNO     [_SB_CHKN]   # If no overflow jump to next test
            BLDX    MSG_003      # Load X reg with addr of message
            JSR     [DISPERR]    # Jump to display error subroutine
                                 # (which terminates the program)

########## Call the CHK2NEG subroutine to test for -32,768
_SB_CHKN:   JSR     [CHK2NEG]    # Call the CHK2NEG subroutine
```

As before, observe that we've deleted nine bytes from our original code and added twelve bytes back in. This means that you need to change the value associated with the Size: Program = 50 bytes comment in the header for the _SUB routine from 50 to 53 bytes.

Testing the Modified _SUB Routine

Use the assembler's **File > Open** command to open up the *lab4c.asm* file you created in Lab 4c, and then use the **File > Save As** command to save this as *lab4e-sub.asm*.

Scroll down the file until you find the start of the original _SUB subroutine, delete this routine in its entirety, and leave the cursor where the routine used to be.

Now use the **Insert > Insert File** command to insert the *int-sub-2-byte-v2.asm* file that contains the new version of our _SUB subroutine. Then use the **Insert > Insert File** command once more to insert the *int-check-32768.asm* file containing the CHK2NEG subroutine we created at the beginning of this lab.

Click the assemble button. As usual, we hope that there won't be any errors, but if there are, track them down and debug them until you receive a "Your file was assembled without any errors" message.

Click the **On/Off** button on the front panel to power-up the calculator, use the **Memory > Load RAM** command to load the *lab4e-sub.ram* file into the calculator's memory, and click the **Run** button to set the program running.

As soon as it's cleared the display, the program loops around, waiting for you to input your first number. As a first test, let's try subtracting a small positive number from a larger one, so click on the following sequence of keys:

"0," "0," "0," "6," "Enter"   "0," "0," "0," "2," "Enter"

The main display should now show the following:

0006 0002 = 0004

And, of course, this is what we expect, because 6 – 2 does equal 4. Now let's see what happens if we subtract a negative number from a positive one as follows. Click any key to clear the display, and then click on the following sequence of keys:

"0," "0," "0," "6," "Enter"   "F," "F," "F," "E," "Enter"

As $FFFE equates to –2 when we're dealing with signed binary numbers, the main display should now show the following:

0006 FFFE = 0008

Again, this is what we expect, because 6 – (–2) equates to 6 + 2, which of course returns a result of 8. Let's get a little more daring, and try subtracting a bigger number from a smaller one. Click any key to clear the display, and then click on the following sequence of keys:

"0," "0," "0," "6," "Enter"   "0," "0," "0," "8," "Enter"

Of course 6 – 8 = –2, and as the 2-byte hexadecimal equivalent to –2 is $FFFE, the main display should now show the following:

0006 0008 = FFFE

As with the _ADD function, this means that our _SUB routine can change the sign of the result without triggering the overflow flag if we perform a legitimate operation. Now let's see what happens if the sign changes as a result of an illegitimate operation. Click any key to clear the display, and then click on the following sequence of keys:

"8," "0," "0," "1," "Enter"   "0," "0," "0," "2," "Enter"

As we know, the $8001 equates to –32,767. If we now attempt to subtract 2 from this value, then as opposed to a result of –32,769 (which cannot be represented in a 2-byte field), we'll actually end up with a result of $7FFF. This equates to +32,767, and the fact that we've performed an illegitimate operation will trigger an overflow condition. When our test in the _SUB routine detects that the O (overflow) flag has been set, it will call the DISPERR subroutine, which will display the following error message, and terminate the program:

ERROR: Overflow

As a final test, rerun the program by clicking the **Run** button and then click on the following sequence of keys:

"8," "0," "0," "1," "Enter"   "0," "0," "0," "1," "Enter"

In this case, $8001 – $0001 = $8000 (which equates to –32,767 – 1 = –32,768 in decimal). In this case, the fact this is a perfectly legitimate operation means that the overflow flag won't be triggered. However, the _SUB routine also calls the CHK2NEG subroutine, and when *it* detects this value it will call the DISPERR subroutine, which will display the following error message and terminate the program:

ERROR: –32,768

Lab 4f

Creating a 16-bit MULTIPLY Subroutine

Objective: To create a general-purpose subroutine that can multiply two 2-byte (16-bit) signed binary number together and return a 2-byte result.

Duration: Approximately 50 minutes.

A Little Thought Experiment

In order to wrap our brains around the way in which our multiplication routine works, we'll first perform a little thought experiment. Let's assume we are working with a really simple computer from the days of yore that is based on a 4-bit data bus and 4-bit memory locations. If we are playing with signed binary numbers, then a 4-bit field (known as a *nybble*) can be used to represent sixteen values in the range –8 through +7 (Figure L4-18).

Now let's assume that we wish to use the shift-and-add technique discussed in Chapter 4 to perform the multiplication operation –3 × +2 = –6. Initially, we can visualize the way in which this takes place as illustrated in Figure L4-19.

As we noted in Chapter 4, the shift-and-add technique works only with positive values, so the first thing we have to do is to negate the negative multiplier (by taking its twos complement) and generate its positive counterpart. (If the multiplicand had been negative, we would have had to invert this also.)

Next, we generate the four partial products. If there is a 0 in the multiplier, the corresponding partial product is all 0s; alternatively, if there is a 1 in the multiplier, the corresponding partial product is a copy of the multiplicand.

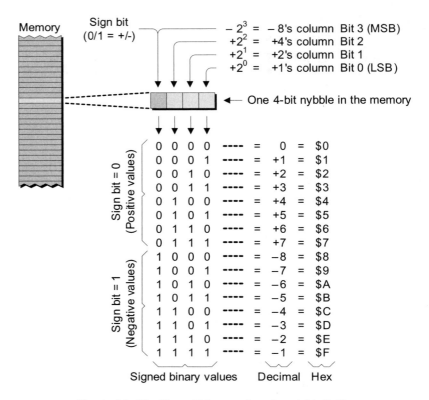

Figure L4-18. Signed binary values in a 4-bit field.

All of the partial products are now added together to give an 8-bit (2-nybble) result. However, due to the fact that we have made the decision to work only with 4-bit values, we need to discard the most-significant nybble (it's our responsibility to check that the result of the multiplication is small enough to fit in a 4-bit field).

Finally, since this particular example involved multiplying a positive value by a negative one, we have to correct the sign by negating the result.

Some cunning tricks

Now let's consider the actions shown in Figure L4-19 in a little more detail. With the exception of the first partial product, the rest are split across two 4-bit nybbles. This means that, if we performed the multiplication as implied in this illustration, we'd have to split the multiplicand

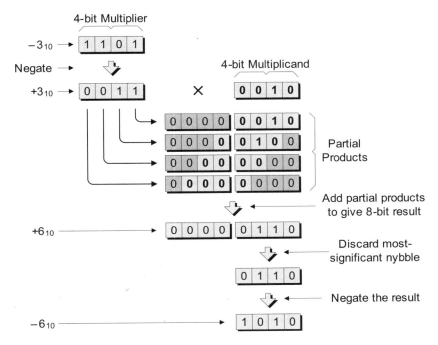

Figure L4-19. Visualizing a simple 4-bit multiplication.

in a different place for each partial product, shift the two resulting fragments by different amounts, and perform two separate additions. We could certainly use this technique, but we would ideally prefer a method that could achieve the same effect with fewer instructions and less hassle.

Fortunately, there are some cunning tricks we can play (that's the really cool thing about working at the assembly language level—there are always interesting ways of doing things). For example, let's assume that we start off by reserving a 4-bit nybble for our multiplicand and two 4-bit nybbles in which to store the 8-bit result from the multiplication. Let's also assume that we initialize the *most-significant* (*MS*) of these result nybbles with 0s, that we initialize the *least-significant* (*LS*) result nybble with a copy of the multiplier (already converted into a positive value), and that we ensure that the C (carry) flag is loaded with 0 [Figure L4-20(a)].

Now we come to perform the first cycle of the multiplication, which involves rotating the entire 2-nybble result field one bit to the right [Figure L4-20(b)]. (Note that the use of bold text and shaded areas in these illustrations is intended only as an aid to understanding the se-

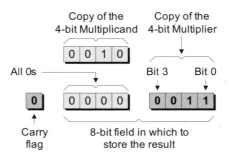

(a) Initial conditions at start of first cycle

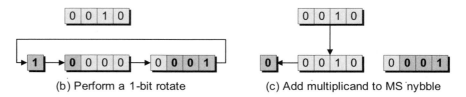

(b) Perform a 1-bit rotate   (c) Add multiplicand to MS nybble

Figure L4-20. First cycle of 4-bit multiplication test case.

quence of events.) Observe that the 0 that used to be in the carry flag [from Figure L4-20(a)] is copied into the MS bit of the MS nybble, the bit that "falls off the end" of the MS nybble is copied into the MS bit of the LS nybble, and the bit that "falls off the end" of the LS nybble is copied back into the carry flag.

This means that the carry flag now contains a copy of bit 0 from the original multiplier. We next perform a test on the carry flag to see if it contains a 0 or a 1. As it contains a 1 in this case, we add a copy of the multiplicand to the current contents of the MS nybble [Figure L4-20(c)]. As a by-product of this addition, the carry flag is reloaded with 0.

Similarly, in the case of the second cycle, we again rotate the 2-nybble result one bit to the right [Figure L4-21(b)]. As usual, the 0 that used to be in the carry flag [from Figure L4-21(a)] is copied into the MS bit of the MS nybble, the bit that "falls off the end" of the MS nybble is copied into the MS bit of the LS nybble, and the bit that "falls off the end" of the LS nybble is copied back into the carry flag.

This means that the carry flag now contains a copy of bit 1 from the original multiplier. Once again, when we perform a test on the carry flag, it contains a 1 in this case, so we add a copy of the multiplicand to the

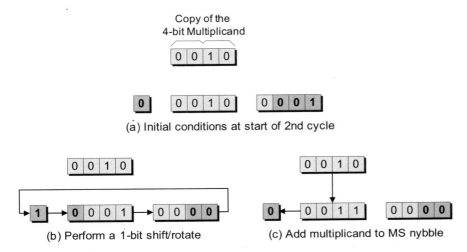

Figure L4-21. Second cycle of 4-bit multiplication test case.

current contents of the MS nybble [Figure L4-21(c)]. And, as before, the carry flag is loaded with 0 as a by-product of this addition.

When we come to the third cycle of our test case, we commence by rotating the 2-nybble result one bit to the right as usual [Figure L4-22(b)]. In this case, the bit that "falls off the end" of the LS nybble into the carry flag—a copy of bit 2 from the original multiplier—is a 0. This

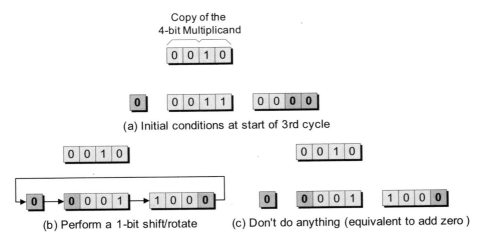

Figure L4-22. Third cycle of 4-bit multiplication test case.

Copy of the
4-bit Multiplicand

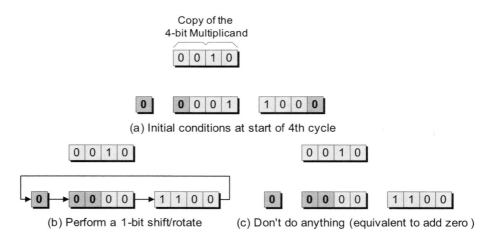

(a) Initial conditions at start of 4th cycle

(b) Perform a 1-bit shift/rotate   (c) Don't do anything (equivalent to add zero)

Figure L4-23. Fourth cycle of 4-bit multiplication test case.

means that we don't actually have to do anything, which is the same as adding zero to the MS nybble [Figure L4-22(c)].

Similarly, with regard to the fourth and final cycle of our test case, when we rotate the 2-nybble result one bit to the right, the bit that "falls off the end" of the LS nybble into the carry flag—a copy of bit 3 from the original multiplier—is a 0 [Figure L4-23(b)]. Once again, this means that we don't have to do anything, which is the same as adding zero to the MS nybble [Figure L4-23(c)].

Last, but not least, we have to perform a final 1-bit rotate so as to leave the penultimate result in the LS nybble (Figure L4-24).

The reason we refer to this as the penultimate result is that, as per the discussions associated with Figure L4-19, we now need to discard the

Copy of the
4-bit Multiplicand

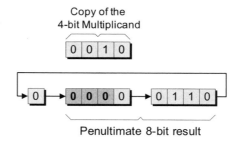

Penultimate 8-bit result

Figure L4-24. Perform one final shift.

MS nybble of the result. Also, due to the fact that our example involved multiplying a positive value by a negative one, we have to correct the sign by negating (taking the twos complement of) the result.

The 16-bit _MULT Subroutine Itself

Of course, when it comes to our real _MULT subroutine, we intend to multiply two 2-byte (16-bit) numbers together. Just to provide a point of reference, the names of the main storage locations used in this routine are as shown in Figure L4-25.

```
###################################################################
# Name:     _MULT                                                 #
#                                                                 #
# Function: Multiply two 16-bit signed binary numbers together and #
#           return a 16-bit result in the range -32,767 to +32,767 #
#                                                                 #
# Entry:    Top of stack                                          #
#           Most-significant byte of return address               #
#           Least-significant byte of return address              #
#           Most-significant byte of second number  (multiplicand) #
#           Least-significant byte of second number (multiplicand) #
#           Most-significant byte of first number   (multiplier)  #
#           Least-significant byte of first number  (multiplier)  #
#                                                                 #
# Exit:     Top of stack                                          #
#           Most-significant byte of result     (product)         #
#           Least-significant byte of result    (product)         #
#                                                                 #
# Modifies: Accumulator and Index Register                        #
#                                                                 #
# Size:     Program = 211 bytes                                   #
#           Data    =   9 bytes                                   #
###################################################################

########## Get return address from stack and store it
_MULT:     POPA                 # Retrieve MS byte of return
           STA    [_ML_RADD+0] #   address from stack and store it
           POPA                 # Retrieve LS byte of return
           STA    [_ML_RADD+1] #   address from stack and store it

########## Get multiplicand from stack and store it
_ML_GMAN:  POPA                 # Retrieve MS byte of multiplicand
           STA    [_ML_MAND+0] #   from stack and store it
           POPA                 # Retrieve LS byte of multiplicand
```

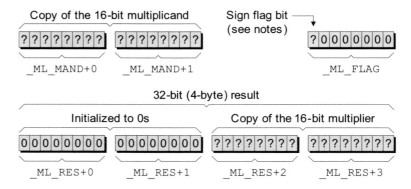

Figure L4-25. Names of primary storage locations in the real subroutine.

```
            STA       [_ML_MAND+1] #   from stack and store it

########## Get multiplier from stack and store it
_ML_GMUL:   LDA       $00          # Load ACC with 0 and save it to
            STA       [_ML_RES+0]  #   the two MS bytes of the
            STA       [_ML_RES+1]  #   4-byte result
            POPA                   # Retrieve MS byte of multiplier
            STA       [_ML_RES+2]  #   from stack and store it
            POPA                   # Retrieve LS byte of multiplier
            STA       [_ML_RES+3]  #   from stack and store it

########## Invert multiplicand if necessary (store sign in flag)
_ML_IMAN:   LDA       [_ML_MAND+0] # Load ACC with MS multiplicand
            STA       [_ML_FLAG]   #   and store a copy in the flag
            JNN       [_ML_IMUL]   # If positive value, jump to next test
                                   #   ... otherwise ...
            LDA       $00          # Load ACC with 0
            SUB       [_ML_MAND+1] #   Subtract (no carry) LS byte of
            STA       [_ML_MAND+1] #   multiplicand and store the result
            LDA       $00          # Load ACC with 0
            SUBC      [_ML_MAND+0] #   Subtract (with carry) MS byte of
            STA       [_ML_MAND+0] #   multiplicand and store the result

########## Invert multiplier if necessary (XOR sign in flag)
_ML_IMUL:   LDA       [_ML_FLAG]   # Load the flag
            XOR       [_ML_RES+2]  #   XOR it with MS byte of multiplier
            AND       %10000000    #   Clear bits (except sign) to 0
            STA       [_ML_FLAG]   #   Store the flag away again
            LDA       [_ML_RES+2]  # Load ACC with MS multiplier
            JNN       [_ML_DUMY]   # If positive value, jump to next part
                                   #   ... otherwise ...
```

```
          LDA       $00              # Load ACC with 0
          SUB       [_ML_RES+3]      #   Subtract (no carry) LS byte of
          STA       [_ML_RES+3]      #   multiplier and store the result
          LDA       $00              # Load ACC with 0
          SUBC      [_ML_RES+2]      #   Subtract (with carry) MS byte of
          STA       [_ML_RES+2]      #   multiplier and store the result

########## Dummy instruction to ensure the carry flag contains 0
_ML_DUMY: ADD       $00              # Add 0 to ACC (will clear C to 0)

########## Hold tight—this is the start of the multiplication loop
          BLDX      17               # Load index reg with the number of
                                     # times we want to go around loop + 1
_ML_DOIT: JNC       [_ML_ROT]        # If carry = 0, jump to next rotate
          LDA       [_ML_RES+1]      #   otherwise add the 16-bit
          ADD       [_ML_MAND+1]     #   multiplicand to the MS 16-bits
          STA       [_ML_RES+1]      #   of the result
          LDA       [_ML_RES+0]      #   :
          ADDC      [_ML_MAND+0]     #   :
          STA       [_ML_RES+0]      #   :

_ML_ROT:  LDA       [_ML_RES+0]      # Rotate the 4-byte (32-bit) result
          RORC                       #   one bit to the right. Start with
          STA       [_ML_RES+0]      #   the MS byte: _ML_RES+0
          LDA       [_ML_RES+1]      # Now do _ML_RES+1
          RORC                       #   :
          STA       [_ML_RES+1]      #   :
          LDA       [_ML_RES+2]      # Now do _ML_RES+2
          RORC                       #   :
          STA       [_ML_RES+2]      #   :
          LDA       [_ML_RES+3]      # Now do _ML_RES+3
          RORC                       #   :
          STA       [_ML_RES+3]      #   :

_ML_ELUP: DECX                       # Decrement the index register
          JNZ       [_ML_DOIT]       # If not zero, then do another loop

########## Check that the result is less than/equal to 32,767
_ML_CHK:  LDA       [_ML_RES+0]      # Load MS byte of result
          JNZ       [_ML_ERR]        #   If not zero, jump to display error
          LDA       [_ML_RES+1]      # Load next byte of result
          JNZ       [_ML_ERR]        #   If not zero, jump to display error
          LDA       [_ML_RES+2]      # Load next byte of result
          JNN       [_ML_SIGN]       #   Jump if MS bit = 0

_ML_ERR:  BLDX      MSG_005          # Load X reg with addr of message
          JSR       [DISPERR]        # Jump to display error subroutine
                                     # (which terminates the program)
```

```
########## Check the flag to see if we have to negate the result
_ML_SIGN:   LDA     [_ML_FLAG]     # Load the flag byte
            JNN     [_ML_SAVE]     # Jump if MS bit = 0
            LDA     $00            #   Otherwise invert LS two bytes
            SUB     [_ML_RES+3]    #   of the result by subtracting
            STA     [_ML_RES+3]    #   them from zero
            LDA     $00            #   :
            SUBC    [_ML_RES+2]    #   :
            STA     [_ML_RES+2]    #   :

########## Save the LS two bytes of the 4-byte result on the stack
_ML_SAVE:   LDA     [_ML_RES+3]    # Load ACC with LS byte of 2-byte
            PUSHA                  #   result and push it onto the stack
            LDA     [_ML_RES+2]    # Load ACC with MS byte of 2-byte
            PUSHA                  #   result and push it onto the stack

########## Return to main program
_ML_RET:    LDA     [_ML_RADD+1]   # Load ACC with LS byte of return
                                   #   address from temp location and
            PUSHA                  #   push it back onto the stack
            LDA     [_ML_RADD]     # Load ACC with MS byte of return
                                   #   address from temp location and
            PUSHA                  #   push it back onto the stack
            RTS                    # That's it, exit the subroutine

########## Reserve temp locations for this subroutine
_ML_RADD:   .2BYTE                 # Reserve 2-byte temp location for
                                   #   the return address
_ML_MAND:   .2BYTE                 # Reserve 2-byte temp location for
                                   #   the multiplicand
_ML_RES:    .4BYTE                 # Reserve 4-byte temp location for
                                   #   the result (product)
_ML_FLAG:   .BYTE                  # Reserve 1-byte to use as a flag
########## This is the end of the _MULT subroutine

# - - - - - - - - - - - - - - - - - - - - - - - - - - - - - - - - - #
```

Don't panic! Just say "There's nothing to fear but fear itself!" to yourself a few times until your heart stops pounding quite so fast. It's true that this routine is a tad bigger than anything we've tackled thus far, but it's really not all that complicated when you start delving into it.

Retrieving and storing the return address

First, at label _MULT, we perform the ritual of popping the return address off the stack and storing it in a temporary 2-byte location called _ML_RADD.

Retrieving and storing the multiplicand

Next, at label _ML_GMAN ("get multiplicand"), we pop the multiplicand off the stack and store it in a temporary 2-byte location called _ML_MAND.

Retrieving and storing the multiplier

When we come to label _ML_GMUL ("get multiplier"), we first load the two MS bytes of our 4-byte result (called _ML_RES) with zeros, and then we pop the multiplier off the stack and store it in the two LS bytes of our 4-byte result.

Inverting the multiplicand (if necessary)

Next, at label _ML_IMAN ("invert multiplicand"), we load the accumulator with the MS byte of the multiplicand and we store a copy of this byte in a temporary location called _MS_FLAG. We then use a JNN ("jump if not negative") instruction to test the MS bit of this MS byte to determine whether it's a 0 or a 1. If this bit is 0, the value is already positive, so we immediately jump to label _ML_IMUL. Otherwise, if this bit is 1, the multiplicand is negative, in which case we subtract it from zero so as to obtain its positive counterpart.

Inverting the multiplier (if necessary)

When we arrive at label _ML_IMUL ("invert multiplier"), we load the accumulator with the MS byte of the multiplier, we XOR this with the contents of our flag, we use an AND instruction to clear the LS seven bits to 0, and we store the result back into the flag. [In fact, due to the way in which we actually use this flag (at label _ML_SIGN), this AND instruction is unnecessary, and serves primarily to remind us that we are interested only in the MS bit of this flag byte.]

The reason we are interested in this bit is that it now contains the XOR of the sign bits from the multiplicand and the multiplier. If you cast your mind back to Chapter 4 and the discussions associated with Figures 4-27 and 4-28, you will recall that a 0 or a 1 in this bit tells us whether the result from our multiplication should be positive or if it needs to be negated, respectively (we'll come back to this point shortly).

Once we've finished playing with the flag, we reload the accumulator with the MS byte of the multiplier and use a JNN ("jump if not negative") to test the MS bit of this MS byte. If this bit is 0, the value is already positive, so we immediately jump to label _ML_DUMY. Otherwise, if this bit is 1, the multiplier is negative, in which case we subtract it from zero so as to obtain its positive counterpart.

Two housekeeping tasks

Before we plunge into the heart of the multiplication, we first need to perform two housekeeping tasks. The ADD at label _ML_DUMY ("dummy") is a dummy instruction whose sole purpose is to ensure that the C (carry) flag contains a 0. This is followed by a BLDX instruction, which we use to load the index register with 17_{10} [the number of times we really wish to go around this loop (16_{10}) plus an extra cycle to ensure that the result is correctly aligned].

The main multiplication loop

The main multiplication loop commences at label _ML_DOIT. First we use a JNC ("jump if not carry") instruction to test the value in the C (carry) flag. If the carry flag contains 0 (which it will do the first time round the loop due to our dummy ADD instruction at label _ML_DUMY), we immediately jump to label _ML_ROT ("rotate"). Alternatively, if the carry flag contains a 1 (which it may well do in future iterations around the loop) we add a copy of our 2-byte multiplicand into the two MS bytes of the result.

When we come to rotate the 4-byte result one bit to the right, we discover the true power of the RORC ("rotate right through the carry flag") instruction that we introduced in Chapter 3. Consider the actions depicted in Figure L4-26.

The values shown in Figure L4-26(a) reflect the initial conditions seen the very first time we go around the loop. Result bytes _ML_RES+0 and _ML_RES+1 contain 0s, wheras result bytes _ML_RES+2 and _ML_RES+2 contain a copy of the multiplier. [As we've already performed any necessary negations, we know that the MS bit of the multiplier will be 0 indicating a positive value; meanwhile, the question mark ("?") characters associated with the remaining multiplier bits indicate that they may be 0s or 1s.]

We commence by loading the MS byte of the result from memory location _ML_RES+0 into the accumulator. We then use an RORC to shift the contents of the accumulator one bit to the right; the original contents of the carry flag are copied into the MS bit of the accumulator and, at the same time, the LS bit that "falls off the end" of the accumulator is loaded

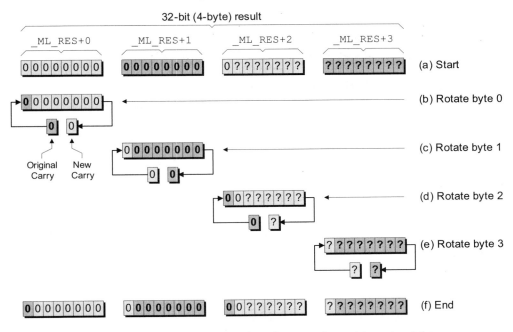

Figure L4-26. Rotating the 4-byte result one bit to the right.

back into the carry flag [Figure L4-26(b)]. We then save the contents of the accumulator back into memory location _ML_RES+0.

Next, we load the contents of memory location _ML_RES+1 into the accumulator and use a new RORC to shift this value one bit to the right. The original contents of the carry flag (which used to be the LS bit from _ML_RES+0) are copied into the MS bit of the accumulator and, at the same time, the LS bit that "falls off the end" of the accumulator is loaded into the carry flag [Figure L4-26(c)]. The contents of the accumulator are then saved back into memory location _ML_RES+1.

And so it goes for the remaining two result bytes [Figures L4-26(d) and L4-26(e)]. After performing all four RORC instructions, the entire 4-byte result has been shifted one bit to the right [Figure L4-26(f)], whereas the LS bit that "fell off the end" of the multiplier ends up in the carry flag.

Once we've performed the rotate, we decrement the index register at label _ML_ELUP ("end of loop?"). As long as the value in the index register is greater than 0, the JNZ ("jump if not zero") instruction will jump back to the beginning of the loop at _ML_DOIT. Of course, the path we take through the loop this time depends on the contents of the carry flag (the bit that "fell off the end" of the multiplier).

Checking the size of the result

When the index register is finally decremented to zero, the JNZ instruction fails, and the program arrives at the _ML_CHK label. This is where we test the result from the multiplication to ensure that it is less than or equal to +32,767.

As shown in Figure L4-27, this means that _ML_RES+0, _ML_RES+0, and the MS bit of _ML_RES+0 must be 0; the remaining bits can be anything from $0000 to $7FFF (which equates to 0 through +32,767 in decimal). If any of these conditions fail, we end up at label _ML_ERR, at which time we load the index register with the start address of the appropriate error message; then we call subroutine DISPERR to display the message and terminate the program.

Correcting the sign of the result

Assuming the result is not too large, we find ourselves at label _ML_SIGN. This is where we test the contents of our flag to see whether its MS bit (which is an XOR of the signs of the original multiplicand and multiplier) is a 0 or a 1. If it's a 1, this means that we have to negate the two LS bytes of the result by subtracting them from 0 (remember that, by this point, we've conceptually "discarded" the two MS bytes).

Saving the result and exiting the subroutine

All that remains now is to push the two LS bytes of the result onto the stack at label _ML_SAVE, to push the return address back onto the stack at label _ML_RET, and to then execute an RTS ("return from subroutine") instruction to return us to the main body of the program.

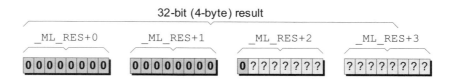

Figure L4-27. Checking that the result is not too large.

Creating and Debugging the _MULT Routine

Invoke the assembler, enter the _MULT subroutine as shown above, and save it in a file called *int-mult-2-byte.asm*.

As for our previous subroutines, in order to perform preliminary testing on this routine, we need a .ORG statement to kick everything off, a .END statement to terminate things, and two labels (MSG_005 and DIS-PERR) to satisfy the external references made by the _MULT routine. Thus, we need to "wrap" our routine with a couple of statements as follows:

```
          .ORG      $4000

{The _MULT subroutine goes here}

MSG_005:  .BYTE
DISPERR:  RTS
          .END
```

Now, try clicking the assemble button and track down and debug any errors until you receive a "Your file was assembled without any errors" message. Once you reach this point, strip out the "wrapping statements" we added above and use the assembler's **File > Save** command to save this file one last time.

Modifying the Testbench to Call the _MULT Routine

Use the assembler's **File > Open** command to open up the *skeleton-lab4.asm* file you created in Lab 4a, and then use the **File > Save As** command to save this as *lab4f.asm*.

Modify the body of the testbench program to call the _MULT routine as follows (the change is highlighted in gray):

```
MAINLOOP:  JSR     [GETNUM]     # Get 1st 16-bit number (NUMA)
           JSR     [GETNUM]     # Get 2nd 16-bit number (NUMB)
           JSR     [_MULT]      # Multiply NUMB and NUMA together
           JSR     [DISPNUM]    # Display 16-bit number from stack
```

Now, scroll down the file until the end of the subroutines to find the comment line that reads as follows:

```
########## This is where the math subroutine will go
```

Delete this comment and leave your cursor at the beginning of this line, then use the assembler's **Insert > Insert File** command, locate the *int-mult-2-byte.asm* source file you just created in the ensuing dialog, and click the **Open** button to insert the contents of that file into this version of the testbench program.

Use the **File > Save** command to save this modified version of the testbench, and then click the assemble button to assemble it.

Using the Testbench to Test the Routine

Click the **On/Off** button on the front panel to power-up the calculator, use the **Memory > Load RAM** command to load your *lab4f.ram* file into the calculator's memory, and click the **Run** button to set the program running.

As usual, once the display has been cleared, the program will loop around, waiting for you to input your first number. As a first test, click on the following sequence of keys:

"0," "0," "0," "8," "Enter"   "0," "0," "0," "8," "Enter"

Now, $8_{10} \times 8_{10} = 64_{10}$, whereas the equivalent in hexadecimal is $8_{16} \times 8_{16} = 40_{16}$. Thus, our program should multiply these two numbers and present the following characters on the calculator's main display:

0008 0008 = 0040

The program now waits for you to click any key, at which point it will clear the main display and wait for you to enter another two numbers. Experiment with different values to make sure that this routine works as expected. Some suggestions are as follows (these are shown in decimal; we'll leave it to you to generate the hexadecimal equivalents):

0×0

0×1

1×0

1×1

1×-1

$$-1 \times 1$$

$$-1 \times -1$$

$$32{,}767 \times 1$$

$$32{,}767 \times -1$$

$$32{,}767 \times 2$$

The last example should, of course, cause the program to issue an error message because the result will be larger than the maximum positive signed integer value we can store in a 2-byte field.

But don't restrict yourself only to the suggestions above, which are primarily intended to test worst-case conditions. Just for giggles and grins, try multiplying a few "interesting" numbers together like $57_{10} \times 42_{10}$, where the former value reflects the last two digits of the year Max (the handsome one of the authors[1]) was born, and the latter is the answer to "Life, the Universe, and Everything" according to *The Hitchhiker's Guide to the Galaxy* by Douglas Adams.

---

[1] By some strange quirk of fate, Max is also the one who is penning these words, whereas Alvin is slaving away at his house on the software.

Lab 4g | Creating a 16-bit DIVIDE Subroutine

Objective: To create a general-purpose subroutine that can divide one 2-byte (16-bit) signed binary number by another and return a 2-byte result.

Duration: Approximately 60 minutes.

Another Little Thought Experiment

As we shall come to see, the way in which we actually perform binary division inside a computer is somewhat tricky to wrap our brains around. It's easy to lose the "thread" here, so let's take things one step at a time. Toward the end of Chapter 4 we showed a simple binary division example, in which we divided 111_2 by 10_2 to give 11_2 with a remainder of 1 (that is, dividing 7_{10} by 2_{10} to give 3_{10} with a remainder of 1 in decimal). The way in which we illustrated this operation is shown in Figure L4-28.

Actually, this is something of a simplification compared to the way in which the computer would carry out the task. In order to see why this should be so, consider the way in which we originally described the first portion of the process in Chapter 4:

> In the case of this particular example, we know that we can't divide 10_2 into 0_2, so we set the first digit in our quotient (result) to 0. Similarly, we know that we can't divide 10_2 into 01_2, so we set the second digit in the quotient to 0. However, we can divide 10_2 into 011_2, so we set the third digit in our quotient to 1.

It all sounds so simple when you say it quickly. For example, when we proclaim things like "we know that we can't divide 10_2 into 01_2," how do

351

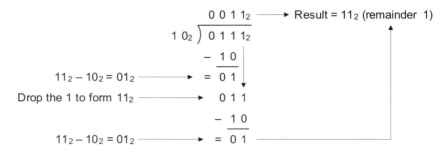

Figure L4-28. Simplified version of performing a long division in binary.

we actually know this? In reality, without even thinking about it we simply looked at the numbers and said to ourselves "10_2 is bigger than 01_2, so we can't divide the former into the latter." But the computer can't do things this way; instead, it has to slog through the calculation one painful step at a time.

As usual, the best way to visualize what's actually happening is by means of a simple thought experiment. Let's once again assume we are working with a really simple computer from the days of yore that is based on a 4-bit data bus and 4-bit memory locations. As we previously discussed, if we are playing with signed binary numbers, then a 4-bit field (known as a *nybble*) can be used to represent sixteen values in the range -8 through $+7$ (Figure L4-29).

Now, let's assume that we wish to perform the division operation $7/2 = 3$ remainder 1; as usual, we are going to employ a few cunning tricks (we can only imagine your surprise).

A few cunning tricks

We start off by reserving a 4-bit nybble for our divisor, and two 4-bit nybbles in which to store the result from the division (one nybble for the quotient and one for the remainder). If the dividend is negative, we must convert this value into its positive counterpart, and similarly for the divisor. Now let's assume that we initialize the *most-significant* (*MS*) result nybble with 0s and that we initialize the *least-significant* (*LS*) result nybble with a copy of the dividend (already converted into a positive value if necessary), as shown in Figure L4-30(a).

The first thing we do is to shift the entire 8-bit result one bit to the left [Figure L4-30(b)]. Note that we are *not* concerned with the contents

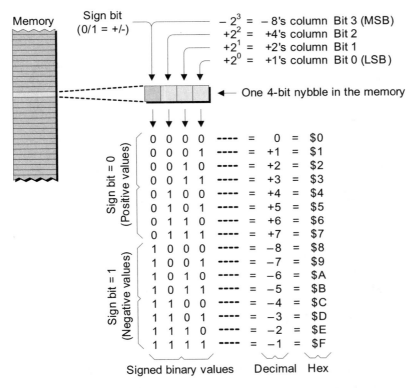

Figure L4-29. Signed binary values in a 4-bit field.

of the carry flag following this operation. Also observe that performing this shift operation automatically causes a 0 to be loaded into the *least-significant* (*LS*) bit of the LS nybble.

Next, we subtract a copy of the divisor from the *most-significant* (*MS*) nybble of the result as shown in Figure L4-30(c) ($0000_2 - 0010_2 = 1110_2$). Now, this is a tricky bit. Although we know that we are actually working with signed binary values, for the purposes of this portion of the algorithm we can "pretend" that they are unsigned quantities. In this case, a 0 in the carry flag (which is acting as a borrow flag in the case of a subtraction) indicates that we just subtracted a larger value from a smaller value. We know that we didn't really wish to do this, so we need to recover from our rash action by adding the divisor back into the MS nybble of the result as shown in Figure L4-30(d) ($1110_2 + 0010_2 = 0000_2$).

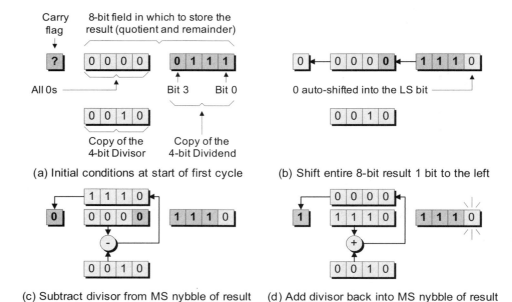

(a) Initial conditions at start of first cycle　　(b) Shift entire 8-bit result 1 bit to the left

(c) Subtract divisor from MS nybble of result　(d) Add divisor back into MS nybble of result

Figure L4-30. First cycle of 4-bit division test case.

In particular, observe the 0 in the LS bit of the LS nybble of the result as highlighted in Figure L4-30(d). This 0, which arrived from our shift operation in Figure L4-30(b), will form the first bit in our quotient.

> **Note**　The fact that we discover that we didn't really wish to perform the subtraction and end up having to add the divisor back into the result means that this is what is known as a *restoring division* algorithm. An alternative would be a *nonrestoring division* algorithm, in which we would first compare the value in the divisor with the value in the MS portion of the result to see whether or not we wish to perform the subtraction.

Now, let's consider the second cycle in our 4-bit division. The initial conditions for this cycle [Figure L4-31(a)] are, of course, the final conditions from the previous cycle [Figure L4-30(d)]. As before, the first thing we do is to shift the entire 8-bit result one bit to the left [Figure L4-31(b)].

Next, we subtract the divisor from the MS nybble of the result as shown in Figure L4-31(c) ($0001_2 - 0010_2 = 1111_2$). Once again, a 0 in the carry (borrow) flag indicates that we have attempted to subtract a larger value from a smaller one, so we have to undo our actions by adding the

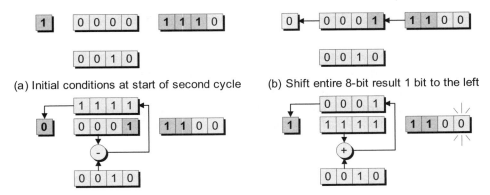

(a) Initial conditions at start of second cycle

(b) Shift entire 8-bit result 1 bit to the left

(c) Subtract divisor from MS nybble of result

(d) Add divisor back into MS nybble of result

Figure L4-31. Second cycle of 4-bit division test case.

divisor back into the MS nybble of the result as shown in Figure L4-31(d) ($1111_2 + 0010_2 = 0001_2$).

As before, observe the 0 in the LS bit of the LS nybble of the result as highlighted in Figure L4-31(d). This 0, which was inserted by our shift operation in Figure L4-31(b), will form the second bit in our quotient.

Things start to become a tad more interesting when we come to consider the third cycle in our 4-bit test case. After we've performed the shift operation as illustrated in Figure L4-32(b), we subtract the divisor

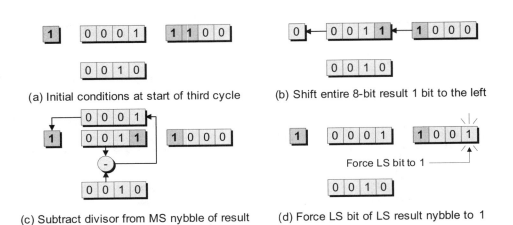

(a) Initial conditions at start of third cycle

(b) Shift entire 8-bit result 1 bit to the left

(c) Subtract divisor from MS nybble of result

Force LS bit to 1

(d) Force LS bit of LS result nybble to 1

Figure L4-32. Third cycle of 4-bit division test case.

from the MS nybble of the result as shown in Figure L4-32(c) ($0011_2 -$
$0010_2 = 0001_2$).

Following this subtraction, the carry (borrow) flag contains a 1,
thereby indicating that the value we subtracted (the divisor) was less than
or equal to the value in the MS nybble of the result. In this case, we keep
the result from the subtraction as-is and we force the LS bit of the LS result nybble to 1, as shown in Figure L4-32(d) (this will form the third bit
in our quotient).

Similarly, in the case of our fourth and final cycle in our 4-bit test
case, once we've performed the shift [Figure L4-33(b)], we again subtract
the divisor from the MS nybble of the result as shown in Figure L4-33(c)
($0011_2 - 0010_2 = 0001_2$).

Once again, the carry (borrow) flag contains a 1 following this subtraction, thereby indicating that the value we subtracted (the divisor) was
less than or equal to the value in the MS nybble of the result. Thus, we
again keep the result from the subtraction as is and we force the LS bit of
the LS result nybble to 1, as shown in Figure L4-33(d) (this will form the
fourth bit in our quotient).

Now let's remind ourselves as to what we were hoping to achieve.
The operation we intended to perform was to divide 0111_2 by 0010_2 to
give 0011_2 with a remainder of 1 (that is, 7_{10} divided by $2_{10} = 3_{10}$ with a
remainder of 1 in decimal). Good grief! If we look at Figure L4-33(d) in

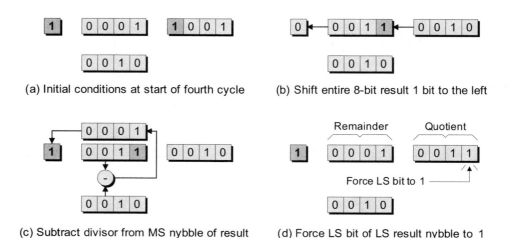

(a) Initial conditions at start of fourth cycle

(b) Shift entire 8-bit result 1 bit to the left

(c) Subtract divisor from MS nybble of result

(d) Force LS bit of LS result nybble to 1

Figure L4-33. Fourth cycle of 4-bit division test case.

more detail, we realize that the LS nybble of our result now contains the quotient ($0011_2 = 3_{10}$), whereas the MS nybble contains the remainder ($0001_2 = 1_{10}$). It's just like magic!

Try it for yourself

Before we proceed, you might wish to try working through a couple of simple test cases as paper exercises just to make sure you have the concepts firmly tied down.

The example we worked through above involved performing the operation $0111_2/0010_2 = 0011_2$ remainder 1 (that is, dividing 7_{10} by 2_{10} to give 3_{10} with a remainder of 1 in decimal). Why don't you start off by performing the corresponding operation $0111_2/0011_2 = 0010_2$ remainder 1 (that is, dividing 7_{10} by 3_{10} to give 2_{10} with a remainder of 1 in decimal)?

As a further example, you might try performing the operation $0101_2/0011_2 = 0001_2$ remainder 10_2 (that is, dividing 5_{10} by 3_{10} to give 1_{10} with a remainder of 2_{10} in decimal). The reason this later example is of interest is that we are going to equip our division subroutine with a *round-half-even* rounding algorithm as discussed in Chapter 4. In this case, we would expect the final result from our division of 5/3 = 1 remainder 2 to be rounded up to 2; that is, 5/3 = 2.

Rounding

Hold onto your hat, because this portion of our discussions may well make your brain go a little "wobbly" around the edges. As we noted above, we intend to equip our division subroutine with a *round-half-even* rounding algorithm. In our discussions in Chapter 4, we made the following statement with regard to this form of rounding (which is also known as "bankers' rounding"):

> In this case, if the digits to be discarded are *greater than* 0.5, then the integer portion of the result will be incremented by 1; or if the digits to be discarded are *less than* 0.5, the integer portion will be left unchanged; otherwise, if the digits to be discarded are equal to 0.5, the integer portion of the result will be rounded so as to make an even number. That is, if the integer portion of the result is already even, then it will be left unaltered, but if the integer portion of the result is odd, it will be incremented by 1 (rounded-up) to form an even number.

This made sense when we were talking about rounding numbers like 27.5, but now we have to decide what we mean by "0.5" in the context of an integer division. In fact, the easiest way to conceptualize this is to think in terms of the divisor and the remainder as follows:

- If the remainder is greater than half of the value of the divisor, then we increment the result by 1.

- If the remainder is less than half of the divisor, then we don't do anything.

- If the remainder is equal to half of the divisor, then we check the quotient; if the quotient is already even, we don't do anything, but if it's odd we round it up to the next even value (which simply means that we add 1 to it).

For example, consider the following operations:

- $29/8 = 3$ remainder 5. In this case, the remainder 5 is greater than half of the divisor ($8/2 = 4$), so round the quotient up to 4.

- $27/8 = 3$ remainder 3. In this case, the remainder 3 is less than half of the divisor ($8/2 = 4$), so leave the quotient as is at 3.

- $28/8 = 3$ remainder 4. In this case, the remainder 4 is exactly equal to half of the divisor ($8/2 = 4$); thus, as the original quotient is odd (3), round it up to the next even value (4).

From this, we see that the key element to our rounding strategy involves comparing the remainder to half of the value of the divisor. In order to achieve this, our "knee jerk" reaction might be to simply divide the divisor by two and then compare it to the remainder; sadly, however, this can lead to errors. For example, suppose we perform the operation $31_{10}/9_{10} = 3_{10}$ remainder 4_{10}. The equivalent in binary is $11111_2/1001_2 = 11_2$ remainder 100_2.

Now, we all know that as the remainder (4_{10}) is less than half of the divisor (9_{10}), we should leave the quotient as is. However, let's see what happens if we use a "divide the divisor by two" approach. As we discussed in Chapter 4, in order to divide a binary value by two, all we need do is to shift it one bit to the right. Thus, if we take our divisor of $1001_2 = 9_{10}$ and shift it one bit to the right, we end up with $100_2 = 4_{10}$. If we now compare the remainder to this value, we find that they are equal, so, due to the fact that the original quotient was an odd value (3_{10}), we would mistakenly round it up to 4_{10}.

The problem is that shifting the divisor one bit to the right caused the least-significant bit to "fall off the end." If the original divisor is an even value, this bit will be 0, so no harm will be done. If the original divisor is an odd value, however, this bit will be 1, in which case we can run into problems, as we've just seen.

Fortunately, there is a very simple solution to this conundrum. As opposed to dividing the divisor by two and then comparing the result to the remainder, we can go about things the other way around. That is, we can leave the divisor as is, multiply the remainder by two, and then perform the comparison. Using this approach, we have to reformulate our rules as follows:

- If two times the remainder is greater than the divisor, then we increment the result by 1.
- If two times the remainder is less than the divisor, then we don't do anything.
- If two times the remainder is equal to the divisor, then we check the quotient; if the quotient is already even, we don't do anything, but if it's odd we round it up to the next even value.

In order to see how this works, let's return to our example operation, $31_{10}/9_{10} = 3_{10}$ remainder 4_{10} (that is, $11111_2/1001_2 = 11_2$ remainder 100_2 in binary). As we discussed in Chapter 4, in order to multiply a binary value by two, all we need do is to shift it one bit to the left. Thus, if we take our remainder of $100_2 = 4_{10}$ and shift it one bit to the left, we end up with $1000_2 = 8_{10}$. If we now compare this value to the divisor, we see that the divisor is the greater, in which case we know to leave the existing quotient as is.

Sign correction

Finally, we need to perform any sign correction on the result. The rules for this are pretty much the same as for multiplication:

+ divided by + equals +
+ divided by − equals −
− divided by + equals −
− divided by − equals +

As you will see in the subroutine described below, the way in which we test for this and perform any necessary sign corrections is the same as for the multiplication routine discussed in the previous lab.

The 16-bit _DIV Subroutine Itself

When it comes to our real _DIV subroutine, of course, we intend to divide one 2-byte (16-bit) number by another. Following rounding, we will discard any remainder and leave the 2-byte quotient on the top of the stack. In order to provide a point of reference, the names of the main storage locations used in this routine are as shown in Figure L4-34.

```
###########################################################################
# Name:      _DIV                                                         #
#                                                                         #
# Function: Divide one 16-bit signed binary number by another and        #
#           return a 16-bit result in the range -32,767 to +32,767       #
#                                                                         #
# Entry:    Top of stack                                                  #
#           Most-significant byte of return address                       #
#           Least-significant byte of return address                      #
#           Most-significant byte of second number  (Divisor)             #
#           Least-significant byte of second number (Divisor)             #
#           Most-significant byte of first number   (Dividend)            #
#           Least-significant byte of first number  (Dividend)            #
#                                                                         #
# Exit:     Top of stack                                                  #
#           Most-significant byte of result          (Quotient)           #
#           Least-significant byte of result         (Quotient)           #
#                                                                         #
# Modifies: Accumulator and Index Register                                #
#                                                                         #
# Size:     Program = 298 bytes                                           #
#           Data    =   9 bytes                                           #
###########################################################################

########## Get return address from stack and store it
_DIV:      POPA                    # Retrieve MS byte of return
           STA     [_DV_RADD+0]    #   address from stack and store it
           POPA                    # Retrieve LS byte of return
           STA     [_DV_RADD+1]    #   address from stack and store it

########## Get divisor from stack and store it
_DV_GDIV:  POPA                    # Retrieve MS byte of divisor
           STA     [_DV_DIV+0]     #   from stack and store it
           POPA                    # Retrieve LS byte of divisor
           STA     [_DV_DIV+1]     #   from stack and store it

########## Check that we're not trying to divide by zero
_DV_CHKZ:  OR      [_DV_DIV+0]     # OR ACC with MS byte of divisor
           JNZ     [_DV_GDND]      # If non-zero, carry on ...
```

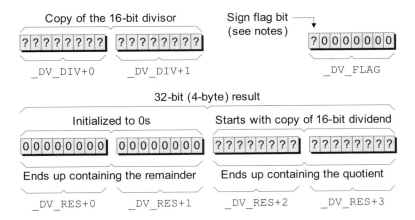

Figure L4-34. Names of primary storage locations in the real subroutine.

```
_DV_ERRZ:   BLDX      MSG_007       # Load X reg with addr of message
            JSR       [DISPERR]     # Jump to display error subroutine
                                    # (which terminates the program)


########## Get dividend from stack and store it
_DV_GDND:   LDA       $00           # Load ACC with 0 and save it to
            STA       [_DV_RES+0]   #   the two MS bytes of the
            STA       [_DV_RES+1]   #   4-byte result
            POPA                    # Retrieve MS byte of dividend
            STA       [_DV_RES+2]   #   from stack and store it
            POPA                    # Retrieve LS byte of dividend
            STA       [_DV_RES+3]   #   from stack and store it


########## Invert divisor if necessary (store sign in flag)
_DV_IDIV:   LDA       [_DV_DIV+0]   # Load ACC with MS divisor
            STA       [_DV_FLAG]    #   and store a copy in the flag
            JNN       [_DV_IDND]    # If positive value, jump to next test
                                    #   ... otherwise ...
            LDA       $00           # Load ACC with 0
            SUB       [_DV_DIV+1]   #   Subtract (no carry) LS byte of
            STA       [_DV_DIV+1]   #   divisor and store the result
            LDA       $00           # Load ACC with 0
            SUBC      [_DV_DIV+0]   #   Subtract (with carry) MS byte of
            STA       [_DV_DIV+0]   #   divisor and store the result


########## Invert dividend if necessary (XOR sign in flag)
_DV_IDND:   LDA       [_DV_FLAG]    # Load the flag
            XOR       [_DV_RES+2]   #   XOR it with MS byte of dividend
            AND       %10000000     #   Clear bits (except sign) to 0
            STA       [_DV_FLAG]    #   Store the flag away again
            LDA       [_DV_RES+2]   # Load ACC with MS dividend
```

```
                JNN       [_DV_LOOP]   # If positive value, jump to next part
                                       #    ... otherwise ...
                LDA       $00          # Load ACC with 0
                SUB       [_DV_RES+3]  #   Subtract (no carry) LS byte of
                STA       [_DV_RES+3]  #   dividend and store the result
                LDA       $00          # Load ACC with 0
                SUBC      [_DV_RES+2]  #   Subtract (with carry) MS byte of
                STA       [_DV_RES+2]  #   dividend and store the result

########## Hold tight—this is the start of the main division loop
_DV_LOOP:       BLDX      16           # Load index reg with the number of
                                       # times we want to go around loop
_DV_DOIT:       LDA       [_DV_RES+3]  # Shift the 4-byte (32-bit) result
                SHL                    #   one bit to the left. Start with
                STA       [_DV_RES+3]  #   the LS byte: _DV_RES+3
                LDA       [_DV_RES+2]  # Now do _DV_RES+2
                ROLC                   #   Use a rotate here
                STA       [_DV_RES+2]  #   :
                LDA       [_DV_RES+1]  # Now do _DV_RES+1
                ROLC                   #   Use a rotate here
                STA       [_DV_RES+1]  #   :
                LDA       [_DV_RES+0]  # Now do _DV_RES+0
                ROLC                   #   Use a rotate here
                STA       [_DV_RES+0]  #   :

########## Subtract the divisor from the two MS bytes of the result
_DV_SUB:        LDA       [_DV_RES+1]  # Load ACC with _DV_RES+1
                SUB       [_DV_DIV+1]  #   Subtract (no carry) LS byte of
                STA       [_DV_RES+1]  #   divisor and store result
                LDA       [_DV_RES+0]  # Load ACC with DV_RES+0
                SUBC      [_DV_DIV+0]  #   Subtract (with carry) MS byte of
                STA       [_DV_RES+0]  #   divisor and store result

########## Check to see if we need to add the divisor back in again
_DV_ADD:        JC        [_DV_SET1]   # Jump if carry (borrow) = 1
                LDA       [_DV_RES+1]  # Load ACC with _DV_RES+1
                ADD       [_DV_DIV+1]  #   Add (no carry) LS byte of
                STA       [_DV_RES+1]  #   divisor and store result
                LDA       [_DV_RES+0]  # Load ACC with DV_RES+0
                ADDC      [_DV_DIV+0]  #   Add (with carry) MS byte of
                STA       [_DV_RES+0]  #   divisor and store result
                JMP       [_DV_ELUP]   # Jump to the end of the loop

########## Force the LS bit of the LS result byte to 1
_DV_SET1:       LDA       [_DV_RES+3]  # Load LS byte of result
                OR        %00000001    #   Force LS bit to 1
                STA       [_DV_RES+3]  #   and store it back again
########## Check to see if we're done
```

```
_DV_ELUP:   DECX                        # Decrement the index register
            JNZ     [_DV_DOIT]  # If not zero, then do another loop

########## Perform a round-half-even algorithm
_DV_RND:    LDA     [_DV_RES+1] # Load LS byte of remainder
            SHL                 #   Multiply by two (shift left one
            STA     [_DV_RES+1] #   bit and store result
            LDA     [_DV_RES+0] # Repeat for MS byte of remainder
            ROLC                #   but use a rotate this time
            STA     [_DV_RES+0] #

                                # Start by comparing MS bytes
_DV_CPMS:   LDA     [_DV_DIV+0] # Load ACC with MS divisor
            CMPA    [_DV_RES+0] # Compare to MS remainder x2
            JC      [_DV_SIGN]  # Do nothing if divisor is bigger
            JNZ     [_DV_FRND]  # Force round up if divisor is smaller

                                # If MS bytes are equal compare LS
_DV_CPLS:   LDA     [_DV_DIV+1] # Load ACC with LS divisor
            CMPA    [_DV_RES+1] # Compare to LS remainder x 2
            JC      [_DV_SIGN]  # Do nothing if divisor is bigger
            JNZ     [_DV_FRND]  # Force round up if divisor is smaller

                                # Do round-half-even if MS&LS are equal
_DV_RHE:    LDA     [_DV_RES+3] # Load LS byte of quotient
            AND     %00000001   # Mask out all but LS bit
            JZ      [_DV_SIGN]  # Do nothing if quotient already even

                                # Otherwise do a round-up
_DV_FRND:   LDA     [_DV_RES+3] # Load ACC with LS quotient
            ADD     $01         #   Add 1 (no carry) to the ACC
            STA     [_DV_RES+3] #   and store the result
            LDA     [_DV_RES+2] # Load ACC with MS quotient
            ADDC    $00         #   Add 0 (but with carry flag)
            STA     [_DV_RES+2] #   and store the result

########## Check the flag to see if we have to negate the result
_DV_SIGN:   LDA     [_DV_FLAG]  # Load the flag byte
            JNN     [_DV_SAVE]  # Jump if MS bit = 0
            LDA     $00         #   Otherwise invert LS two bytes
            SUB     [_DV_RES+3] #   of the result by subtracting
            STA     [_DV_RES+3] #   them from zero
            LDA     $00         #   :
            SUBC    [_DV_RES+2] #   :
            STA     [_DV_RES+2] #   :

########## Save the LS two bytes of the 4-byte result on the stack
_DV_SAVE:   LDA     [_DV_RES+3] # Load ACC with LS byte of 2-byte
            PUSHA               #   quotient and push it onto the stack
```

```
                LDA     [_DV_RES+2]  # Load ACC with MS byte of 2-byte
                PUSHA                #    quotient and push it onto the stack
########## Return to main program
_DV_RET:        LDA     [_DV_RADD+1] # Load ACC with LS byte of return
                                     #    address from temp location and
                PUSHA                #    push it back onto the stack
                LDA     [_DV_RADD]   # Load ACC with MS byte of return
                                     #    address from temp location and
                PUSHA                #    push it back onto the stack
                RTS                  # That's it, exit the subroutine

########## Reserve temp locations for this subroutine
_DV_RADD: .2BYTE                     # Reserve 2-byte temp location for
                                     #    the return address
_DV_DIV:  .2BYTE                     # Reserve 2-byte temp location for
                                     #    the divisor
_DV_RES:  .4BYTE                     # Reserve 4-byte temp location for
                                     #    the result (remainder + quotient)
_DV_FLAG: .BYTE                      # Reserve 1-byte to use as a flag
########## This is the end of the _DIV subroutine

# - - - - - - - - - - - - - - - - - - - - - - - - - - - - - - - - #
```

Wow! This is even bigger than the multiply routine from the previous lab. But, as usual, it's really not all that scary when you start delving into it.

Retrieving and storing the return address

First, at label _DIV, we perform the standard task of popping the return address off the stack and storing it in a temporary 2-byte location called _DV_RADD.

Retrieving and storing the divisor

Next, at label _DV_GDIV ("get divisor"), we pop the divisor off the stack and store it in a temporary 2-byte location called _DV_DIV.

Testing for a "divide by zero"

Following the previous operation, the accumulator already contains a copy of the LS byte of the divisor. Thus, when we come to label _DV_CHKZ ("check for zero"), all we have to do is OR in a copy of the MS byte of the divisor. If the accumulator still contains zero following this operation, then we know we have a problem, in which case we will display the appropriate error message and terminate the program.

Retrieving and storing the dividend

When we come to label _DV_GDND ("get dividend"), we first load the two MS bytes of our 4-byte result (called _DV_RES) with zeros, and then we pop the dividend off the stack and store it in the two LS bytes of our 4-byte result.

Inverting the divisor (if necessary)

Next, at label _DV_IDIV ("invert divisor"), we load the accumulator with the MS byte of the divisor and we store a copy of this byte in a temporary location called _DV_FLAG. We then use a JNN ("jump if not negative") instruction to test the MS bit of this MS byte to determine whether it's 0 or a 1. If this bit is 0, the value is already positive, so we immediately jump to label _DV_IDND. Otherwise, if this bit is 1, the divisor is negative, in which case we subtract it from zero so as to obtain its positive counterpart.

Inverting the dividend (if necessary)

When we arrive at label _DV_IDND ("invert dividend"), we load the accumulator with the MS byte of the dividend, we XOR this with the contents of our flag, we use an AND instruction to clear the LS seven bits to 0, and we store the result back into the flag.

As before, the reason we are interested in this bit is that it now contains the XOR of the sign bits from the divisor and the dividend. A 0 or a 1 in this bit tells us whether the result from our division should be positive or if it will need to be negated, respectively (we'll return to this point in a little while).

Once we've finished playing with the flag, we reload the accumulator with the MS byte of the dividend and use a JNN ("jump if not negative") to test the MS bit of this MS byte. If this bit is 0, the value is already positive, so we immediately jump to label _DV_LOOP. Otherwise, if this bit is 1, the dividend is negative, in which case we subtract it from zero so as to obtain its positive counterpart.

The main division loop

Just before we commence the main division loop, we use a BLDX ("big load index register") instruction to load the index register with 16_{10}, which is the number of times we wish to go around the loop.

The main division loop itself commences at label _DV_DOIT. Our first task is to shift the 4-byte result field one bit to the left. This is where we take advantage of the ROLC ("rotate left through the carry flag") instruction that we introduced in Chapter 3. Consider the actions depicted in Figure L4-35.

The values shown in Figure L4-35(a) reflect the initial conditions seen the very first time we go around the loop. Result bytes _DV_RES+0 and _DV_RES+1 contain 0s, whereas result bytes _DV_RES+2 and _DV_RES+3 contain a copy of the dividend. [As we've already performed any necessary negations, we know that the MS bit of the dividend will be 0 indicating a positive value; meanwhile, the question mark ("?") characters associated with the remaining dividend bits indicate that they may be 0s or 1s.]

We commence by loading the LS byte of the result from memory location _DV_RES+3 into the accumulator. We then use a SHL ("shift left") instruction to shift the contents of the accumulator one bit to

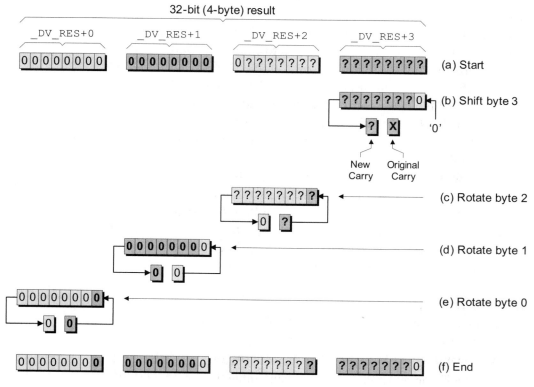

Figure L4-35. Shifting the 4-byte result one bit to the left.

the left [Figure L4-35(b)]. As part of this shift operation, a 0 is automatically loaded into the LS bit and, at the same time, the MS bit
that "falls off the end" of the accumulator is loaded into the carry flag.
(The "X" character associated with the original value in the carry flag
is used to indicate that we don't have any interest in this value.) We
then save the contents of the accumulator back into memory location
_DL_RES+3.

Next, we load the contents of memory location _DL_RES+2 into
the accumulator and use a ROLC to shift this value one bit to the left [Fig-
ure L4-35(c)]. The original contents of the carry flag (which used to be
the MS bit from _DV_RES+3) are copied into the LS bit of the accumulator, and, at the same time, the MS bit that "falls off the end" of the accumulator is loaded into the carry flag. The contents of the accumulator are
then saved back into memory location _DV_RES+2.

And so it goes for the remaining two result bytes [Figures L4-35(d)
and L4-35(e)]. The combination of the SHL instruction and the three
ROLC instructions means that the entire 4-byte result ends up shifted
one bit to the left with a 0 inserted into the LS bit of the LS byte [Figure
L4-35(f)].

Once we've performed this shift, we subtract the divisor from the
MS two bytes of the result at label _DV_SUB. Next, at label _DV_ADD,
we check the state of the carry flag. If the carry flag is 0, we know that
we've tried to subtract a larger value from a smaller one, in which
case we add the divisor back into the MS two bytes of the result and
then jump to the end of the loop. Otherwise, we jump to label
_DV_SET1, at which point we load the LS bit of the LS byte of the
result with a 1.

When we eventually arrive at label DV_ELUP ("end of loop") we
decrement the index register. So long as the index register still contains a
nonzero value, we jump back to the beginning of the loop to perform
another cycle.

Performing a round-half-even

When the index register is finally decremented to zero, the JNZ instruction fails, and the program arrives at the _DV_RND ("round") label. We
start by multiplying the remainder by two, which involves shifting the
MS two bytes of the result (the two bytes containing the remainder) one
bit to the left.

Next, at label _DV_CPMS ("compare MS bytes"), we load the accumulator with the MS byte of the divisor. We then use a CMPA ("compare accumulator") instruction to compare this value with the MS byte of the remainder (which we just multiplied by two). We now have three possibilities:

a) The contents of the accumulator (the MS byte of the divisor) are the larger. This will be indicated by the fact that the carry flag contains 1 (and the zero flag contains 0). In this case, we don't have to do anything and can jump immediately to label _DV_SIGN.

b) The contents of the accumulator are the smaller. This will be indicated by the fact that both the carry and zero flags contain 0. In this case, we definitely wish to force a round operation (add 1 to the quotient) by jumping to label _DV_FRND ("force a round").

c) The two values are equal. This will be indicated by the fact that the zero flag contains 1 (and the carry flag contains 0). In this case, we drop through to label _DV_CPLS ("compare LS bytes") as discussed below.

If we arrive at label _DV_CPLS, we know that the MS bytes of the divisor and the remainder (multiplied by two) are equal. This means that we now have to compare their LS bytes, so we load the accumulator with the LS byte of the divisor and use a new CMPA instruction to compare this value with the LS byte of the remainder. As before, there are three possibilities:

d) The contents of the accumulator (the LS byte of the divisor) are the larger. In this case, we don't have to do anything and can jump immediately to label _DV_SIGN.

e) The contents of the accumulator are the smaller. In this case, we definitely wish to force a round operation (add 1 to the quotient) by jumping to label _DV_FRND ("force a round").

f) The two values are equal. This is the only case in which we may need to perform the round-half-even operation. We start by loading the LS byte of the quotient and masking out all but the LS bit. If this bit is a zero, the quotient is already an even number, in which case we don't have to do anything and can jump immediately to label _DV_SIGN. Otherwise, the quotient is an odd number, so we fall through to label

_DV_FRND, at which point we force a round operation (that is, we add 1 to the quotient).

Correcting the sign of the result

Following the round operation, we find ourselves at label _DV_SIGN. This is where we test the contents of our flag to see whether its MS bit (which is an XOR of the signs of the original divisor and dividend) is a 0 or a 1. If it's a 1, this means that we have to negate the quotient (the two LS bytes of the result) by subtracting them from 0.

Saving the result and exiting the subroutine

All that remains now is to push the quotient onto the stack at label _DV_SAVE, push the return address back onto the stack at label _DV_RET, and execute an RTS ("return from subroutine") instruction to return us to the main body of the program.

Creating and Debugging the _DIV Routine

Invoke the assembler, enter the _DIV subroutine as shown above, and save it in a file called *int-div-2-byte.asm*.

As usual, in order to perform preliminary testing on this routine, we need a .ORG statement to kick everything off, a .END statement to terminate things, and two labels (MSG_007 and DISPERR) to satisfy the external references made by the _DIV routine. Thus, we need to "wrap" our routine with a couple of statements as follows:

```
          .ORG      $4000

{The _DIV subroutine goes here}

MSG_007:  .BYTE
DISPERR:  RTS
          .END
```

Now, try clicking the assemble button and track down and debug any errors until you receive a "Your file was assembled without any errors" message. Once you reach this point, strip out the "wrapping statements" we added above and use the assembler's **File > Save** command to save this file one last time.

Modifying the Testbench to Call the _DIV Routine

Use the assembler's **File > Open** command to open up the *skeleton-lab4.asm* file you created in Lab 4a, and then use the **File > Save As** command to save this as *lab4g.asm*.

Modify the body of the testbench program to call the _DIV routine as follows (the change is highlighted in gray):

```
MAINLOOP:   JSR      [GETNUM]    # Get 1st 16-bit number (NUMA)
            JSR      [GETNUM]    # Get 2nd 16-bit number (NUMB)
            JSR      [_DIV]      # Divide NUMA by NUMB
            JSR      [DISPNUM]   # Display 16-bit number from stack
```

Now, scroll down the file until the end of the subroutines to find the comment line that reads as follows:

```
########## This is where the math subroutine will go
```

Delete this comment and leave your cursor at the beginning of this line, then use the assembler's **Insert > Insert File** command, locate the *int-div-2-byte.asm* source file you just created in the ensuing dialog, and click the **Open** button to insert the contents of that file into this version of the testbench program.

Use the **File > Save** command to save this modified version of the testbench, and then click the assemble button to assemble it.

Using the Testbench to Test the Routine

Click the **On/Off** button on the front panel to power-up the calculator, use the **Memory > Load RAM** command to load your *lab4g.ram* file into the calculator's memory, and click the **Run** button to set the program running.

As usual, once the display has been cleared, the program will loop around, waiting for you to input your first number. As a first test, click on the following sequence of keys:

"0," "0," "0," "7," "Enter"   "0," "0," "0," "3," "Enter"

Now 7_{10} divided by $3_{10} = 2_{10}$, remainder 1. As this remainder is less than half of the divisor, the result will not be rounded, so our program should present the following characters on the calculator's main display:

$$0007\ 0003 = 0002$$

The program now waits for you to click any key, at which point it will clear the main display and wait for you to enter another two numbers. Now try clicking the following sequence of keys:

"0," "0," "0," "7," "Enter"   "0," "0," "0," "2," "Enter"

In this case, 7_{10} divided by $2_{10} = 3_{10}$, remainder 1. However, as this remainder is equal to half of the divisor, and as the original quotient is odd and we're using a round-half-even algorithm, the result *will* be rounded. Thus, our program should present the following characters on the calculator's main display:

$$0007\ 0002 = 0004$$

Experiment with different values to make sure that this routine works as expected. Some suggestions are as follows (these are shown in decimal; we'll leave it to you to generate the hexadecimal equivalents):

1/1

1/–1

–1/1

–1/–1

32,767/1

32,767/–1

32,767/2

1/0

The last example should, of course, cause the program to issue an error message and terminate itself, because the result of dividing anything by zero is undefined in this implementation of our calculator.

As usual, don't restrict yourself only to the suggestions above, which are primarily intended to test worst-case conditions. Just for giggles and grins, try dividing a few "interesting" numbers like $1984_{10}/42_{10}$ to ensure that everything functions as planned.

LABS FOR CHAPTER 5

CREATING AN INTEGER CALCULATOR

Lab 5a

Creating the Calculator Framework

Objective: To create a "framework" program that can be used as the basis to implement an extensible calculator.

Duration: Approximately 40 minutes.

Creating the Framework Program

After the complexities of the previous labs, you may be a tad worried that creating a full-fledged calculator program is going to be mind-bogglingly complex. But there's no need to fret because we've already set the scene and done most of the drudgework in our earlier experiments. All we really have to do at this stage is to gather our existing functions and add a bit of "glue code" to stick everything together.

Our first task—the purpose of this lab—is to create a framework program that we can use as the basis to implement a very simple four-function calculator. Furthermore, we want to create this framework in such a way that we can easily add enhanced capabilities and additional functions in the future.

Building a skeleton framework

Just to make sure that we're all marching to the same drumbeat, we are going to construct our framework program from the ground up. As a starting point, let's invoke the assembler and enter the following comments and directive statements:

```
##############################################################
## Start of constant declarations                          ##
##############################################################

## Constant declarations will go here
```

375

```
##########################################################################
## End of constant declarations                                       ##
##########################################################################

          .ORG      $4000         # Set program origin

##########################################################################
## Start of initialization                                            ##
##########################################################################

## Initialization statements will go here

##########################################################################
## End of initialization                                              ##
##########################################################################


##########################################################################
## Start of main program body                                         ##
##########################################################################

## Main body will go here

##########################################################################
## End of main program body                                           ##
##########################################################################


##########################################################################
## Start of subroutines                                               ##
##########################################################################

## Subroutines will go here

##########################################################################
## End of subroutines                                                 ##
##########################################################################


##########################################################################
## Start of global data                                               ##
##########################################################################

## Global data will go here

##########################################################################
## End of global data                                                 ##
##########################################################################

          .END                    # That's all folks
```

Use the assembler's **File > Save As** command to save this skeleton framework as Lab5a.asm (as always, you should never fail to take an opportunity to save your work). Next, use the assembler's **File > Assemble** command to make sure that everything is as it should be (the sooner we detect any errors, the easier they are to identify and rectify).

Adding constant declarations

Scroll your way to the "Constant Declarations" portion of the skeleton; delete the comment that says "## Constant declarations will go here," and replace it with the following statements:

```
MAINDISP:   .EQU    $F031       # Address of out port for main display
SIXLEDS:    .EQU    $F032       # Address of out port for six LEDs
KEYPAD:     .EQU    $F011       # Address of input port for keypad

CLRCODE:    .EQU    $10         # Special code to clear main display
BELLCODE:   .EQU    $11         # Special code to "beep" the display
BACKCODE:   .EQU    $12         # Code to delete last character
BINMODE:    .EQU    %00000100   # LED code to indicate binary mode
DECMODE:    .EQU    %00000010   # LED code to indicate decimal mode
HEXMODE:    .EQU    %00000001   # LED code to indicate hexadecimal mode

CLRKEY:     .EQU    $10         # Code associated with the "Clear" key
CEKEY:      .EQU    $11         # Code associated with the "CE" key
BACKKEY:    .EQU    $12         # Code associated with the "Back" key
ENTERKEY:   .EQU    $13         # Code associated with the "Enter" key
```

As with our previous programs, the first block of these statements is used to associate labels with the calculator's input and output ports.

The second block defines the codes used to clear the main display, to cause the display to emit an annoying "beep", and to delete the last character to have been displayed.

The third block specifies the different binary patterns of 0s and 1s required to light the binary ("Bin"), decimal ("Dec"), and hexadecimal ("Hex") lights on the main display.

The fourth block is used to associate labels with the keypad's four main "control" keys. We may or may not decide to implement the functions associated with all of these keys, but if we do, then the tasks they perform will be as follows:

Clear: Clear the entire calculation and start again

CE: Clear the current data entry (but keep the rest of the calculation as is)

Back: Delete the last digit from the display (clicking this key multiple times should delete multiple digits)

Enter: Indicates the end of a numerical value

Once again, use the assembler's **File > Assemble** command to save these changes and check that we haven't introduced any errors (we won't keep harping on about this—it's your responsibility to stay on top of this sort of thing henceforth).

Adding the initialization statements

Now scroll your way to the "Initialization" portion of the skeleton, delete the comment that says "## Initialization statements will go here," and replace it with the following statements:

```
INIT:     LDA     CLRCODE       # Load accumulator with clear code
          STA     [MAINDISP]    #   and write it to the main display
          LDA     DECMODE       # Load accumulator with dec mode code
          STA     [SIXLEDS]     #   and write it to port driving LEDs
          BLDSP   $EFFF         # load stack pointer with initial value
```

Reserving placeholders for subroutines

We don't actually want to enter any subroutines at this moment in time, but we do want to reserve some placeholders for them so as to make our lives easier. So, scroll your way to the "Subroutines" portion of the skeleton, delete the comment that says "## Subroutines will go here," and replace it with the following "stub" statements:

```
GETSTUFF:  RTS                 # This will be the raw input routine
GETNUM:    RTS                 # This will get a decimal number
DISPNUM:   RTS                 # This will display a decimal number
DISPERR:   RTS                 # This will display an error message
CHK2NEG:   RTS                 # This will check for -32,768

_NOTYET:   RTS                 # Catch-all for unimplemented functions
_ADD:      RTS                 # This will be our 16-bit addition
_SUB:      RTS                 # This will be our 16-bit subtraction
_MULT:     RTS                 # This will be our 16-bit multiply
_DIV:      RTS                 # This will be our 16-bit divide
_NEG:      RTS                 # This will be our 16-bit negation
```

> **Note** In the context of programming, the term "stub" refers to a dummy function, procedure, or subroutine. The stub routine need not contain any code and is primarily used to prevent "undefined label" errors during assembly, compilation, or linking.

Reserving global temporary data locations and messages

Now we're going to reserve some locations in which to store temporary data values as required. Scroll your way to the "Global Data" portion of the skeleton, delete the comment that says "## Global data will go here," and replace it with the following statements:

```
INSTRING: .BYTE  *10                # Reserve 10 bytes to store a string
TEMPX:    .2BYTE $0000              # Reserve 2-byte location for X reg

########## Start of message strings

########## Start of cunning re-direction
```

Next, place your cursor at the beginning of the blank line following the "Start of message strings" comment; then, use the assembler's **Insert > String** command to access its associated dialog window and use this tool to insert the following message strings (this utility was introduced in Lab 2f):

Label	Message
MSG_000:	ERROR:
MSG_001:	Carry = 1
MSG_002:	Borrow = 0
MSG_003:	Overflow
MSG_004:	Underflow
MSG_005:	Too big
MSG_006:	Too small
MSG_007:	Divide by 0
MSG_008:	Out of range
MSG_009:	-32,768
MSG_010:	Not implemented

As we previously discussed when we did this in Lab 4a, with regard to message MSG_000, make sure you include a space following the colon (that is, the string should be "ERROR: ").

Now, place your cursor at the beginning of the blank line following the "Start of cunning redirection" comment and insert the following statements (we'll discuss what this is all about shortly):

```
KEY_H14:  .2BYTE _NEG            # +/-   Negate
KEY_H15:  .2BYTE _NOTYET         # .     Decimal point
KEY_H16:  .2BYTE _ADD            # +     Add
KEY_H17:  .2BYTE _SUB            # -     Subtract
KEY_H18:  .2BYTE _MULT           # *     Multiply
KEY_H19:  .2BYTE _DIV            #/      Divide
KEY_H1A:  .2BYTE _NOTYET         # =     Equals
KEY_H1B:  .2BYTE _NOTYET         # (     Left parenthesis
KEY_H1C:  .2BYTE _NOTYET         # Pi    Constant Pi
KEY_H1D:  .2BYTE _NOTYET         # Mod   Modulus
KEY_H1E:  .2BYTE _NOTYET         #       Unassigned
```

```
KEY_H1F:    .2BYTE _NOTYET        # )     Right parenthesis
KEY_H20:    .2BYTE _NOTYET        # F-S   Scientific on/off
KEY_H21:    .2BYTE _NOTYET        # Exp   Exponential
KEY_H22:    .2BYTE _NOTYET        #       Unassigned
KEY_H23:    .2BYTE _NOTYET        #       Unassigned
KEY_H24:    .2BYTE _NOTYET        #       Unassigned
KEY_H25:    .2BYTE _NOTYET        #       Unassigned
KEY_H26:    .2BYTE _NOTYET        #       Unassigned
KEY_H27:    .2BYTE _NOTYET        #       Unassigned
KEY_H28:    .2BYTE _NOTYET        #       Unassigned
KEY_H29:    .2BYTE _NOTYET        #       Unassigned
KEY_H2A:    .2BYTE _NOTYET        #       Unassigned
KEY_H2B:    .2BYTE _NOTYET        #       Unassigned
KEY_H2C:    .2BYTE _NOTYET        #       Unassigned
KEY_H2D:    .2BYTE _NOTYET        #       Unassigned
KEY_H2E:    .2BYTE _NOTYET        #       Unassigned
KEY_H2F:    .2BYTE _NOTYET        #       Unassigned
KEY_H30:    .2BYTE _NOTYET        #       Unassigned
KEY_H31:    .2BYTE _NOTYET        #       Unassigned
KEY_H32:    .2BYTE _NOTYET        #       Unassigned
KEY_H33:    .2BYTE _NOTYET        #       Unassigned
KEY_H34:    .2BYTE _NOTYET        #       Unassigned
KEY_H35:    .2BYTE _NOTYET        #       Unassigned
KEY_H36:    .2BYTE _NOTYET        # n!    Factorial
KEY_H37:    .2BYTE _NOTYET        # Log   logarithm
KEY_H38:    .2BYTE _NOTYET        # Tan   Tangent
KEY_H39:    .2BYTE _NOTYET        # Cos   Cosine
KEY_H3A:    .2BYTE _NOTYET        # Sin   Sine
KEY_H3B:    .2BYTE _NOTYET        # 1/x   Reciprocal
KEY_H3C:    .2BYTE _NOTYET        # Rx    Square root
KEY_H3D:    .2BYTE _NOTYET        # x^2   X squared
KEY_H3E:    .2BYTE _NOTYET        # x^3   X cubed
KEY_H3F:    .2BYTE _NOTYET        # y^x   Y to the power of X
KEY_H40:    .2BYTE _NOTYET        # Hex   Switch to hex
KEY_H41:    .2BYTE _NOTYET        # Dec   Switch to decimal
KEY_H42:    .2BYTE _NOTYET        # Bin   Switch to binary
KEY_H43:    .2BYTE _NOTYET        #       Unassigned
KEY_H44:    .2BYTE _NOTYET        #       Unassigned
KEY_H45:    .2BYTE _NOTYET        #       Unassigned
```

Actually, this is a little tricky to explain and the real power of this scheme won't become fully apparent until we implement the body of the calculator in Lab 5e. For the moment, let's simply note that, by means of these statements, we've managed to associate target subroutine addresses with a series of labels that will eventually correspond to the various keys on the calculator's keypad.

In order to visualize how this all hangs together, first use the assembler's **File > Assemble** command to save this latest version of the skeleton program and assemble it. Next, use the assembler's **Window > View List File** command to look at the list file associated with this current version of the program.

The format and use of the list file was discussed in Lab 2f. If you scroll down to the "Subroutines" section of the list file, you should see something like the following:

```
00055 ####################################################################
00056 ## Start of subroutines                                          ##
00057 ####################################################################
00058
00059 400D CF         GETSTUFF: RTS
00060 400E CF         GETNUM:   RTS
00061 400F CF         DISPNUM:  RTS
00062 4010 CF         DISPERR:  RTS
00063 4011 CF         CHK2NEG:  RTS
00064
00065 4012 CF         _NOTYET:  RTS
00066 4013 CF         _ADD:     RTS
00067 4014 CF         _SUB:     RTS
00068 4015 CF         _MULT:    RTS
00069 4016 CF         _DIV:     RTS
00070 4017 CF         _NEG:     RTS
00071
00072 ####################################################################
00073 ## End of subroutines                                            ##
00074 ####################################################################
```

For the purposes of these discussions, we will refer to the line numbers and memory addresses in our list file (your line numbers and other aspects of this file may be slightly different, depending on how closely you've followed our instructions, but the overall "feel" should be the same).

As an example, observe that the first instruction in the _ADD subroutine occurs at address $4013 (on line 00066). Now scroll further down the list file to the redirection statements we entered in the "Global Data" area of the program:

```
00097 ########## Start of cunning re-direction
00098 4097 40 17    KEY_H14:   .2BYTE _NEG
00099 4099 40 12    KEY_H15:   .2BYTE _NOTYET
00100 409B 40 13    KEY_H16:   .2BYTE _ADD
00101 409D 40 14    KEY_H17:   .2BYTE _SUB
00102 409F 40 15    KEY_H18:   .2BYTE _MULT
00103 40A1 40 16    KEY_H19:   .2BYTE _DIV
    :
  etc.
```

Observe the label KEY_H16 ("the key whose code is hexadecimal $16") located on line 00100. If you look at the calculator's keypad (as illustrated in Lab 2d), you'll see that this key corresponds to the "+" function. This label occurs at address $409B, but that's of little interest to us at the moment. What *is* of interest is the fact that the assembler has automatically loaded the 2-byte data value associated with this label with the start address of the _ADD subroutine at address $4013.

The way this works is that, while analyzing the source file, the assembler automatically replaces any labels with their corresponding values. Thus, when the assembler sees the following statement:

```
KEY_H16:   .2BYTE _ADD
```

it automatically replaces the _ADD label with the start address of the _ADD subroutine. We could have achieved exactly the same effect by explicitly entering this address in our source code as follows:

```
KEY_H16:   .2BYTE $4013
```

The problem with doing things explicitly is that we don't know what this address is going to be until we've assembled the program. Furthermore, as we add more lines of code to the program, the start addresses of the various subroutines will wander around all over the place. Thus, the cool thing about our scheme is that, irrespective of any changes we make to the source code, whenever we reassemble our program, the 2-byte data value associated with label KEY_H16 will always point to the first instruction in the _ADD subroutine.

Spend a few moments pondering this, but don't worry unduly if you find it a little tricky to wrap your brain around this concept because all will be revealed in Lab 5e. When you are ready, use the assembler's **Window > View Source File** command to return to the source file and continue to build the framework program as discussed in the following labs.

Lab 5b

Adding Some Low-Level Utility Routines

Objective: To augment our calculator framework program with some low-level utility subroutines.

Duration: Approximately 30 minutes.

Adding the GETSTUFF Routine

The first of the simple utility routines we are going to need is one that will accept and store a sequence of keystrokes from the calculator's keypad.

For the purposes of our first-pass calculator, we are going to adhere to the KISS principle ("Keep it simple, stupid!"). The idea here is that we are going to create a low-level input routine called GETSTUFF whose sole task in life is to read keystrokes from the calculator's keypad and store them as a series of codes in the 10-byte temporary global variable called INSTRING we reserved in the previous lab.

It's important to note that our GETSTUFF routine is not going to perform any serious error checks on the key codes as they come in; instead, it will blindly store any codes it receives until the **Enter** key is pressed.

Before you do anything else, use the assembler to open the *Lab5a.asm* file you created in the previous lab and save it out as *Lab5b.asm*. Now scroll down to the "Subroutines" section, delete the line containing the statement that says "GETSTUFF: RTS" and replace it with the following routine:

```
########## Start of GETSTUFF subroutine
########## Read a series of codes from the keypad until
########## the "Enter" key is pressed
```

```
########## Initialize the input string with dummy values
GETSTUFF:  BLDX     10             # Load the index register with 10
           LDA      $FF            # Load ACC with $FF
_GS_ISTR:  DECX                    # Decrement the index reg
           STA      [INSTRING,X]   # Store ACC in string
           JNZ      [_GS_ISTR]     # Jump back if index reg not zero

########## Initialize the "first" flag with 1
_GS_IFST:  LDA      $01            # Load ACC with 1 (meaning "first")
           STA      [_GS_FST]      # Store this value into the flag

########## This is the main loop where we load a series of key codes
########## into 'INSTRING' until the enter key is pressed
           BLDX     0              # Load the index register with 0

########## Wait for a key to be pressed
_GS_LOOP:  LDA      [KEYPAD]       # Load accumulator from keypad
           JN       [_GS_LOOP]     # Jump back if no key pressed
           STA      [_GS_TEMP]     # Store this code in temp location

########## If this is the first key, then clear the display
_GS_TFST:  LDA      [_GS_FST]      # Retrieve the "first" flag
           JZ       [_GS_DOIT]     # Jump if 0 (not the first key)
           LDA      CLRCODE        # Otherwise load ACC with clear code
           STA      [MAINDISP]     #    store to main display
           LDA      $00            #    then load ACC with 0
           STA      [_GS_FST]      #    and set "first" flag to "not first"

########## Process the key
_GS_DOIT:  LDA      [_GS_TEMP]     # Retrieve the key code from temp
           CMPA     ENTERKEY       # Compare to code for 'Enter" key
           JNZ      [_GS_STOR]     #    if not the same store it
           RTS                     #    else return from this routine


_GS_STOR:  STA      [INSTRING,X]   # Else store this code in string
           INCX                    # Increment the index register
           CMPA     $09            # Compare code to that for number '9'
           JC       [_GS_LOOP]     # If code is bigger, don't display
           STA      [MAINDISP]     #    else copy code to main display
           JMP      [_GS_LOOP]     #    then wait for next key

_GS_FST:  .BYTE                    # Flag to indicate first character
_GS_TEMP: .BYTE                    # Just a temp location
########## End of GETSTUFF subroutine

# - - - - - - - - - - - - - - - - - - - - - - - - - - - - - - - - - - #
```

Unlike most of our subroutines, in this case we don't have to start by popping a return address off the top of the stack, because this routine doesn't actually use the stack per se.

Thus, the first thing we do when we enter the routine is to load the index register with 10 and the accumulator with a dummy (non-key-code) value of $FF. We then enter a short loop at label _GS_ISTR ("initialize string") in which we load our 10-byte INSTRING with a series of these $FF codes.

Next, at label _GS_IFST ("initialize first"), we load the accumulator with 1 and store this value in a temporary location called _GS_FST ("first"). As we shall see, we're going to use this flag to tell us whether or not we're processing the first key in a sequence. The reason for this is that we wish to leave whatever value is on the display until we start to enter a new value, at which point we will need to clear the display.

When we reach the main loop at label _GS_LOOP, we cycle around, reading from the keypad and waiting for a key to be pressed. When a key *is* pressed, we immediately squirrel that code away in a temporary location.

Next, at label _GS_TFST ("test for first"), we reload our "first" flag. If this flag already contains a 0, we know this is not the first key in the sequence, so we jump to label _GS_DOIT. Otherwise, if the flag contains 1, then this is the first key in the sequence, in which case we clear the main display and load the "first" flag with 0.

When we arrive at label _GS_DOIT, we retrieve the original key code from its temporary location and compare it to the code for the **Enter** key. If the codes are the same, we immediately return from this subroutine; otherwise we store the code in the appropriate position in INSTRING. Furthermore, so long as the key code is less than or equal to $09 (which means it corresponds to a decimal digit in the range "0" to "9"), we copy this value back to the calculator's main display.

Implications with regard to the usage model

The simplicity of the GETSTUFF routine presented above is going to dramatically affect our calculator's usage model, because it will force the user to press the **Enter** key after every number and every operator. For example, consider the following equation:

$$30 * 3 + 8 = 92 \qquad \text{\{Infix notation\}}$$

We've shown this using a conventional infix notation in order to get the idea across. As we discussed in Chapter 5, however, our first-pass calculator is going to be based on a postfix notation known as *reverse Polish notation (RPN)*. In this case, the operands and operators may be entered as follows:

8 30 3 * +   {Postfix (RPN) notation}

Based on our discussions above, the way in which our GETSTUFF routine works means that the user will have to click the **Enter** key after each number and each operator, so the actual keystrokes required to execute this calculation will be as follows:

[8] [Enter] [3] [0] [Enter] [3] [Enter] [*] [Enter] [+] [Enter]

Some of the very early calculators employed just such a usage model. However, this is obviously somewhat unwieldy, so at some stage in the future, you may decide to enhance our input routines so as to:

a) Not require the **Enter** key to be pressed after each operator

b) Allow operators like +, −, *, and/to automatically indicate the end of a number without the need to press the **Enter** key.

This will increase the complexity of the code, but it will also make the calculator much easier to use because our equation could now be entered using fewer keystrokes as follows:

[8] [Enter] [3] [0] [Enter] [3] [*] [+]

But, once again, such enhancements are for the future. At the moment, we will just have to learn to be satisfied with the limitations of our simple input routines.

Adding the DISPERR Routine

This is going to be exactly the same DISPERR ("display error") routine that we used in Lab 4a, so we won't discuss it in detail here. All we need say is to scroll down to the "Subroutines" section of our framework program, delete the line containing the statement that says "DISPERR: RTS" and replace it with the following routine:

```
########## Start of DISPERR subroutine
########## Display an error message and then quit the program

########## First clear the display
DISPERR:    LDA       CLRCODE       # Load ACC with clear code
            STA       [MAINDISP]    #   and copy it to the main display
```

```
##########  Display the word "Error: " (note the space)
_DE_ERR:    BSTX    [TEMPX]     # Store X reg in temp location
            BLDX    MSG_000     # Load X reg with start of msg 000
_DE_LUPA:   LDA     [0,X]       # Load a character from the msg
            JZ      [_DE_MSG]   # If it's a NUL jump to next bit
            STA     [MAINDISP]  #   otherwise copy it to main display
            INCX                # Increment the index register
            JMP     [_DE_LUPA]  # Jump back for next character

##########  Now display the main error message then terminate
_DE_MSG:    BLDX    [TEMPX]     # Reload X reg from temp location
_DE_LUPB:   LDA     [0,X]       # Load a character from the msg
            JZ      [$0000]     # If it's a NUL terminate the program
            STA     [MAINDISP]  #   otherwise copy it to main display
            INCX                # Increment the index register
            JMP     [_DE_LUPB]  # Jump back for next character
##########  End of DISPERR subroutine
# - - - - - - - - - - - - - - - - - - - - - - - - - - - - - - #
```

As we previously noted in Lab 4a, we are not concerned with storing the return address in this case because once the DISPERR routine has finished displaying the required message it will terminate the program.

Adding the _NOTYET Routine

At last, an easy one! This is the routine that will be called if the user clicks on a function key that hasn't been implemented yet, in which case we simply wish to display an appropriate error message and terminate the program.

Scroll down to the "Subroutines" section of our framework program, delete the line containing the statement that says "_NOTYET: RTS" and replace it with the following routine:

```
##########  Start of _NOTYET subroutine (for non-implemented functions)
##########  Display an error message and then quit the program

_NOTYET:    BLDX    MSG_010     # Load X reg with addr of message
            JSR     [DISPERR]   # Jump to display error subroutine

##########  End of _NOTYET subroutine

# - - - - - - - - - - - - - - - - - - - - - - - - - - - - - - #
```

Adding the CHK2NEG Routine

And last, but not least, we need to add the routine that will be used by the _ADD and _SUB routines to check that they don't end up generating a value of –32,768. As you may recall, we created and used this CHK2NEG routine in Lab 4e.

Furthermore, as we were planning ahead (we love it when a plan comes together), as part of that lab we saved a copy of this routine in the file *int-check-32768.asm.* Thus, all you have to do is to scroll down to the "Subroutines" section of our framework program and delete the line containing the statement that says "CHK2NEG: RTS."

Leave the cursor at the beginning of this now-blank line and use the assembler's **Insert > Insert File** command to select and insert the contents of *int-check-32768.asm.*

Finally, use the assembler's **File > Assemble** command to save this latest copy of your work and ensure that we still haven't introduced any glaring errors.

Lab 5c

Creating a Decimal GETNUM ("Get Number") Routine

Objective: To augment our calculator framework program with a routine that converts a string of key codes into a 2-byte (16-bit) binary integer.

Duration: Approximately 30 minutes.

Introducing the GETNUM Routine

When this routine is called, it assumes that our 10-byte INSTRING contains one or more key codes corresponding to the decimal digits "0" through "9." The task of this routine is to convert such a sequence of codes into an equivalent 2-byte (16-bit) binary integer.

For example, let's assume that, while under the control of the GETSTUFF routine discussed in Lab 5b, the user entered the following sequence of characters:

[2] [6] [3] [Enter]

Not surprisingly, this means that the contents of INSTRING will be as shown in Figure L5-1.

Now, we know that the first three codes in INSTRING are intended to represent the decimal number 263_{10} ("two hundred and sixty-three"). (The remaining $FF codes are just dummy values of no interest to us here.) This equates to 0107_{16} in hexadecimal or $0000\ 0001\ 0000\ 0111_2$ in binary.

The trick, of course, is to manipulate the individual key codes in such a way as to form the equivalent binary number. There are a number of ways in which we might set about achieving this, the simplest of which involves multiplication.

The idea here is that we start with a 2-byte temporary value containing all zeros. We next enter a loop in which, for each key code in se-

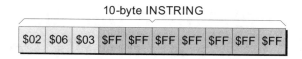

Figure L5-1. Example contents of INSTRING.

quence, we multiply the temporary value by ten and then add in the value of the key code. The easiest way to understand how this works is to consider the progression shown in Figure L5-2.

	Initial value = 0	=	0_{10}	= 0000_{16}	= 0000 0000 0000 0000
Key code = 2	Value = (Value x 10) + 2	=	2_{10}	= 0002_{16}	= 0000 0000 0000 0010
Key code = 6	Value = (Value x 10) + 6	=	26_{10}	= $001A_{16}$	= 0000 0000 0001 1010
Key code = 3	Value = (Value x 10) + 3	=	263_{10}	= 0107_{16}	= 0000 0001 0000 0111

Figure L5-2. Building up the binary value.

Another cunning trick

Well, this seems easy enough, doesn't it? All we need is some way to multiply our 2-byte temporary value by ten. Also, we will require some way to add the resulting value to our key code.

We know what you're thinking. We already have the _MULT and _ADD subroutines that we created during the course of the labs associated with Chapter 4. You are right; we could use these if we really wanted to, but multiplication involves performing a lot of instructions. Surely, we can come up with something just a little more chic.

You've hit the nail on the head. It's time for yet another cunning trick. As we've noted, we wish to multiply our temporary value by ten. But hold hard, because this is equivalent to multiplying the value by eight and by two and adding the results together; for example:

Starting with:	Value = Value * 10
This is the same as saying:	Value = Value * (8 + 2)
Which is the same as saying:	Value = (Value * 8) + (Value * 2)

The point of all of this is that, from Chapter 4, we know that multiplying a binary value by two can be achieved by shifting it one bit to the left. Similarly, multiplying a binary value by eight can be achieved by shifting it three bits to the left.

The GETNUM routine itself

Using the cunning stratagem discussed above means that we can create a reasonably elegant decimal GETNUM routine as follows:

```
########## Start of decimal GETNUM subroutine
########## Assumes INSTRING contains one or more decimal digits

########## Pop the return address off the stack and save it
GETNUM:    POPA                    # Pop the MS byte of return address
           STA     [_GN_RADD]      #   off the stack and store it
           POPA                    # Pop the LS byte of return address
           STA     [_GN_RADD+1]    #   off the stack and store it

########## Initialize temp 2-byte value and index register
_GN_INIT:  BLDX    0               # Load the index register with 0
           LDA     0               # Load ACC with 0
           STA     [_GN_TMPA]      # Clear MS byte of temp location
           STA     [_GN_TMPA+1]    # Clear LS byte of temp location

########## Start of main loop
_GN_DOIT:  LDA     [INSTRING,X]    # Load the next key code
           JN      [_GN_STA]       # If code = $FF we're done

########## Multiply by 2 and store in two locations
_GN_ML2:   LDA     [_GN_TMPA+1]    # Load LS byte of temp value
           SHL                     #   shift left by 1 bit
           STA     [_GN_TMPA+1]    #   store it in LS 1st temp value
           STA     [_GN_TMPB+1]    #   also store in LS 2nd temp value

           LDA     [_GN_TMPA]      # Load MS byte of temp value
           ROLC                    #   rotate left by 1 bit
           STA     [_GN_TMPA]      #   store it in MS 1st temp value
           STA     [_GN_TMPB]      #   also store in MS 2nd temp value

########## Multiply main location by 4 and store it
_GN_ML4:   LDA     [_GN_TMPA+1]    # Load LS byte of temp value
           SHL                     #   shift left 1 bit
           STA     [_GN_TMPA+1]    #   store it
           LDA     [_GN_TMPA]      # Load MS byte of temp value
           ROLC                    #   rotate left by 1 bit
           STA     [_GN_TMPA]      #   store it
```

```
          LDA      [_GN_TMPA+1] # Load LS byte of temp value
          SHL                   #   shift left 1 bit
          STA      [_GN_TMPA+1] #   store it
          LDA      [_GN_TMPA]   # Load MS byte of temp value
          ROLC                  #   rotate left by 1 bit
          STA      [_GN_TMPA]   #   store it

########## Add the value multiplied by 2 to the value multiplied by 8
_GN_AMUL: LDA      [_GN_TMPA+1] # Load LS byte of x8
          ADD      [_GN_TMPB+1] #   add (no carry) LS byte x2
          STA      [_GN_TMPA+1] #   and store the result
          LDA      [_GN_TMPA]   # Load MS byte of x8
          ADDC     [_GN_TMPB]   #   add (with carry) MS byte x2
          STA      [_GN_TMPA]   #   and store the result

########## Add the value of the key code in
_GN_AKEY: LDA      [INSTRING,X] # Load ACC with key code
          ADD      [_GN_TMPA+1] #   add (no carry) the LS temp value
          STA      [_GN_TMPA+1] #   and store the result
          LDA      $00          # Load ACC with 0
          ADDC     [_GN_TMPA]   #   add (with carry) the MS temp value
          STA      [_GN_TMPA]   #   and store the result

########## Increment the index register and do it all again
_GN_INCX: INCX                  # Increment the index register
          JMP      [_GN_DOIT]   # Jump back to look at the next code

########## Store the 2-byte binary number on the stack
_GN_STA:  LDA      [_GN_TMPA+1] # Load ACC with LS byte of number
          PUSHA                 # Push it into the stack
          LDA      [_GN_TMPA]   # Load ACC with MS byte of number
          PUSHA                 # Push it into the stack

########## Return gracefully from this subroutine
_GN_RET:  LDA      [_GN_RADD+1] # Get LS byte of return address from
          PUSHA                 #   temp location and push onto stack
          LDA      [_GN_RADD]   # Get MS byte of return address from
          PUSHA                 #   temp location and push onto stack
          RTS                   # Return from this subroutine

_GN_RADD: .2BYTE                # 2-byte temp location used to store
                               #   the return address for this routine
_GN_TMPA: .2BYTE                # One 2-byte temp location
_GN_TMPB: .2BYTE                # Another 2-byte temp location
########## End of GETNUM subroutine

# - - - - - - - - - - - - - - - - - - - - - - - - - - - - - - - - #
```

We can race through this routine really quickly. The first thing we do when we enter the routine at label GETNUM is to pop the return address off the stack and store it away.

Next, at label _GN_INIT ("initialize") we load the index register with 0 (to point it at the first key code in INSTRING) and we load a 2-byte temporary location called _GN_TMPA with zeros.

And as quickly as that we've reached the start of this routine's main loop at label _GN_DOIT. We commence by loading the key code pointed to by the index register and check it to see if it's a digit or a dummy ($FF) value. If it's a dummy value we're finished, in which case we jump directly to label _GN_STA ("store the number"). Otherwise, we fall through to label _GN_ML2 ("multiply by 2"), where we shift the value in our 2-byte location 1 bit to the left, thereby multiplying it by two. Observe that we store the result in our main _GN_TMPA temporary location and also in an additional temporary location called _GN_TMPB.

Next, at label _GN_ML4 ("multiply by 4"), we shift the value in our main temporary location left by two more bits, thereby multiplying it by four. This means that, as we previously multiplied it by two and now we've further multiplied it by four, the main location has now been multiplied by $2 \times 4 = 8$.

When we reach label _GN_AMUL ("add the multiplied values together"), we add the contents of our two 2-byte temporary locations together and store the result in the main location at _GN_TMPA. Thus, this location now contains its original value multiplied by ten.

And then we come to label _GN_AKEY, where we add the binary code associated with the key into our main temporary location. Of course, the key code itself occupies only 1 byte, whereas the temporary location is two bytes, but we can get around this by adding $00 to the MS byte of the temporary location (along with the carry from the addition of the LS bytes).

All that happens at label _GN_INCX is that we increment the contents of the index register to point to the next key code and then we jump back to the beginning of the loop.

Once we've processed the final key code, we find ourselves at label _GN_STA ("store the number"), at which point we push our 2-byte integer onto the top of the stack. Finally, at label _GN_RET, we push the return address back onto the stack and exit from this subroutine.

Adding the GETNUM routine into the framework

Use the assembler to open the *Lab5b.asm* file you created in the previous lab and save it as *Lab5c.asm*. Now scroll down to the "Subroutines" section, delete the line containing the statement that says "GETNUM: RTS" and replace it with the GETNUM routine described above.

You know the drill by now. Use the assembler's **File > Assemble** command to save this latest copy of your masterpiece and ensure that everything remains error-free.

Lab 5d

Creating a Decimal DISPNUM ("Display Number") Routine

Objective: To augment our calculator framework program with a routine that pops a 2-byte (16-bit) signed binary integer off the top of the stack and displays it as a decimal value.

Duration: Approximately 30 minutes.

Introducing the DISPNUM Routine

When this routine is called, it assumes that there is at least one 2-byte signed binary number on the top of the stack. The task of the DISPNUM routine is to pull a copy of this value off the top of the stack, convert it into an equivalent series of decimal character codes, and present these codes to the calculator's main display.

There are two key points to remember here:

a) The DISPNUM routine needs to leave a copy of the original number on the top of the stack, because this number may well be required by subsequent routines.

b) The value pulled from the top of the stack will be a signed binary number, which means that it may be positive or negative. If the value is negative, the GETNUM routine must first display a minus ("–") character; then it will convert *its* copy of the number into its positive counterpart; and finally it will translate this positive value into the decimal character codes for display.

As usual, there are a number of techniques we might use to convert a binary number into a series of decimal characters. For example, we could use a divide-by-ten approach. In this case, we would take our original binary value and divide it by ten; then the remainder from this operation would represent the least-significant decimal digit to be displayed.

395

We would then divide our new binary value by ten, and the remainder from *this* operation would represent the next least-significant digit to be displayed. We could proceed in this vein until we had extracted all of the digits. The only twist here is that we would eventually have to display the decimal digits in the reverse order to that in which we extracted them. In order to visualize what we mean by this, consider the progression shown in Figure L5-3.

In this case, we start with a binary value of $0000\ 0001\ 0000\ 0111_2$, which equates to 263_{10} in decimal. When we divide this value by ten, we end up with a new binary value of $0000\ 0000\ 0001\ 1010_2$ (which equates to 26_{10} in decimal), along with a remainder of 0011_2 (which equates to 3_{10} in decimal).

By continuing to perform this operation until the main value is zero, we end up extracting remainders of 3_{10}, 6_{10}, and 2_{10}. We would then need to offer these values to the calculator's display in their reverse order so as to present a final value of 263.

Yet another cunning ploy

The above technique seems relatively easy, doesn't it? We've already created a 2-byte _DIV ("divide") routine in Lab 4g. In the case of that routine, we used the remainder to perform a rounding operation and then threw it away. But it wouldn't take a vast amount of effort to create a modified version of the routine that omitted the rounding function and returned a 1-byte remainder.

Once again, however, functions such as multiplication and division are expensive in terms of the large number of instructions they

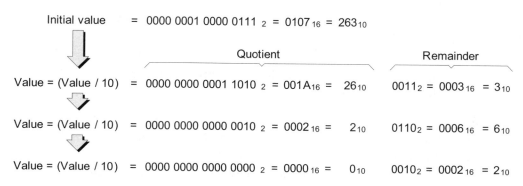

Figure L5-3. Extracting the decimal digits.

have to perform. Could we possibly come up with yet another clever ruse?

Actually, the solution we are going to use here really isn't all that cunning, because it simply involves performing a series of subtractions. Due to the fact we're working with 2-byte (16-bit) values, we know that the largest positive number we're going to have to handle is $32,767_{10}$. This means that we can begin by subtracting $10,000_{10}$ from our original value and keeping track of how many times we can do this before the value goes negative.

When the value *does* go negative, we add $10,000_{10}$ back into the main number and display the number of times we successfully subtracted the $10,000_{10}$. We then repeat this process for values of $1,000_{10}$, 100_{10}, 10_{10}, and 1_{10}.

For example, in the case of a number like $26,847_{10}$, repeatedly subtracting $10,000_{10}$, then $1,000_{10}$, then 100_{10}, then 10_{10} will result in the following (as we shall discover, we don't actually need to perform any subtractions in the case of the final units value):

$$2_{10} \times 10,000_{10}$$

$$6_{10} \times 1,000_{10}$$

$$8_{10} \times 100_{10}$$

$$4_{10} \times 10_{10}$$

$$7_{10} \times 1_{10}$$

Then, all we need to do is to present the codes 2, 6, 8, 4, and 7 to the main display, which will therefore show the following:

$$26847$$

There is one slight "gotcha" we need to keep in mind with regard to any zero values. In the case of a number such as $6,847_{10}$, for example, the algorithm described above will result in the following:

$$0_{10} \times 10,000_{10}$$

$$6_{10} \times 1,000_{10}$$

$$8_{10} \times 100_{10}$$

$$4_{10} \times 10_{10}$$

$$7_{10} \times 1_{10}$$

In this case, we do not wish to present the codes 0, 6, 8, 4, and 7 to the main display, because this would result in the following:

06847

This means that we have to introduce a rule that we will not display any leading 0s. That is, we will not display a 0 unless we've previously displayed a nonzero value. However, "there is an exception to every rule," as they say, and the exception to this one occurs in the units (1s) column. In this case, we will always display the quantity of 1s, even if that quantity is zero. The reason for doing this is that the value we are trying to display may well be 0_{10}, in which case we will wish to see a single 0 appear on the calculator's main display.

The DISPNUM routine itself

Using the technique discussed above means that we can create a reasonably simple decimal DISPNUM routine that uses relatively few instructions. All we need to keep in mind are the following decimal to hexadecimal to binary equivalent values:

Decimal	Hexadecimal	Binary
$10,000_{10}$	2710_{16}	$0010\ 0111\ 0001\ 0000_2$
$1,000_{10}$	$03E8_{16}$	$0000\ 0011\ 1110\ 1000_2$
100_{10}	0064_{16}	$0000\ 0000\ 0110\ 0100_2$
10_{10}	$000A_{16}$	$0000\ 0000\ 0000\ 1010_2$

```
##########  Start of decimal DISPNUM subroutine
##########  Assumes a 2-byte integer is on top of the stack

##########  Pop the return address off the stack and save it
DISPNUM:    POPA                 # Pop the MS byte of return address
            STA    [_DN_RADD]    #   off the stack and store it
            POPA                 # Pop the LS byte of return address
            STA    [_DN_RADD+1]  #   off the stack and store it

##########  Clear the main display
_DN_CDIS:   LDA    CLRCODE       # Load ACC with clear code
            STA    [MAINDISP]    #   and copy to main display

##########  Initialize the "first" flag with 1
_DN_IFST:   LDA    $01           # Load ACC with 1 (meaning "first")
            STA    [_DN_FST]     # Store this value into the flag
```

```
########## Pop the number off the top of the stack
_DN_GNUM:   POPA                    # Pop the MS byte of number
            STA     [_DN_TMP]       #   off the stack and store it
            POPA                    # Pop the LS byte of number
            STA     [_DN_TMP+1]     #   off the stack and store it


########## Push a copy of the number back onto the stack
_DN_PNUM:   PUSHA                   # Push LS byte of number onto stack
            LDA     [_DN_TMP]       # Retrieve MS byte of number
            PUSHA                   #   and push it onto the stack


########## Display '-' and invert number if necessary
_DN_INV:    JNN     [_DN_DOIT]      # Jump if the value isn't negative
            LDA     $2D             #   else load ASCII code for minus sign
            STA     [MAINDISP]      #   and store to main display
            LDA     $00             # Load ACC with 0
            SUB     [_DN_TMP+1]     #   subtract (no carry) LS byte of
            STA     [_DN_TMP+1]     #   number and store it
            LDA     $00             # Load ACC with 0
            SUBC    [_DN_TMP]       #   subtract (with carry) MS byte of
            STA     [_DN_TMP]       #   number and store it


########## Subtract the different powers of ten
_DN_DOIT:   BLDX    $2710           # Load X reg with $2710 (10,000)
            JSR     [_DN_NEST]      #   and call nested subroutine
            BLDX    $03E8           # Load X reg with $03E8 (1,000)
            JSR     [_DN_NEST]      #   and call nested subroutine
            BLDX    $0064           # Load X reg with $0064 (100)
            JSR     [_DN_NEST]      #   and call nested subroutine
            BLDX    $000A           # Load X reg with $000A (10)
            JSR     [_DN_NEST]      #   and call nested subroutine


########## Display the number of 1s irrespective of its value
_DN_01:     LDA     [_DN_TMP+1]     # Load ACC with LS number (1s)
            STA     [MAINDISP]      #   and display it


########## Return gracefully from the main DISPNUM subroutine
_DN_RET:    LDA     [_DN_RADD+1]    # Get LS byte of return address from
            PUSHA                   #   temp location and push onto stack
            LDA     [_DN_RADD]      # Get MS byte of return address from
            PUSHA                   #   temp location and push onto stack
            RTS                     # Return from this subroutine


########## This is the start of the nested subroutine that repeatedly
########## subtracts the power of 10 (passed in via the X reg)
_DN_NEST:   BSTX    [_DN_TMPX]      # Store contents of index reg
            BLDX    0               # Load index register with 0
```

```
########## Subtract the power of 10 from the number
_DN_SUB:    LDA    [_DN_TMP+1]   # Load ACC with LS number
            SUB    [_DN_TMPX+1]  #   subtract (no carry) LS power of 10
            STA    [_DN_TMP+1]   #   and store result
            LDA    [_DN_TMP]     # Load ACC with MS number
            SUBC   [_DN_TMPX]    #   subtract (w carry) MS power of 10
            STA    [_DN_TMP]     #   and store it
            JNC    [_DN_ADD]     # If carry = 0 then recover
            INCX                 #     else increment index reg
            JMP    [_DN_SUB]     #     and go for it again

########## We've gone too far, so add power of 10 back in again
_DN_ADD:    LDA    [_DN_TMP+1]   # Load ACC with LS number
            ADD    [_DN_TMPX+1]  #   add (no carry) LS power of 10
            STA    [_DN_TMP+1]   #   and store result
            LDA    [_DN_TMP]     # Load ACC with MS number
            ADDC   [_DN_TMPX]    #   add (w carry) MS power of 10
            STA    [_DN_TMP]     #   and store it

########## Test to see if this is the first digit to be displayed
_DN_TSTF:   BSTX   [_DN_TMPX]    # Store the index register
            LDA    [_DN_FST]     # Load the first flag
            JNZ    [_DN_TSTZ]    # If 1 (first) then jump to test for 0
            LDA    [_DN_TMPX+1]  #   else load ACC with LS X
            STA    [MAINDISP]    #     store to main display
            RTS                  #     and return to main routine

########## If this is the first digit, test for a non-zero value
_DN_TSTZ:   LDA    [_DN_TMPX+1]  # Load ACC with LS byte of X reg
            JNZ    [_DN_DFST]    #   if non-0 jump to display
            RTS                  #     else return to main routine
_DN_DFST:   STA    [MAINDISP]    # Store code to main display
            LDA    $00           #   then load ACC with 0
            STA    [_DN_FST]     #   and store it to "first" flag
            RTS                  #   and return to main routine
########## This is the end of the nested subroutine

_DN_RADD:   .2BYTE               # 2-byte temp location used to store
                                 #   the return address for this routine
_DN_TMP:    .2BYTE               # A 2-byte temp location
_DN_TMPX:   .2BYTE               # A 2-byte temp location for the X reg
_DN_FST:    .BYTE                # A 1-byte temp location
########## End of DISPNUM subroutine

# - - - - - - - - - - - - - - - - - - - - - - - - - - - - - - - - - - #
```

Fear not! We know the above looks a bit "hairy," but, as usual, it's nowhere near as bad as it seems. We commence by performing a few simple housekeeping tasks. When we enter the routine at label DISPNUM, we pop the return address off the stack and store it away. Next, at label _DN_CDIS ("clear display"), we write a clear code to the display to remove any value that is already on display. And when we get to label _DN_IFST ("initialize first flag") we load a flag called _DN_FST ("first") with a 1 to indicate that we have not, thus far, displayed any characters (we'll return to this point in a moment).

At label _GN_GNUM ("get number") we pop a copy of the 2-byte number off the top of the stack; then, at label _GN_PNUM ("put number") we push a copy back onto the stack for future use by other subroutines.

When we arrive at label _DN_INV ("invert"), we still have a copy of the MS byte of our number stored in the accumulator, so we test to see if this value is negative. In the case of a positive value, we immediately jump to label _DN_DOIT; otherwise, we write the ASCII code for a minus ("−") character to the main display and convert the number into its positive counterpart by subtracting it from zero.

The fun really starts when we arrive at label _GN_DOIT. Now, pay attention and don't blink, because if you're not careful this will be a case of "the quickness of the hand deceives the eye" (note, however, that our hands never leave the ends of our arms).

In a moment we're going to introduce a nested subroutine called _DN_NEST. As we shall see, we need some way to pass a 2-byte value into this subroutine, and we've decided to use the index register (in a somewhat unconventional role) for this purpose. All we need know at this point is that the _DN_NEST routine will take whatever value is passed in via the index register. It will loop around, subtracting this value from the main number; it will maintain a count of how many times the value can be subtracted; and, ultimately, it will present a decimal digit corresponding to this count value to the calculator's main display. (Oh yes, it will also make sure not to display any leading zeros and it will make our smiles whiter and brighter!)

Thus, as we see at label _DN_DOIT, we load the index register with $2710 (which corresponds to $10,000_{10}$ in decimal) and call our nested _DN_NEST routine. We then repeat this process for values of $03E8 (or $1,000_{10}$ in decimal), $0064 (or 100_{10} in decimal), and $000A (or 10_{10} in decimal).

Once we've used the _DN_NEST routine to display the digits corresponding to multiples of $10,000_{10}$, $1,000_{10}$, 100_{10}, and 10_{10}, we arrive at label _DN_01, at which point we display the digit corresponding to multiples of 1_{10}. Finally, at label _DN_RET, we push the return address for the main DISPNUM routine back onto the stack and exit from this routine.

The nested _DN_NEST routine

We shouldn't boast, but this really is a surprisingly efficacious little routine. We start by taking the value in the index register (this is the value we wish to subtract from the main number) and storing it in a temporary location called _DN_TMPX. Once we've stored this value, the index register is free to be used as a counter, so we initialize it to contain 0.

At label _DN_SUB, we loop around, subtracting the 2-byte value in the _DN_TMPX location from our 2-byte number in the _DN_TMP location. Every time we successfully perform this subtraction without our main value going negative, we increment the contents of the index register.

When the main value *does* go negative (indicated by a 0 in the carry (borrow) flag), we fall through to label _DN_ADD, at which point we add the 2-byte value in the _DN_TMPX location back into the 2-byte number in the _DN_TMP location.

When we reach label _DN_TSTF ("test for first"), we store the count value in the index register into our temporary location. Next we load the _DN_FST flag and test to see if it contains 1 (indicating that this is the first character to be displayed) or 0 (indicating that we've already displayed a character).

If we have already displayed another character, then we immediately display this character and return to the main routine. If this is the first character to be displayed, however, we jump to label _DN_TSTZ ("test for zero"). If the value is zero, we return to the main routine without displaying it; otherwise, we display the value, load the _DN_FST flag with 0 to indicate that we have now displayed a character, and *then* return to the main routine.

Adding the DISPNUM routine into the framework

Use the assembler to open the *Lab5c.asm* file you created in the previous lab and save it as *Lab5d.asm*. Now scroll down to the "Subroutines" section, delete the line containing the statement that says "DISPNUM: RTS" and replace it with the DISPNUM routine described above.

We're almost there! We're poised on the edge of excitement beyond our wildest dreams (it's been a long day). In the next, and final (hurray!), lab we're going to put the finishing touches to our program and actually perform some calculations. But before that happy event occurs, use the assembler's **File > Assemble** command to save this penultimate copy of the little scamp and ensure that the term "error" is not one that sullies our vocabulary.

Lab 5e

Implementing a Four-Function Integer Calculator

Objective: To pull our existing math subroutines together to implement a simple four-function integer calculator.

Duration: Approximately 30 minutes.

Inserting the Math Subroutines

We're almost ready to create the body of the calculator, but before we proceed, we need to incorporate our math subroutines into the framework program.

Use the assembler to open the *Lab5d.asm* file you created in the previous lab and save it as *Lab5e.asm*. Now scroll down to the "Subroutines" section and delete the following lines:

```
_ADD:      RTS          # This will be our 16-bit addition
_SUB:      RTS          # This will be our 16-bit subtraction
_MULT:     RTS          # This will be our 16-bit multiply
_DIV:      RTS          # This will be our 16-bit divide
_NEG:      RTS          # This will be our 16-bit negation
```

Leave the cursor at the beginning of these now-blank lines and use the assembler's **Insert > Insert File** command to select and insert the contents of the following files:

int-add-2-byte-v2.asm This is the _ADD routine that we originally created in Lab 4b and then "tweaked" in Lab 4e.

int-sub-2-byte-v2.asm This is the _SUB routine that we originally created in Lab 4c and then "tweaked" in Lab 4e.

int-mult-2-byte.asm This is the _MULT routine that we originally created in Lab 4f.

405

int-div-2-byte.asm	This is the _DIV routine that we originally created in Lab 4g.
int-neg-2-byte.asm	This is the _NEG routine that we originally created in Lab 4d.

Creating the Body of the Calculator Program

Yes! This is it! We've traveled a long hard road together, but now we're poised on the very brink of success. All we need do is to add the "secret squirrel sauce" that will enable our calculator perform its machinations. Can you hear the buzz of excitement and feel the tingle of anticipation?

So, without further delay, scroll down to the "Body" section of the framework, delete the comment that says "## Main body will go here," and replace it with the following statements:

```
MAINLOOP:  JSR     [GETSTUFF]    # Call routine to get some input
           LDA     [INSTRING]    # Load ACC with first code in INSTRING
           CMPA    $09           # Compare value in ACC with $09
           JC      [DOFUNC]      # If value in ACC > $09 then jump
           JSR     [GETNUM]      #   else it's a number so get it
           JMP     [MAINLOOP]    #   then jump back for more input

########## Call the appropriate math/other function/subroutine
DOFUNC:    SUB     $14           # Subtract offset from ACC
           SHL                   # Shift left (multiply by 2)
           STA     [TEMPX+1]     # Store in LS byte of temp X reg
           BLDX    [TEMPX]       # Load index register
           JSR     [[KEY_H14,X]] # Call appropriate subroutine
           JSR     [DISPNUM]     # Call routine to display result
           JMP     [MAINLOOP]    # Jump back for more input
```

"Surely this can't be all there is to it!" you cry. But yes, it is! Let's go through this from the top (as you can see, it won't take long). We start at label MAINLOOP where we call our GETSTUFF subroutine. As we know, this routine reads a series of key codes from the calculator's keypad and stores them in our 10-byte INSTRING until it sees the "Enter" key.

Next, we load the accumulator with the first code from INSTRING and compare it to a value of $09 to determine whether we are dealing with a number or a function key. If the value in the accumulator is greater than $09, it's a function key, in which case we jump to the DO-FUNC ("do the function") label. Otherwise, we call subroutine GETNUM, which converts the string of decimal codes stored in INSTRING into a 2-byte binary integer and stores this value on the stack. Once we've converted this number, we jump back to the MAINLOOP label to wait for the user to enter another number or function key.

The clever part occurs when the user clicks on a function key and we jump to label DOFUNC. In order to understand what happens next, we need to cast our minds way back through the mists of time. As you may recall, in those days of yore we used to refer to as Lab 5a, we included the following statements toward the end of our framework program:

```
KEY_H14:    .2BYTE _NEG          # +/-   Negate
KEY_H15:    .2BYTE _NOTYET       # .     Decimal point
KEY_H16:    .2BYTE _ADD          # +     Add
KEY_H17:    .2BYTE _SUB          # -     Subtract
KEY_H18:    .2BYTE _MULT         # *     Multiply
KEY_H19:    .2BYTE _DIV          #/   Divide
KEY_H1A:    .2BYTE _NOTYET       # =     Equals
KEY_H1B:    .2BYTE _NOTYET       # (     Left parenthesis
    :
  etc.
```

The first of these 2-byte locations is identified by the KEY_H14 label. This corresponds to the first function key in which we are interested—the negate (+/−) key—whose associated code is $14. As we previously discussed, the assembler will automatically load this 2-byte location with the start address of the _NEG subroutine.

Similarly, in the case of the KEY_H15 label, which is associated with the decimal point key whose code is $15, the assembler will automatically load this 2-byte location with the start address of the _NOTYET subroutine. And so it goes.

The point here is that, so long as we know the address of the first of these labels (the KEY_H14 label), we can easily calculate the address of any of the other labels. That is, the address of the KEY_H15 label is equal to the address of the KEY_H14 label + 2, the address of the KEY_H16 label (which is associated with the _ADD subroutine) is equal to the address of the KEY_H14 label + 4, and so forth.

Now let's return to the DOFUNC label in the main body of the program. When we reach this label, we know that the accumulator contains the code associated with one of our function keys. The first thing we do is to subtract $14 from this value. This means that if the user clicked the

"+/−" key whose code is $14, the accumulator will now contain $00; if the user clicked the decimal point "." key whose code is $15, the accumulator will now contain $01; if the user clicked the "+" key whose code is $16, the accumulator will now contain 2; and so on.

Stay with us now. The next thing we do is to shift the contents of the accumulator left by one bit, which is the same as multiplying this value by two. Thus, if the user originally clicked the "+/−" key, the accumulator will now contain $00 × 2 = $00; if the user clicked the decimal point "." key, the accumulator will now contain $01 × 2 = $02; if the user clicked the "+" key, the accumulator will now contain $02 × 2 = $04; and so on.

Now, you may recall that when we originally created our framework, we reserved a 2-byte temporary variable called TEMPX and we loaded this with $0000. We now store the contents of the accumulator into the LS byte of this variable at address TEMPX+1. This means that if the user originally clicked the "+/−" key, the TEMPX location will now contain $0000; if the user clicked the decimal point "." key, TEMPX will now contain $0002; if the user clicked the "+" key, TEMPX will now contain $0004; and so on.

Next, we load the index register with the contents of TEMPX, and then we use the following statement:

```
JSR     [[KEY_H14,X]]# Call appropriate subroutine
```

This JSR ("jump to subroutine") is using a *preindexed indirect* addressing mode. In this case, the CPU will first add the address of the KEY_H14 label to the value in the index register. The CPU will then look at the 2-byte location pointed to by this address, and it will use the contents of *this* location as the target address.

> **Note**   The ways in which the various addressing modes work are detailed in Appendix B. Also, the addressing modes supported by the different instructions are summarized in Appendix C.

For example, as we've already discussed, if the user clicks the "+" key whose code is $16, the index register will end up containing ($16 − $14) * 2 = $0004. The CPU will then read the 2-byte value stored at the location whose address is equal to KEY_H14 + $0004. And this 2-byte value (which corresponds to the KEY_H16 label) is the address of the _ADD subroutine.

This does take a little thought to wrap your brain around things, but once it "clicks" you will say "wow, how capriciously cunning" (and you'll be right!).

And that's pretty much it. As soon as we return from performing the appropriate math function, which will leave its 2-byte result on the top of the stack, we call the DISPNUM subroutine to display this value, and then jump back to MAINLOOP to wait for the next number or operator to be keyed in.

Testing the Calculator Program

Hold onto your socks, because here we go. Use the assembler's **File > Assemble** command to assemble your masterpiece. Click the **On/Off** button on the front panel to power-up the calculator, use the **Memory > Load RAM** command to load the *lab5e.ram* file into the calculator's memory, and click the **Run** button to set the program running.

The display clears and the calculator sits there waiting for us to perform our first calculation. Remembering that our calculator is based on a *reverse Polish notation* (*RPN*) usage model, let's try entering the following equation:

$$123\ 456\ 111 - +$$

This is, of course, equivalent to entering $456 - 111 + 123$ on a standard pocket calculator using infix-based input routines, but we digress. In order to enter the above equation, we need to use the following keystrokes:

[1] [2] [3] [Enter] [4] [5] [6] [Enter] [1] [1] [1] [Enter]

[−] Enter [+] [Enter]

Let's take this step by step to see what happens. We'll start with the following keystrokes:

[1] [2] [3] [Enter]

As soon as we press the **Enter** key, the calculator places a 2-byte binary number containing the equivalent of 123_{10} onto the top of the stack. Now continue with the following keystrokes:

[4] [5] [6] [Enter]

Observe how the previous number remains on the display until you click the first of these number keys. This time, pressing the **Enter** key causes the calculator to place a 2-byte binary number containing the equivalent of 456_{10} onto the top of the stack. Now key in the third value:

[1] [1] [1] [Enter]

On this occasion, pressing the **Enter** key causes the calculator to place a 2-byte binary number containing the equivalent of 111_{10} onto the top of the stack. Thus, the stack currently contains three values [Figure L5-4(a)].

Now, let's perform the first operation by entering the following keystrokes:

[–] [Enter]

This causes our _SUB routine to subtract 111_{10} from 456_{10} and leave the resulting 345_{10} on the top of the stack [Figure L5-4(b)]. This is the value that is now presented to the display by our DISPNUM routine.

Similarly, when we perform the second operation by entering the following keystrokes:

[+] [Enter]

This causes our _ADD routine to add 345_{10} to 123_{10} and leave the resulting 468_{10} on the top of the stack. And *this* is the value that is now presented to the display by our DISPNUM routine [Figure L5-4(c)].

Now, let's try negating this value by entering the following keystrokes:

[+/–] [Enter]

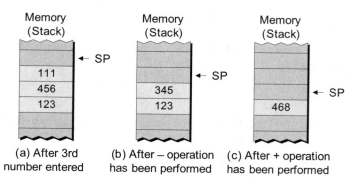

(a) After 3rd number entered

(b) After – operation has been performed

(c) After + operation has been performed

Figure L5-4. The stack during the first example (values are shown in decimal).

This causes our _NEG routine to pop the 2-byte value off the top of the stack, negate it by subtracting it from 0, and push the result back onto the stack. The ensuing -468_{10} value is now presented to the display by our DISPNUM routine.

OK, you're on your own. Run wild and free experimenting with different equations to ensure that the calculator functions as planned.

Enhancing the Calculator Program

A simple error check

Let's start off with a simple suggestion. We know that a math operator like "+" requires two numbers to be on the top of the stack, for example:

[1] [2] [3] [Enter] [4] [5] [6] [Enter] [+] [Enter]

So what will happen if there's only one number on top of the stack; for example, what would occur if we were to start the calculator program running and enter the following:

[1] [2] [3] [Enter] [+] [Enter]

Not surprisingly, the results will be undefined (but they will doubtless not be anything we would like to see). Thus, one enhancement you might consider would be to modify the program to keep a count of exactly how many numbers are on the stack at any particular time. When a unary operator like the "+/−" (negate) is selected, the program should first check to see that there's at least one number on the stack. Similarly, when a binary operator like "+", "−," "*," or "/" is selected, the program should make sure that there are at least two numbers on the stack.

Activating the Back, CE, and Clear keys

Another relatively simple enhancement would be to modify the input routines to support the use of the "Back," "CE," and "Clear" keys. As we discussed in Lab 5a, these keys could be used to perform the following functions:

Back: Delete the last digit from the display (clicking this key multiple times should delete multiple digits)

CE: Clear the current data entry (but keep the rest of the calculation as is)

Clear: Clear the entire calculation and start again

Adding binary and hexadecimal input and display

If we wished to get a tad more sophisticated, we could activate the binary (**Bin**), decimal (**Dec**), and hexadecimal (**Hex**) buttons on the keypad. This would involve modifying our input and output routines to accept and display values in each of these three modes (our core math routines would, of course, remain unchanged).

Changing the usage model

It won't take long before you start chaffing at the restrictions imposed by our simple usage model. Surely we could modify the input routines such that:

a) We don't need to press the **Enter** key after each math function key.

b) We don't need to press the **Enter** key to terminate a number if that number is followed by a math function key.

As we discussed in Chapter 5, implementing these two changes alone would make our lives a lot simpler, For example, instead of entering our original equation using the following keystrokes:

$$[1]\ [2]\ [3]\ [\text{Enter}]\ [4]\ [5]\ [6]\ [\text{Enter}]\ [1]\ [1]\ [1]\ [\text{Enter}]$$
$$[-]\ \text{Enter}\ [+]\ [\text{Enter}]$$

We could now use the following:

$$[1]\ [2]\ [3]\ [\text{Enter}]\ [4]\ [5]\ [6]\ [\text{Enter}]\ [1]\ [1]\ [1]\ [-]\ [+]$$

But why stop here? Why not completely redesign the input routines to support a standard infix usage model. This would allow us to enter the equation as:

$$[4]\ [5]\ [6]\ [-]\ [1]\ [1]\ [1]\ [+]\ [1]\ [2]\ [3]\ [=]$$

Yes, yes, YES!!! You can do it! So have at it with gusto and abandon (and don't hesitate to let us know how you get on)!

Note   For your delectation and delight, a whole slew of additional functions, enhancements, and experiments are presented in Chapter 6.

APPENDIX

A INSTALLING YOUR DIY CALCULATOR

Installing the DIY Calculator on your IBM-compatible PC is very easy, as discussed below. Note that you need at least 7 megabytes of free space on your hard disk drive; also note that your computer should be running Microsoft® Windows® 2000, ME, XP (home or professional), or later.

1) Place the CD-ROM accompanying this book into your computer's CD/DVD drive. For the purpose of these discussions, we will assume that the letter associated with this drive is D: (substitute the letter used by your system if this is different).

2) Double-click the My Computer icon on your desktop or use the **Start** > **My Computer** command to see the drives associated with your computer.

3) Double-click the CD/DVD drive to access the contents of the CD-ROM, which will appear as follows:

/Altium	(This is a folder containing other files)
/Databook	(This is a folder containing other files)
/Educators	(This is a folder containing other files)
/History	(This is a folder containing other files)
/Xnumbers	(This is a folder containing other files)
setup.exe	(This is the setup program)

The contents of the various folders are discussed in more detail in Appendix D.

4) Double-click the *setup.exe* file to launch the DIY Calculator Setup Wizard and initiate the setup process, then click the **Next** button to continue.

5) For the purpose of these discussions, we will assume that the letter associated with your main system drive is C: (substitute the letter used by your computer if this is different). By default, the DIY Calculator will be installed in a folder called *C:\DIY Calculator*.

It is recommended that you use this default location, but you may use the **Browse** button to select a different target location if you wish. When you are ready, click the **Next** button to continue.

6) By default, the installation will insert a shortcut called DIY Calculator in your **Start** > **Programs** menu. It is recommended that you use this shortcut name, but you may specify a different name if you wish. When you are ready, click the **Next** button to continue.

7) If you wish to have an icon associated with the DIY Calculator placed on your desktop (this is recommended), ensure that the appropriate tick-box is selected, and then click the **Next** button to continue.

8) At this point, you will be presented with a dialog window showing a summary of the installation options you have selected. If you agree with these options, click the Install button to complete the installation.

9) The final dialog in the setup wizard allows you to use a tick-box to specify whether or not you wish to launch the DIY Calculator at this time. Make your selection and then click the **Finish** button when you are ready to continue.

10) In order to launch the DIY Calculator in the future, you may use the **Start** > **Programs** > **DIY Calculator** > **DIY Calculator** command, or you may double-click the DIY Calculator icon on your desktop [this assumes that you opted to have this icon placed on your desktop as described in point (7) above].

Additional Points Worth Noting

Before you plunge into the fray, there are a couple of points that are well worth noting, because these can save you oodles of time when you are "up to your ears in alligators" trying to create or debug a program.

Fonts and terminology used in this book

- *Italics* font is used for emphasis, book titles, commands, and path and file names.

- **Bold** may be used for emphasis.

- **Menu** > **Command** identifies the path used to select a menu command.

- `Courier` font is used for program listings and for any text messages that the software displays on the screen.
- "Select" means click the left mouse button on the indicated item.
- "Click-left" (or just "click") means click the left mouse button on the indicated item.
- "Click-middle" means click the middle mouse button on the indicated item.
- "Click-right" means click the right mouse button on the indicated item.
- "Double-click" means click twice consecutively with the left mouse button.
- "Drag-left" (or just "drag") means press-and-hold the left mouse button on the indicated item, then move the cursor (pointer) to the destination and release the button.
- "Shift-click-left" means press-and-hold the <Shift> key and then click the left mouse button on the indicated item.
- "Ctrl-click-left" means press-and-hold the <Ctrl> key and then click the left mouse button on the indicated item.

Screen resolution

Although the DIY Calculator can run on a screen resolution as low as 800 × 600, you will quickly discover that things get a little crowded on the display. Thus, it is strongly recommended that you set your resolution to 1024 × 768 or higher.

The quickest and easiest way to change your screen resolution is to point your mouse cursor at a blank area of your desktop (the background screen image), away from any applications or icons, click-right on the desktop, and then select the **Properties** item from the ensuing pop-up menu. Next, click the **Settings** tab, change the **Screen Resolution** setting, and click the **OK** button.

Arranging the various display utilities

As soon as you invoke one of the utilities such as the Memory Walker, CPU Register, or I/O Ports displays, it is recommended that you immediately drag the display (using its blue title bar) away from the main calculator front panel and locate it in a clear area of the interface. (Note that you can also resize the Memory Walker in a vertical direction so as to make more rows of data visible.)

The reason we suggest this is that, if you inadvertently click the front panel when one of the other utilities is "on top," the front panel will "come forward" and may completely obscure the other utility.

Should this occur, you can drag the calculator front panel around the screen with its blue title bar until you see a portion of the hidden display. Clicking on any part of that display will bring it to the foreground, at which point you can drag the display to a clear area of the interface.

"Losing" and "recovering" the assembler

This is similar in concept to the previous point. The assembler is an independent executable application (program) that is launched from within the main DIY Calculator interface. When you click on any part of the DIY Calculator interface after you have launched the assembler, then the main interface will "come forward" and may completely obscure the assembler.

Should this occur, you can click the Assembler item in the main Windows® taskbar to bring this application to the foreground.

APPENDIX B

ADDRESSING MODES

Introduction

The term *addressing modes* refers to the way in which the CPU determines or resolves the addresses of any data to be used in the execution of its instructions. Different computers can support a wide variety of addressing modes; the selection of such modes depends both on the computer's architecture and the whims of the designer. The seven addressing modes supported by the DIY Calculator's CPU are as follows:

- Implied
- Immediate
- Absolute
- Indexed (or absolute-indexed)
- Indirect
- Preindexed indirect
- Indirect postindexed

Implied Addressing (imp)

Implied addressing refers to instructions that comprise only an opcode without an operand; for example, the INCA ("increment accumulator") instruction. In this case, any data required by the instruction and the destination of any result from the instruction are *implied* by the instruction itself (Figure B-1).

An implied sequence commences when the PC reaches the opcode for an implied instruction (a), loads that opcode into the IR (b), and increments the PC (c). Recognizing that this is an implied instruction, the CPU executes it and continues on to the next instruction.

Instructions that use implied addressing are: CLRIM, DECA, DECX, HALT, INCA, INCX, NOP, POPA, POPSR, PUSHA, PUSHSR, ROLC, RORC, RTI, RTS, SETIM, SHL, and SHR.

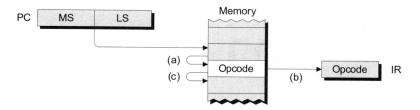

Figure B-1. Implied addressing.

Standard Immediate Addressing (imm)

An instruction using standard immediate addressing has one data operand byte following the opcode; for example, ADD $03 ("add $03 to the contents of the accumulator) (Figure B-2).

The sequence commences when the PC reaches the opcode for an immediate instruction (a), loads that opcode into the IR (b), and increments the PC (c). Recognizing that this is an immediate instruction, the CPU reads the data byte pointed to by the PC, executes the instruction using this data, stores the result in the accumulator (d), and increments the PC to look for the next instruction (e).

Instructions that use standard immediate addressing are: ADD, ADDC, AND, CMPA, LDA, OR, SUB, SUBC, and XOR.

Big Immediate Addressing (imm)

The big immediate addressing mode is very similar to the standard mode, but it refers to instructions that are used to load the 16-bit X, SP, and IV registers. An instruction using big immediate addressing has two

Registers		Flags		Addressing Modes		Other	
ACC	= Accumulator	Z	= Zero	imp	= Implied	LS	= Least-significant
PC	= Program Counter	N	= Negative	imm	= Immediate	MS	= Most-significant
IR	= Instruction Register	C	= Carry	abs	= Absolute	Addr	= Address
X	= Index Register	O	= Overflow	abs-x	= Indexed		
SP	= Stack Pointer	I	= Interrupt Mask	ind	= Indirect		
IV	= Interrupt Vector			x-ind	= Preindexed indirect		

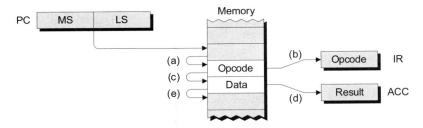

Figure B-2. Standard immediate addressing.

data operand bytes following the opcode; for example, BLDSP $01C4 ("load $01C4 into the stack pointer") (Figure B-3).

The sequence commences when the PC reaches the opcode for an immediate instruction (a), loads that opcode into the IR (b), and increments the PC (c). Recognizing that this is a big immediate instruction, the CPU reads the MS data byte from memory, stores it in the MS byte of the target register (d), and increments the PC (e). The CPU then reads the LS data byte from memory, stores it in the LS byte of the target register (f), and increments the PC to look for the next instruction (g).

Instructions that use big immediate addressing are: BLDSP, BLDX, and BLDIV.

Standard Absolute Addressing (abs)

An instruction using standard absolute addressing has two address operand bytes following the opcode, and these two bytes are used to point to a byte of data (or to a byte in which to store data); for example,

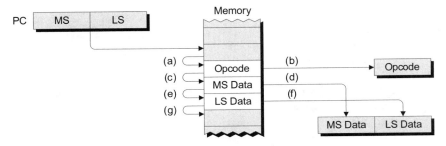

Figure B-3. Big immediate addressing (using BLDSP as an example).

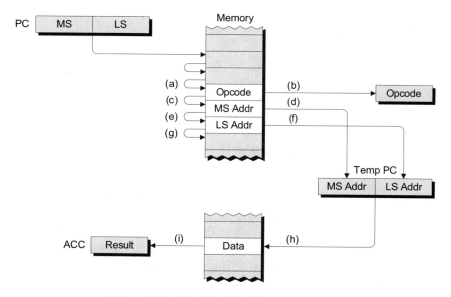

Figure B-4. Standard absolute addressing.

ADD [$4B06] ("add the data stored in location $4B06 to the contents of the accumulator) (Figure B-4).

The sequence commences when the PC reaches the opcode for an absolute instruction (a), loads that opcode into the IR (b), and increments the PC (c). Recognizing that this is a standard absolute instruction, the CPU reads the MS address byte from memory, stores it in the MS byte of one of the temporary PCs (d), and increments the main PC (e). The CPU then reads the LS address byte from memory, stores it in the LS byte of the temporary PC (f), and increments the main PC (g).

Registers		Flags		Addressing Modes		Other	
ACC =	Accumulator	Z =	Zero	imp =	Implied	LS =	Least-significant
PC =	Program Counter	N =	Negative	imm =	Immediate	MS =	Most-significant
IR =	Instruction Register	C =	Carry	abs =	Absolute	Addr =	Address
X =	Index Register	O =	Overflow	abs-x =	Indexed		
SP =	Stack Pointer	I =	Interrupt Mask	ind =	Indirect		
IV =	Interrupt Vector			x-ind =	Preindexed indirect		

The main PC is now "put on hold" while the CPU uses the temporary PC to point to the target address containing the data (h). The CPU executes the original instruction using this data, stores the result into the accumulator (i), and returns control to the main PC to look for the next instruction.

Instructions that use standard absolute addressing are: ADD, ADDC, AND, CMPA, LDA, OR, STA, SUB, SUBC, and XOR. Note that, in the case of a STA ("store accumulator"), the contents of the accumulator would be copied (stored) *into* the data byte in memory. Also note that the jump instructions JMP, JC, JNC, JN, JNN, JO, JNO, JZ, JNZ, and JSR can use absolute addressing. In this case, however, the address operand bytes point to the target address which will be loaded into the main PC.

Big Absolute Addressing (abs)

The big absolute addressing mode is very similar to the standard mode, but it refers to instructions that affect our 16-bit X, SP, and IV registers. An instruction using big absolute addressing has two address operand bytes following the opcode, and these two bytes are used to point to a *pair* of bytes from which to load or store data; for example, BLDSP [$4B06] ("load the two bytes of data starting at location $4B06 into the stack pointer) (Figure B-5).

The sequence commences when the PC reaches the opcode for an absolute instruction (a), loads that opcode into the IR (b), and increments the PC (c). Recognizing that this is a big absolute instruction, the CPU reads the MS address byte from memory, stores it in the MS byte of one of our temporary PCs (d), and increments the main PC (e). The CPU then reads the LS address byte from memory, stores it in the LS byte of the temporary PC (f), and increments the main PC (g).

The main PC is now "put on hold" while the CPU uses the temporary PC to point to the target address containing the MS data byte (h) and store it in the MS byte of our target register (i). The CPU then increments the temporary PC so as to point to the LS data byte (j) and store it in the LS byte of our target register (k). The CPU now returns control to the main PC to look for the next instruction.

Remember that the above sequence described a "big load" of one of our 16 bit registers (the stack pointer in this example). In the case of a

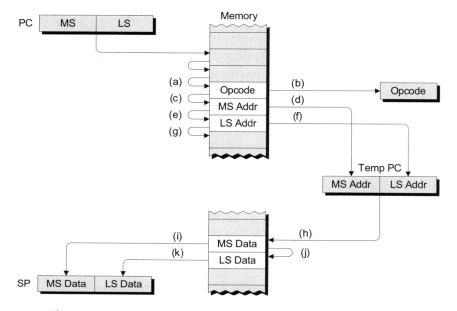

Figure B-5. Big absolute addressing (using BLDSP as an example).

"big store," the contents of the 16-bit register in question would be copied (stored) *into* the two data bytes in memory.

Instructions that use big absolute addressing are: BLDSP, BLDX, BLDIV, BSTSP, and BSTX.

Indexed Addressing (abs-x)

An indexed instruction is very similar to its absolute counterpart, in that it has two address operand bytes following the opcode. However, these

Registers		Flags		Addressing Modes		Other	
ACC	= Accumulator	Z	= Zero	imp	= Implied	LS	= Least-significant
PC	= Program Counter	N	= Negative	imm	= Immediate	MS	= Most-significant
IR	= Instruction Register	C	= Carry	abs	= Absolute	Addr	= Address
X	= Index Register	O	= Overflow	abs-x	= Indexed		
SP	= Stack Pointer	I	= Interrupt Mask	ind	= Indirect		
IV	= Interrupt Vector			x-ind	= Preindexed indirect		

two bytes are added to the contents of the index register (X), and the result is used to point to a byte of data (or to a byte in which to store data); for example, ADD [$4B06,X] ("add the data stored in location ($4B06 + X) to the contents of the accumulator) (Figure B-6).

The sequence commences when the PC reaches the opcode for an indexed instruction (a), loads that opcode into the IR (b), and increments the PC (c). Recognizing that this is an indexed instruction, the CPU reads the MS address byte from memory, stores it in the MS byte of one of our temporary PCs (d), and increments the main PC (e). The CPU then reads the LS address byte from memory, stores it in the LS byte of the temporary PC (f), and increments the main PC (g).

The main PC is now "put on hold" while the CPU adds the contents of the temporary PC to the contents of the index register and uses the result to point to the target address containing the data (h). The CPU now executes the original instruction using this data and stores the result into the accumulator (i). Finally, the CPU returns control to the main PC to look for the next instruction. (Note that the act of adding the temporary PC to the index register does not affect the contents of the index register. Also note that the index register must be loaded with a valid value prior to the first indexed instruction.)

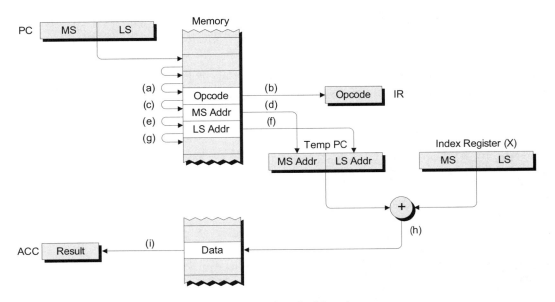

Figure B-6. Indexed addressing.

Instructions that use indexed addressing are: ADD, ADDC, AND, CMPA, LDA, OR, STA, SUB, SUBC, and XOR. Note that, in the case of a STA ("store accumulator"), the contents of the accumulator would be copied (stored) *into* the data byte in memory. Also note that the jump instructions JMP and JSR can use indexed addressing; in this case, however, the result of adding the contents of the temporary program counter to the index register forms the target jump address, which is loaded into the main PC.

Indirect Addressing (ind)

As for an absolute instruction, an indirect instruction has two address operand bytes following the opcode. However, these two bytes do not point to the target data themselves, but instead point to the first byte of another pair of address bytes, and it is *these* address bytes that point to the data (or to a byte in which to store data). Thus, an indirect instruction is so-named because it employs a level of indirection. For example, consider an LDA [[$4B06]] ("load the accumulator with the data stored in the location pointed to by the address whose first byte occupies location $4B06) (Figure B-7).

When the PC reaches an indirect opcode (a), the CPU loads that opcode into the IR (b), and increments the PC (c). Next, the CPU reads the MS address byte from memory, stores it in the MS byte of temporary PC A (d), and increments the main PC (e). Next the CPU reads the LS address byte from memory, stores it in the LS byte of temporary PC A (f), and increments the main PC (g).

The CPU now employs temporary PC A to read the MS byte of the second address (h), store it in the MS byte of temporary PC B (i), and

Registers		Flags		Addressing Modes		Other	
ACC	= Accumulator	Z	= Zero	imp	= Implied	LS	= Least-significant
PC	= Program Counter	N	= Negative	imm	= Immediate	MS	= Most-significant
IR	= Instruction Register	C	= Carry	abs	= Absolute	Addr	= Address
X	= Index Register	O	= Overflow	abs-x	= Indexed		
SP	= Stack Pointer	I	= Interrupt Mask	ind	= Indirect		
IV	= Interrupt Vector			x-ind	= Preindexed indirect		

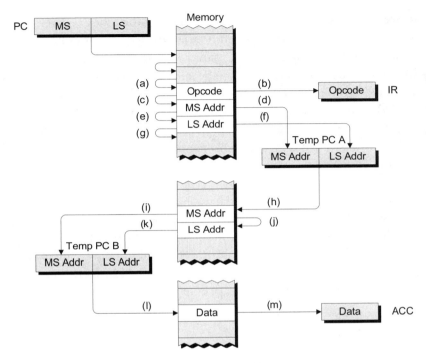

Figure B-7. Indirect addressing.

increment temporary PC A (j). Next the CPU reads the LS byte of the second address, stores it in the LS byte of temporary PC B (k), uses temporary PC B to point to the target data (l), and loads this data into the accumulator (m). Finally, the CPU returns control to the main PC to look for the next instruction.

Instructions that use indirect addressing are LDA and STA. Note that, in the case of a STA ("store accumulator"), the contents of the accumulator would be copied (stored) *into* the data byte in memory. Also, the jump instructions JMP and JSR can use indirect addressing; in this case, however, the second address is the target jump address which is loaded into the main PC.

Preindexed Indirect Addressing (x-ind)

Preindexed indirect addressing is a combination of the indexed and indirect modes. This form of addressing is so-named because the address in the opcode bytes is first added to the contents of the index register, and the result points to the first byte of the second address. For example, consider an LDA [[$4B06,X]] ["load the accumulator with the data stored in the location pointed to by the address whose first byte occupies location ($4B06 + X)] (Figure B-8).

When the PC reaches a preindexed indirect opcode (a), the CPU loads that opcode into the IR (b), and increments the PC (c). Next, the CPU reads the MS address byte from memory, stores it in the MS byte of temporary PC A (d), and increments the main PC (e). Now the CPU reads the LS address byte from memory, stores it in the LS byte of temporary PC A (f), and increments the main PC (g).

The CPU now adds the contents of temporary PC A to the contents of the index register, uses the result to point to the MS byte of the second address (h), and stores this byte in the MS byte of temporary PC B (i). The CPU then points to the LS byte of the second address (j), stores it in the LS byte of temporary PC B (k), uses temporary PC B to point to the

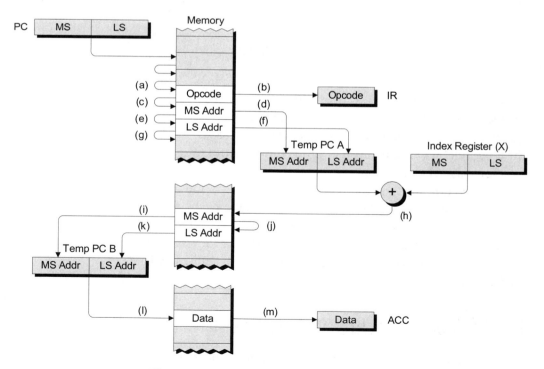

Figure B-8. Preindexed indirect addressing.

target data (l), and loads this data into the accumulator (m). Finally, the CPU returns control to the main PC to look for the next instruction.

Instructions that use preindexed indirect addressing are LDA and STA. Note that, in the case of a STA ("store accumulator"), the contents of the accumulator would be copied (stored) *into* the data byte in memory. Also, the jump instructions JMP and JSR can use this form of addressing; in this case, however, the address pointed to by the combination of temporary PC A and the index register is the target jump address that is loaded into the main PC.

Indirect Postindexed Addressing (ind-x)

Indirect postindexed addressing is similar in concept to preindexed indirect addressing. In this case, however, the address in the opcode bytes points to a second address, and it is this second address that is added to the contents of the index register to generate the address of the target data. For example, consider an LDA [[$4B06],X] (Figure B-9).

When the PC reaches an indirect postindexed opcode (a), the CPU loads that opcode into the IR (b), and increments the PC (c). Next, the CPU reads the MS address byte from memory, stores it in the MS byte of temporary PC A (d), and increments the main PC (e). Now the CPU reads the LS address byte from memory, stores it in the LS byte of temporary PC A (f), and increments the main PC (g).

The CPU uses the contents of temporary PC A to point to the MS byte of the second address (h), and stores this byte in the MS byte of temporary PC B (i). The CPU then increments temporary PC A to point to the LS byte of the second address (j), and stores this byte in the LS byte of temporary PC B (k). Now the CPU adds the contents of tempo-

Registers		Flags		Addressing Modes		Other	
ACC	= Accumulator	Z	= Zero	imp	= Implied	LS	= Least-significant
PC	= Program Counter	N	= Negative	imm	= Immediate	MS	= Most-significant
IR	= Instruction Register	C	= Carry	abs	= Absolute	Addr	= Address
X	= Index Register	O	= Overflow	abs-x	= Indexed		
SP	= Stack Pointer	I	= Interrupt Mask	ind	= Indirect		
IV	= Interrupt Vector			x-ind	= Preindexed indirect		

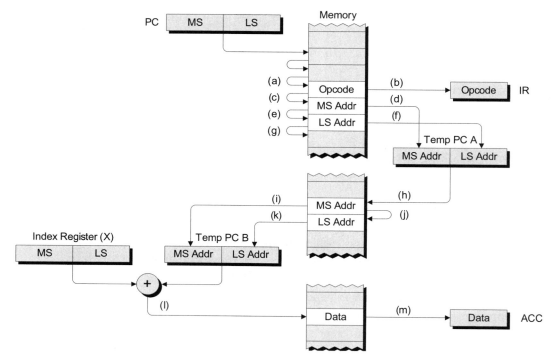

Figure B-9. Indirect postindexed addressing.

rary PC B to the contents of the index register, uses the result to point to the target data (l), and loads this data into the accumulator (m). Finally, the CPU returns control to the main PC to look for the next instruction.

Instructions that use indirect postindexed addressing are LDA and STA. Note that, in the case of a STA ("store accumulator"), the contents of the accumulator would be copied (stored) *into* the data byte in memory. Also, the jump instructions JMP and JSR can use this form of addressing; however, in this case, the address pointed to by the combination of temporary PC B and the index register is the target jump address that is loaded into the main PC.

APPENDIX C

INSTRUCTION SET SUMMARY

Table C-1. Instructions by categories

Control	NOP	No-operation, CPU doesn't do anything.
	HALT	Generate internal NOPs until an interrupt occurs.
	SETIM	Set the interrupt mask flag in the status register.
	CLRIM	Clear the interrupt mask flag in the status register.
Arithmetic	ADD	Add data in memory to the accumulator.
	ADDC	Like an ADD, but include contents of the carry flag.
	SUB	Subtract data in memory from the accumulator.
	SUBC	Like a SUB, but include contents of the carry flag.
Logical	AND	AND data in memory to the accumulator.
	OR	OR data in memory to the accumulator.
	XOR	XOR data in memory to the accumulator.
Comparison, Shifts, and Rotates	CMPA	Compare data in memory to the accumulator.
	SHL	Shift the accumulator left 1 bit (arithmetic shift).
	SHR	Shift the accumulator right 1 bit (arithmetic shift).
	ROLC	Rotate the accumulator left 1 bit (through carry flag).
	RORC	Rotate the accumulator right 1 bit (through carry flag).
Increments and Decrements	INCA	Increment the accumulator.
	DECA	Decrement the accumulator.
	INCX	Increment the index register.
	DECX	Decrement the index register.
Loads and Stores	LDA	Load data in memory into the accumulator.
	STA	Store data in the accumulator into memory.
	BLDX	Load data in memory into the index register.
	BSTX	Store data in the index register into memory.
	BLDSP	Load data in memory into the stack pointer.
	BSTSP	Store data in the stack pointer into memory.
	BLDIV	Load data in memory into the interrupt vector.
Push and Pop	PUSHA	Push the accumulator onto the stack.
	POPA	Pop the accumulator from the stack.
	PUSHSR	Push the status register onto the stack.
	POPSR	Pop the status register from the stack.
Jumps	JMP	Jump to a new memory location.
	JSR	Jump to a subroutine.
	JZ	Jump if the result was zero.
	JNZ	Jump if the result wasn't zero.
	JN	Jump if the result was negative.

(continued)

Table C-1. *Continued*

Jumps (*cont.*)	JNN	Jump if the result wasn't negative.
	JC	Jump if the result generated a carry.
	JNC	Jump if the result didn't generate a carry.
	JO	Jump if the result generated an overflow.
	JNO	Jump if the result didn't generate an overflow.
Returns	RTS	Return from a subroutine.
	RTI	Return from an interrupt.

Table C-2. Instruction set summary

	imp		imm		abs		abs-x		ind		x-ind		ind-x		flags				
	op	#	op	#	op	#	op	#	op	#	op	#	op	#	I	O	N	Z	C
ADD			$10	2	$11	3	$12	3							—	O	N	Z	C
ADDC			$18	2	$19	3	$1A	3							—	O	N	Z	C
AND			$30	2	$31	3	$32	3							—	—	N	Z	—
BLDIV			$F0	3	$F1	3									—	—	—	—	—
BLDSP			$50	3	$51	3									—	—	—	—	—
BLDX			$A0	3	$A1	3									—	—	—	—	—
BSTSP					$59	3									—	—	—	—	—
BSTX					$A9	3									—	—	—	—	—
CLRIM	$09	1													0	—	—	—	—
CMPA			$60	2	$61	3	$62	3							—	—	—	≥	≥
DECA	$81	1													—	—	N	Z	—
DECX	$83	1													—	—	—	Z	—
HALT	$01	1													—	—	—	—	—
INCA	$80	1													—	—	N	Z	—
INCX	$82	1													—	—	—	Z	—
JC					$E1	3									—	—	—	—	—
JMP					$C1	3	$C2	3	$C3	3	$C4	3	$C5	3	—	—	—	—	—
JN					$D9	3									—	—	—	—	—
JNC					$E6	3									—	—	—	—	—
JNN					$DE	3									—	—	—	—	—
JNO					$EE	3									—	—	—	—	—
JNZ					$D6	3									—	—	—	—	—
JO					$E9	3									—	—	—	—	—
JSR					$C9	3	$CA	3	$CB	3	$CC	3	$CD	3	—	—	—	—	—
JZ					$D1	3									—	—	—	—	—

Table C-2. *Continued*

	imp		imm		abs		abs-x		ind		x-ind		ind-x		flags				
	op	#	op	#	op	#	op	#	op	#	op	#	op	#	I	O	N	Z	C
LDA			$90	2	$91	3	$92	3	$93	3	$94	3	$95	3	—	—	N	Z	—
NOP	$00	1													—	—	—	—	—
OR			$38	2	$39	3	$3A	3							—	—	N	Z	—
POPA	$B0	1													—	—	N	Z	—
POPSR	$B1	1													Φ	Φ	Φ	Φ	Φ
PUSHA	$B2	1													—	—	—	—	—
PUSHSR	$B3	1													—	—	—	—	—
ROLC	$78	1													—	—	N	Z	↔
RORC	$79	1													—	—	N	Z	↔
RTI	$C7	1													Φ	Φ	Φ	Φ	Φ
RTS	$CF	1													—	—	—	—	—
SETIM	$08	1													1	—	—	—	—
SHL	$70	1													—	—	N	Z	↔
SHR	$71	1													—	—	N	Z	↔
STA	$99	3			$99	3	$9A	3	$9B	3	$9C	3	$9D	3	—	—	—	—	—
SUB			$20	2	$21	3	$22	3							—	O	N	Z	C
SUBC			$28	2	$29	3	$2A	3							—	O	N	Z	C
XOR			$40	2	$41	3	$42	3							—	—	N	Z	—

Legend

op = Opcode
$ = Hexadecimal value
# = Number of bytes
— = No change
≥ = Magnitude comparison
↔ = Shift or rotate thru carry bit
Φ = Restored by popping status register

Flags

Z = Zero
N = Negative
C = Carry
O = Overflow
I = Interrupt Mask

Addressing Modes

imp = Implied
imm = Immediate
abs = Absolute
abs-x = Indexed
ind = Indirect
x-ind = Preindexed indirect
ind-x = Indirect postindexed

APPENDIX

D

ADDITIONAL RESOURCES

Resources on the CD-ROM

When you open up the CD-ROM accompanying this book, you will see a number of files and folders (directories) as shown below:

/Altium   (This is a folder containing other files)
/Databook   (This is a folder containing other files)
/Educators (This is a folder containing other files)
/History   (This is a folder containing other files)
/Xnumbers   (This is a folder containing other files)
setup.exe   (This is the setup program)

As is discussed in Appendix A, the *setup.exe* file is used to install the DIY Calculator software on your computer. Meanwhile, the contents of the remaining folders are as follows:

/Altium

The term *LiveDesign* refers to an integrated digital system design methodology that allows you to easily and interactively develop hardware or processor-based applications implemented inside a *field-programmable gate array* (*FPGA*). It does this by allowing you to "see" inside the FPGA during development and to communicate in real time with FPGA-based virtual instruments and processors running within the FPGA-based design. You can develop and debug your design interactively, without the need for simulation at the system level.

Altium's LiveDesign-enabled product range provides a seamlessly integrated set of design technologies for capture through to final PCB production. Altium's DXP 2004-based products, including Nexar 2004 and Protel 2004, interface to Altium's NanoBoard-NB1 (a device-independent, FPGA-based development board) or to existing third-party development boards to provide a complete LiveDesign experience.

This folder contains a collection of Web-based multimedia presentations. These interactive product demonstrations highlight some of the

433

advanced features in Altium's Nexar and Protel electronic design software solutions, as well as profiling some of Altium's available hardware that supports interactive FPGA development, such as the Nanoboard-NB1 and a range of LiveDesign Evaluation kits (featuring either a low-cost development board or a Universal JTAG Interface).

In order to view these presentations, use your Web browser to launch the top-level HTML (web) page, called *Start.html*, that you will find in this folder.

/Databook

For your delectation and delight, the authors have created a tome called *The Official DIY Calculator Data Book*. This magnum opus contains a lot of useful information, including a detailed breakdown of the CPU's architecture and an in-depth summary of all aspects of our assembly language.

A complete copy of this databook is to be found as an Adobe® Acrobat® PDF file in this folder on the CD-ROM.

> **Note** If you don't already have an Adobe Acrobat Reader installed on your computer, you can download a free copy from the www.Adobe.com website.

/Educators

Inside this folder, teachers and lecturers will find Adobe Acrobat PDF files for each of the labs in the book. These files can be printed out, copied, and used as handouts for students.

Furthermore, all of the images used throughout the book (in the chapters, labs, and appendices) are included in this folder in the form of Microsoft® PowerPoint® presentations. These images may be used as is, or they can be employed as the basis to form additional presentations and handouts so long as appropriate credits are given to this tome.

/History

Over the course of writing this and other books, the authors have run across all sorts of interesting nuggets of history. Over time, these tidbits of trivia have grown to almost form a small book in their own right. For your amusement, they are gathered together and presented in this folder as a PDF document called *The History of Calculators, Computers, and Other Stuff.*

/Xnumbers

As was discussed in Chapter 6, calculator enthusiast James Redin holds U.S. Patent 5,623,433 on what he refers to as *Structured Numbers* or *Verbal Numerals.*

In this folder, James has kindly provided a suite of web pages that include an extremely interesting article (presented as a PDF document) on language and numbers, a selection of sample calculators created using JavaScript that feature James' Structured Number input techniques, and a link to an in-depth *History of Calculators* located on James' website.

In order to view this information, use your web browser to launch the top-level HTML (Web) page, called *Index.html,* that you will find in this folder.

Resources on the DIY Calculator Website

The DIY Calculator website at www.DIYCalculator.com contains a plethora of additional resources. These include, but are not limited to, the following:

Updates

As and when we make any changes to the DIY Calculator, we will make new versions of the core software available for free download.

Tools

The authors have ideas for the creation of a number of support tools such as code coverage and profiler applications. As such utilities are created (either by the authors or by readers), we will make them available via the DIY Calculator website.

File Formats

In order to facilitate the creation of additional utilities, the formats of the various files that are input and output by the DIY Calculator (and its support tools such as the assembler) are documented on the DIY Calculator website.

Articles

The DIY Calculator website features a number of articles on a variety of topics. Some examples are as follows.

Alternative Architectures. The virtual microprocessor featured in the DIY Calculator contains a single accumulator, an index register, and a stack pointer. Over the years, computer architects have experimented with a variety of different architectures, such as multiple accumulators, additional general-purpose registers, and so forth. This article summarizes the various architectures used in some of the early microprocessors.

Floating-point. The math routines presented in this book are designed to work only with integer values. In reality, of course, this is very limiting, because it means that we can't represent real numbers like 3.142. This article introduces the concepts of floating-point representations and defines a simple binary floating-point format. Also described are the creation of associated floating-point subroutines for the fundamental math operations (+, −, *, and /).

Binary coded decimal (BCD). Many folks belittle the BCD format, in which each digit in a decimal number is represented by four binary bits. However, there are a lot of interesting aspects to this approach of which one should at least be aware. This article introduces core BCD concepts and describes the creation of associated BCD subroutines for the fundamental math operations (+, −, *, and /).

Subroutines

Over time, the authors (and, we hope, some of our readers) will create additional integer, floating-point, and BCD-based subroutines. These routines will be made available via the DIY Calculator website for other readers to download, peruse, and ponder.

Further Reading

The various facts and tidbits of trivia with which this tome is festooned were gleaned from a wide variety of sources. In many cases, different references present contradictory information, so one has to weigh the various offerings in the hope of achieving something of a reasonable consensus. Taking all of this into account, the authors have found the following sources to be of particular use and interest.

Books

Bebop to the Boolean Boogie (An Unconventional Guide to Electronics), Second Edition, by Clive "Max" Maxfield, Newnes ISBN: 0-7506-7543-8. A highly understandable introduction as to how transistors can be used to implement simple logic gates, how these gates can be combined to form more complex functions, and how silicon chips are fabricated.

CODE (The Hidden Language of Computer Hardware and Software), by Charles Petzold, Microsoft Press, ISBN: 0-7356-1131-9. A nicely presented, easy-to-understand introduction to the ways in which computers represent and manipulate data.

Dictionary of Word Origins, by John Ayto, Arcade, ISBN: 1-55970-214-1. Described as ". . . perhaps the best inexpensive etymological dictionary available today. . . ," this useful tome offers insights into the origins and histories of 8000-plus English words and the connections to their non-English ancestors.

Dimboxes, Epopts, and other Quidams, by David Grambs, Workman, ISBN: 0-89480-155-4. A compendium of words that can be used to describe life's indescribable people, from the most innocuous creep at the office party ("pigwidgeon") to the motormouth who tyrannizes the car pool ("blatteroon").

Latin For All Occasions, by Henry Beard, Villark, ISBN: 0-394-58660-3. If you have ever found yourself in the position of needing to know the Latin for such everyday phrases as: "God, look at the time! My wife will kill me!" or "Darn, there goes my beeper!" or even "I have a catapult. Give me all the money or I will fling an enormous rock at your head," then this is the book for you!

The New Hacker's Dictionary, Third Edition, by Eric Raymond, MIT Press, ISBN: 0-262-68092-0. A great source for tracking down the origin and historical development of computer jargon and terms.

The Timetables of Science, by Alexander Hellemans and Bryan Bunch, Simon & Schuster, ISBN: 0-671-73328-1. An incredibly useful quick-reference chronology of the most important people and events in the history of science (astronomy, biology, chemistry, earth science, mathematics, medicine, physics, and technology) from the earliest verifiable manufacture of stone tools to the present time.

The Universal History of Computing, by Georges Ifrah, Wiley, ISBN: 0-471-44147-3. In this well-researched tome, the author traces the conceptual, scientific, and technical achievements that made modern computing a reality; from the abacus to the quantum computer, and from the invention of the binary number system three hundred years ago to the recent discovery of "surreal numbers."

The Universal History of Numbers, by Georges Ifrah, Wiley, ISBN: 0-471-39340-1. Not an easy read by any stretch of the imagination, but this book is an absolute treasure-trove of information. Described as ". . . a mind-boggling and enriching experience . . ." by the British newspaper *The Guardian.*

Zero to Lazy Eight: The Romance of Numbers, by Alexander Humez, Nicholas Humez, and Joseph Maguire, Touchstone, ISBN: 0-671-74281-7. Did you ever wonder why "a stitch in time saves nine" and not, say, four? Or why the number seven is considered to be so lucky? Or to what number the word "googol" refers? Well, all is explained in this delightfully wacky and entertaining offering.

Websites

www.Ask.com. Of all of the search engines available, this "Ask Jeeves" website almost invariably proves to be particularly useful when it comes to finding simple explanations for tricky concepts.

www.Dictionary.com. In addition to coming into play for simply checking the spelling of a word, this is an incredibly useful site for tracking down the origins of words, different ways of using the same word, and alternative words by means of its Thesaurus function.

http://members.aol.com/jeff570/mathsym.html. Maintained by Jeff Miller, a teacher at Gulf High School in New Port Richey, Florida, these pages provide a wealth of information as to the earliest usage of various mathematical symbols.

We need your help!

The great thing about computer-based math is that there are a myriad ways of doing everything. So if you are aware of any interesting algorithms for any applicable math routines, then please feel free to share these with us and we'll make them available to other readers by means of the DIY Calculator website (our contact details are to be found on the website at www.DIYCalculator.com).

Furthermore, if you are an educator and come up with your own laboratories, lesson plans, and tests (in the form of question-and-answer or multiple-choice papers), we would very much like to make these available to other lecturers via our website (except for the answers to any questions, of course, which we would make available via email only to teachers and not to their students <grin>).

ABOUT THE AUTHORS

Clive "Max" Maxfield and Alvin Brown are both eminent in the field of electronics, although Alvin was already eminent when Max's eminence was merely imminent.

Max is 6'1" tall, outrageously handsome, English and proud of it. In addition to being a hero, trendsetter, and leader of fashion, he is widely regarded as an expert in all aspects of electronics and computers (at least by his mother). After receiving his B.Sc. in Control Engineering in 1980 from Sheffield Polytechnic (now Sheffield Hallam University), England, Max began his career as a designer of central processing units (CPUs) for mainframe computers. To cut a long story short, he now finds himself president of TechBites Interactive Inc. (www.techbites.com), a high-tech marketing consultancy where he gets to pen his musings about technology. To occupy his spare time (Ha!), Max is coeditor and copublisher of the Web-delivered electronics and computer hobbyist magazine *EPE Online* (www.epemag.com). In addition to numerous technical articles and papers appearing in magazines and at conferences around the world, Max is also the author of *Bebop to the Boolean Boogie (An Unconventional Guide to Electronics)*, coauthor of *Bebop BYTES Back (An Unconventional Guide to Computers)*, author of *Designus Maximus Unleashed*, coauthor of *EDA: Where Electronics Begins*, and author of *The Design Warrior's Guide to FPGAs (Devices, Tools, and Flows)*. On the off chance that you're still not impressed, Max was once referred to as an "industry notable" and a "semiconductor design expert" by someone famous who wasn't prompted, coerced, or remunerated in any way!

Alvin is a well-traveled English gentleman, born in Assam, India where his father managed several tea plantations. After enjoying a lifestyle most of us can only dream of, he returned to England at the age of nine to suffer the rigors of the British educational system. Upon leaving college, Alvin spent ten years acquiring an extensive knowledge of electronics computer-aided design while working for a UK-based defense contractor. During this time, Alvin devoted himself to developing thick- and thin-film hybrid microelectronics circuits and writing computer-aided design software. He subsequently held a number of managerial positions in software development and consultative services groups. Alvin moved to the United States in 1989, and he currently holds a senior management position with a prominent EDA vendor. Alvin has also presented papers at international conferences, has published a number of technical articles, and is the coeditor and copublisher of EPE *Online* magazine (www.epemag.com).

Acknowledgments

It's only when one looks back over the course of a large project, like the writing of this tome, that one realizes just how many people have made significant contributions in one way or another.

The authors would especially like to thank the folks at Altium (www.Altium.com) for sponsoring the writing of our magnum opus and for providing the supporting materials for inclusion in the Altium folder on the CD-ROM accompanying the book.

With regard to the technical aspects of the manuscript in general and, in particular, to the additional materials presented on the www.DIYCalculator.com website relating to the concepts of floating-point and binary-coded-decimal (BCD) representations and operations, the authors would especially like to thank Alon Kfir, Charles Abzug, Malcolm Wiles, Mike Cowlishaw, Peter Hemsley, Robert Munafo, Cyrille de Brebisson, and Steve Heinrich for their stimulating input.

Many are the e-mails that have winged their way across the Internet over the course of the last few years in which the authors posed "But what if . . ." type questions to the above-named experts. Before long, detailed "Ah, in this case . . ." answers would come racing back to clear everything up.

The authors are also grateful to James Redin for his insights into the concept of structured numbers and for providing the supporting materials for inclusion in the Xnumber folder on the CD-ROM accompanying the book.

Thanks also go to Steve Pesto for reviewing the early drafts of the manuscript, to Fred Hudson for wending his weary way through the laboratories to find and fix those little "gotchas" that can be so annoying if they go undetected, and to Joan Doggrell for her painstaking copy editing and myriad helpful suggestions.

Last but certainly not least, the authors would like to thank Bob Sallee for creating the DIY Calculator cartoon image; and Denis Crowder for creating the toolbar icons and the DIY Calculator website.

As usual, we are happy to take full credit for everything that came out right, whereas any errors that may have slithered in can only be attributed to cosmic rays and spurious events of unknown origin.

Index